Discovering Chemistry
With Natural Bond
Orbitals

Discovering Chemistry With Natural Bond Orbitals

Frank Weinhold
Clark R. Landis
Theoretical Chemistry Institute
and Department of Chemistry
University of Wisconsin Madison
Wisconsin

WILEY

A JOHN WILEY & SONS, INC., PUBLICATION

Published by John Wiley & Sons, Inc., Hoboken, New Jersey
Published simultaneously in Canada

For general information on our other products and services or for technical support, please contact our Customer Care Department within the United States at (800) 762-2974, outside the United States at (317) 572-3993 or fax (317) 572-4002.

Wiley also publishes its books in a variety of electronic formats. Some content that appears in print may not be available in electronic formats. For more information about Wiley products, visit our web site at www.wiley.com.

Library of Congress Cataloging-in-Publication Data:

Weinhold, Frank, 1941–
   Discovering chemistry with natural bond orbitals / by Frank Weinhold, Clark R. Landis.
      p. cm.
     Includes index.
     ISBN 978-1-118-11996-9 (pbk.)
  1.   Chemical bonds. 2.   Molecular orbitals. I.   Landis, Clark R., 1956– II.   Title.
     QD461.W45 2012
     541'.28–dc23

                                  2011047575

100989114
Paper ISBN: 9781118119969

Printed in the United States of America

10 9 8 7 6 5 4 3 2 1

The nature of the chemical bond is the problem at the heart of all chemistry.

Bryce Crawford

If anybody says he can think about quantum problems without getting giddy, that only shows he has not understood the first thing about them.

Neils Bohr

It is nice to know that the computer understands the problem, but I would like to understand it too.

Eugene Wigner

Contents

vii

Preface

Recent advances in computers, networking, and electronic structure software now make it feasible for practically every student of chemistry to gain access to powerful computational tools for solving Schrödinger's equation, the ultimate oracle of chemical knowledge. With proper guidance, students having but little quantum mechanical background can undertake creative explorations of modern bonding and valency concepts that often surpass common textbook expositions in accuracy and sophistication. The goal of this book is to provide a practical "how to" guide for such chemical explorers, giving nuts and bolts examples of how chemical questions can be addressed with the help of modern wavefunction or density functional technology, as translated into familiar chemical language through the "Rosetta stone" of Natural Bond Orbital analysis.

The "natural" orbital concept, as originally formulated by Per-Olov Löwdin, refers to a mathematical algorithm by which *best possible* orbitals (optimal in a certain maximum-density sense) are determined from the system wavefunction itself, with no auxiliary assumptions or input. Such orbitals inherently provide the most compact and efficient numerical description of the many-electron molecular wavefunction, but they harbor a type of residual multicenter indeterminacy (akin to that of Hartree–Fock molecular orbitals) that somewhat detracts from their chemical usefulness.

However, a localized adaptation of the natural orbital algorithm allows one to similarly describe *few*-center molecular subregions in optimal fashion, corresponding to the localized lone pairs (one-center) and bonds (two-center) of the chemist's Lewis structure picture. The "Natural Bond Orbitals" (NBOs) that emerge from this algorithm are intrinsic to, uniquely determined by, and optimally adapted to localized description of, the system wavefunction. The compositional descriptors of NBOs map directly onto bond hybridization, polarization, and other freshman-level bonding concepts that underlie the modern electronic theory of valency and bonding.

The NBO mathematical algorithms are embedded in a well-tested and widely used computer program (currently, *NBO 5.9*) that yields these descriptors conveniently, and is attached (or attachable) to many leading electronic structure packages in current usage. Although the student is encouraged to "look under the hood" (Appendix A), the primary goal of this book is to enable students to gain proficiency in *using* the NBO program to re-express complex many-electron wavefunctions in terms of intuitive chemical concepts and orbital imagery, with minimal distractions from underlying mathematical or programming details. "NBO analysis" should be considered a *strategy* as well as a collection of keyword *tools*. Successful usage of the NBO toolkit involves intelligent visualization of the blueprint as well as mastery of individual tools to construct a sound explanatory framework.

This book owes an obvious debt to Foresman and Frisch's useful supplementary manual, *Exploring Chemistry with Electronic Structure Methods* (2nd ed., Gaussian Inc., Pittsburgh, PA, 1996), which provides an analogous how to guide for the popular *Gaussian*™ electronic structure program. Combined with popular utilities such as those made available on the WebMO website, the Gaussian program often makes calculating a wavefunction as simple as a few mouse-clicks, and many such choices of electronic structure system (ESS) are now widely available. The current Gaussian version, *Gaussian 09* (G09), is still the most widely used ESS in the chemical literature, and it includes an elementary NBO module (the older "NBO 3.1" version) that lets the student immediately perform many of the exercises described in this book. However, the NBO program is indifferent to which ESS provided the wavefunction, or even what *type* of wavefunction or density was provided, and the current book is largely independent of such choices. For options that involve intricate interactions with the host ESS and are implemented in only a select set of ESS packages, the Gaussian/NBO form of input file will be used for illustrative purposes. However, the present book has no specific association with the Gaussian program or the Foresman–Frisch guidebook, and the only requirement is that the chosen host ESS can pass wavefunction information to an NBO program (linked or stand-alone) that allows the ESS wavefunction to be analyzed in chemically meaningful terms with the help of the procedures and keywords described herein.

This book also serves as a complementary companion volume to the authors' research monograph, *Valency and Bonding: A Natural Bond Orbital Donor–Acceptor Perspective* (Cambridge University Press, 2005). The latter is theory- and applications-dominated, offering little or no practical know-how for coaxing the NBO program to yield the displayed numerical tables or graphical images. However, the instructions and examples given in this book should allow the student to easily reproduce any of the results given in *Valency and Bonding*, or to extend such treatment to other chemical systems or higher levels of approximation. For complete consistency with the numerical values and graphical orbital displays of *Valency and Bonding*, we employ the same B3LYP/6-311 $++$ G** density functional theoretic (DFT) methodology in this work. However, the student is encouraged to pursue independent explorations of other computational methodologies (correlated or uncorrelated, perturbative or variational, DFT or wavefunction-based, etc.) and other chemical systems after mastering the illustrative examples of this book.

We thank Franklin Chen, Ken Fountain, John Harriman, J. R. Schmidt, Peter Tentscher, and Mark Wendt for comments and suggestions on earlier drafts, with special thanks to Mohamed Ayoub for reviewing Problems and Exercises throughout the book.

We wish all readers of this book success on the path to discovery of enriched chemical understanding from modern electronic structure calculations.

FRANK WEINHOLD AND CLARK R. LANDIS
Madison, May, 2011

Chapter 1

Getting Started

1.1 TALKING TO YOUR ELECTRONIC STRUCTURE SYSTEM

In order to begin natural bond orbital (NBO) analysis of a wavefunction, you first need to establish communication between a chosen electronic structure system (ESS) that calculates the wavefunction and the NBO program that performs the analysis. Many ESS programs in common usage have integrated NBO capability or a convenient interface with the most recent version of the NBO program [currently *NBO 5.9* (NBO5)]. We assume you have access to such a program.

In favorable cases, the ESS and NBO programs may already be integrated into a linked ESS/NBO module (such as G09/NBO of current *Gaussian 09*™ distributions). In this case, communication between the ESS and NBO programs only requires appending the $NBO keylist (see below) to the end of the usual ESS input file that performs the desired wavefunction calculation. [Instructions for creating the ESS input file and appending the $NBO keylist are generally included in the ESS program documentation; see, for example, J. B. Foresman and A. Frisch, *Exploring Chemistry with Electronic Structure Calculations: A Guide to Using Gaussian* (Gaussian Inc., Pittsburgh, PA, 1996) for the Gaussian program.] Such an integrated ESS/NBO program module allows the ESS and NBO programs to interactively cooperate on certain complex tasks that are unavailable in the unlinked stand-alone configurations described in the following paragraph. Optimally, the combined module will incorporate the latest NBO5 capabilities (ESS/NBO5), allowing the greatest possible range of analysis options; however, even older NBO versions (such as the older "NBO 3.1" incorporated in binary G09W Gaussian for Windows) can correctly perform most of the core NBO analysis options of Chapters 1–4. Ask your System Manager to upgrade the ESS to the latest NBO5-compatible form if a source-code version of the ESS is available. (Those fortunate readers with access to a full-featured ESS/NBO5 installation may skip to Section 1.2.)

Users of unlinked ESS hosts (including G09W users who wish to gain access to NBO5-level options) may use a stand-alone version of NBO5 (e.g., *GENNBO 5.0W* for PC-Windows users), but the process is a little trickier. In this case, the ESS program must first be instructed to produce the NBO "archive" file for the calculated

Discovering Chemistry With Natural Bond Orbitals, First Edition. Frank Weinhold and Clark R. Landis.
© 2012 John Wiley & Sons, Inc. Published 2012 by John Wiley & Sons, Inc.

——————————————————— *I/O-1.1* ———————————————————

```
$GENNBO  NATOMS=1  NBAS=8  UPPER  BODM  $END
$NBO  $END
$COORD
H atom test job
     1     1      0.000000      0.000000      0.000000
$END
  :
  :
```

wavefunction (see Sidebar 1.1 for Gaussian users). This file normally has the extension .47 following the chosen job filename (e.g., JOBNAME.47) and will be found to contain an empty $NBO keylist ("$NBO $END") as the second line of the file, as illustrated in the sample I/O-1.1 listing.

You can use any text editor to add desired keyword entries to the $NBO keylist, specifying the analysis options to be performed by the ensuing GENNBO5 processing. You can also insert a new keylist after the $NBO keylist, just as though you were appending the keylist to the end of the input file for an integrated ESS/NBO5 program.

The JOBNAME.47 archive file becomes the input file for your GENNBO5 job, which performs the actual NBO analysis. With the PC-Windows *GENNBO5.0W* version, you merely launch the program by mouseclick and select the JOBNAME.47 job from the displayed menu selections. Alternatively, if the GENNBO5 program has been set up as a binary executable (gennbo5.exe) on your system, you can launch the job by a command of the form

```
gennbo5 < JOBNAME.47 > JOBNAME.OUT
```

that pipes the analysis output to a chosen "JOBNAME.OUT" file. Details of interfacing the ESS with GENNBO5 may have been set up differently on your particular installation or website, but logically this is what is going on.

No matter whether you are working with a linked or stand-alone NBO configurations, the manner of controlling NBO analysis through the keyword entries of the $NBO keylist (the subject of this book) is the same for all setups. Although different ESS hosts boast somewhat different capabilities, the implementation of $NBO keylist commands is consistent across all ESS platforms. We shall ignore further ESS-specific details as far as possible.

SIDEBAR 1.1 *HOW GAUSSIAN USERS OBTAIN THE NBO ARCHIVE FILE FOR NBO5-LEVEL PROCESSING*

For Gaussian G09W (Windows binary) users wishing to bypass the limitations of the integrated NBO 3.1, the "trick" is to include the ARCHIVE keyword (and suitable FILE name) in the $NBO keylist that follows ordinary Gaussian input. As an example, for a simple H-atom calculation, the input file takes the form

```
#B3LYP/6-311++G** POP=NBOREAD

H atom test job

0 2
H

$NBO ARCHIVE FILE=H_atom $END
```

This produces the "H_atom.47" archive file that serves as input to GENNBO5, as described above.

Several points should be particularly noted:

(1) The Gaussian route card should include the "POP=NBOREAD" keyword to read and process the $NBO keylist (or the "POP=NBODEL" keyword to process a $DEL keylist). Follow the instructions of the Gaussian manual or Foresman–Frisch supplementary manual for further details of NBO-specific keyword options.

(2) Keyword input in both Gaussian and NBO is generally case-insensitive, except for literals such as the FILE specification.

(3) Certain keyword options that superficially appear to "work" in NBO 3.1 are obsolete or erroneous with respect to more recent NBO versions. This applies particularly to the PLOT keyword, where the files produced by NBO 3.1 are incompatible with the *NBOView* orbital viewer (Appendix B). Significant algorithmic differences between NBO3 and NBO5 are particularly apparent in details of natural population analysis for transition metals and rare-earth species. In addition, NBO5-level methodological improvements often result in significant numerical discrepancies between NBO3-level and NBO5-level output, particularly in cases of near-linear dependence (e.g., large basis sets including diffuse functions). NBO5 also includes numerous keyword options (e.g., NRT, STERIC, NEDA, NCS, NJC, and numerous checkpointing and matrix output options) with no counterpart in NBO3. Gaussian users are therefore advised to use the NBO3-level program only to generate the necessary ARCHIVE file for accessing higher NBO5-level analysis whenever possible.

1.2   HELPFUL TOOLS

The reader should be aware of three important resources that complement the present book and provide additional useful details on many topics:

(1) The *NBO 5.0 Program Manual* (which accompanies every authorized copy of the *NBO 5*-level program) is an essential resource for every serious NBO user. In addition to documentation of all program keywords, sample output, and background references, the manual contains (Section C, pp. C1–C72) extensive documentation of the Fortran source program itself, including brief descriptions of each SUBROUTINE and FUNCTION. For those so determined (presumably a small fraction of readers of this book!), it thereby becomes possible to locate the source code and program comments that connect back to the original description of the program algorithm in the research literature. Together with the documentation within the NBO source code itself, the NBO Manual should be relied upon as the ultimate authority on many points of details beyond the scope of the present book.

(2) The *NBO website* [www.chem.wisc.edu/~nbo5] contains a variety of important resources for both novice and accomplished NBO users, including tutorials, interactive "self-explaining" output samples for all major program options, FAQ (frequently asked questions), comprehensive literature references to recent NBO applications, and much else. The NBO website also contains program documentation for the *NBOView* orbital viewer program that is used extensively throughout this book (see Appendix B).

(3) The authors' companion research monograph *Valency and Bonding: A Natural Bond Orbital Donor–Acceptor Perspective* (Cambridge University Press, Cambridge, 2005) describes applications of NBO analysis to a broad variety of chemical problems spanning the periodic table. This monograph also provides extensive theoretical background (*V&B*, Chapter 1) on the physical and mathematical concepts that underlie NBO program options, allowing the interested student to trace calculated NBO descriptors back to fundamental quantum mechanical principles.

While the goal of this book is to facilitate the student's entry into the ranks of accomplished NBO users with minimal prerequisites or assumed background, we shall freely include cross-references to *NBO Manual* pages, *NBO website* URLs, or *V&B* content where appropriate.

1.3 GENERAL $NBO KEYLIST USAGE

The entryway to communication with your NBO program is the $NBO keylist, which allows you to include desired keywords between initial $NBO and final $END delimiters, namely,

```
$NBO (chosen keywords) $END
```

Other NBO keylists to be described below (such as the $GENNBO . . . $END and $COORD . . . $END keylists shown in I/O-1.1) are similarly opened by an identifying $KEY identifier and closed by a matching $END delimiter, so it is important that these delimiters be correctly located and spelled. A given keylist may extend over multiple lines, for example,

```
$NBO
(chosen keywords)
$END
```

but no two keylists (or portions thereof) may occur on the same line. (In some non-U.S. installations, the "$" identifier of keylist delimiters may be replaced by a more convenient keyboard character.)

The keywords appearing between $NBO...$END delimiters may generally occur in any order, and both keywords and keylist delimiters are case-insensitive (though we generally write them in upper case in this book). Keywords can be separated by a comma or any number of spaces. A keyword may also include a single parameter PARM in the form

```
KEYWORD=PARM
```

or a set of parameters PARM1, PARM2, ..., PARMn in "bracket-list" format

```
KEYWORD < PARM1/PARM2/.../PARMn>
```

Bracket-list syntax rules are summarized in Sidebar 1.2.

The $NBO keylist may contain any assortment of plain, parameterized, and bracket-listed keywords, such as

```
$NBO FILE=tryout
FNBO < 13,27/8,34>
STERIC=0.4 < 16,22/8,24/17,6 > PLOT NRT $END
```

Each input keyword will be echoed near the top of the NBO output file (as shown in I/O-1.2 for the above keylist), allowing you to check that the program "understands" your input commands.

The listing includes some extra keywords that were automatically activated as prerequisites for requested options. If a requested keyword fails to appear in this list, you may find it (perhaps misspelled?) in a list of "Unrecognized keywords" that appears before any other NBO output. The *NBO website* gives many other illustrations of $NBO keylist entry for main program keyword options (www.chem.wisc. edu/~nbo5/mainprogopts.htm).

In preparing an NBO input file, it is important to use an ordinary text editor (rather than *Word* or other word processor) in order to scrupulously avoid tabs or other control characters embedded in the plain-ASCII text file. Unseen control characters, corresponding to ASCII characters outside the printable range 32–126, cause unpredictable errors in processing the input file. Check also that *text-file format* is consistent between the platform on which the input file was prepared and that under which the

---

I/O-1.2

```
/NLMO    / : Form Natural Localized Molecular Orbitals
/NRT     / : Natural Resonance Theory Analysis
/FNBO    / : Print the NBO Fock matrix
             for user-requested matrix elements
/STERIC / : Print NBO/NLMO steric analysis
             with user-requested <NBO indices>
             Print threshold set to  0.40
/LFNPR   / : set to 21
/PLOT    / : Write information for the orbital plotter
/FILE    / : Set to tryout
```

NBO program will run; a particularly exasperating inconsistency is the different choice of CR/LF versus CR "end-line" markers in PC-Windows versus Macintosh or linux text files. When in doubt, use a file-transfer protocol (ftp) or file-conversion utility (dos2unix, etc.) to transfer text files from one platform to another.

SIDEBAR 1.2   *BRACKET-LIST SYNTAX*

Several NBO keywords can be modified by inclusion of parameters (PARM1, PARM2, ..., PARM*n*) of numerical or text content. In such cases, the parameters are enclosed in a "bracket-list" that is associated with the keyword through an input entry of the form

```
KEYWORD <PARM1/PARM2/.../PARMn>
```

The bracket-list "<", ">" terminators must be separated by at least one space from the preceding keyword, as well as from any following keyword. Bracket-lists may be broken up onto separate lines following any "/" separator,

```
KEYWORD <PARM1/
PARM2/
.../
PARMn>
```

The entries of the bracket-list vary considerably according to the keyword they modify. A common usage is to specify selected index pairs (i, j) of an array to be printed; for example, the command

```
FNBO <13 27/8 24>
```

specifies that only the $F_{13,27}$ and $F_{8,24}$ elements of the NBO Fock matrix ("FNBO" array) should be printed, rather than the entire array. A bracket-list may also follow a parameterized keyword (separated, as always, by at least one space at either end); for example, the command

```
STERIC=0.4 <16 22/ 8 24/ 17 6>
```

resets the STERIC output threshold to 0.4 kcal/mol and restricts printout of pairwise steric interactions to the NBO pairs (16, 22), (8, 24), and (17, 6). In case of text entries, each "/" separator should be set off by at least one blank (on each side) from text characters of the entry. Consult the *NBO Manual* for further details of allowed bracket-list options for each keyword.

1.4   PRODUCING ORBITAL IMAGERY

In many cases, the key to developing effective chemical intuition about NBOs is accurate *visualization* of their shapes and sizes. For this purpose, it is important to gain access to a suitable graphical utility for displaying images of NBOs and other orbitals. NBO graphical output can be exported to many popular orbital-viewing programs, such as *Gaussview, Jmol, Molden, Spartan, Molekel*, and *ChemCraft*, each offering distinctive features or limitations with respect to other programs. Sidebar 1.3 summarizes some details of how NBO "talks" to such programs and provides links to their further description.

The orbital images of this book are produced by the *NBOView 1.0* program, whose usage is briefly described in Appendix B. *NBOView* is specifically adapted to flexible display of the entire gamut of localized NBO-type (NAO, NHO, NBO, NLMO, and preorthogonal PNAO, PNHO, PNBO, and PNLMO "visualization orbitals") as well as conventional AO/MO-type orbitals in a variety of 1D (profile), 2D (contour), and 3D (view) display forms. The *NBOView Manual* link on the NBO website (http://www.chem.wisc.edu/~nbo5/v_manual.htm) provides full documentation and illustrative applications of *NBOView* usage.

SIDEBAR 1.3 *EXPORTING NBO OUTPUT TO ORBITAL VIEWERS*

Most orbital viewers are designed to import orbital data from the checkpoint file of the host ESS program or to directly read NBO "PLOT" (.31–.46) or "ARCHIVE" (.47) files. Communication with a chosen orbital viewer will therefore depend on details of its interface to the host ESS or NBO program.

For programs that read from a Gaussian or GAMESS checkpoint file, such as

Gaussview (http://www.gaussian.com/g_prod/gv5.htm)

Molden (http://www.cmbi.ru.nl/molden/)

Molekel (http://molekel.cscs.ch/wiki/pmwiki.php/Main/DownloadBinary)

Chemcraft (http://www.chemcraftprog.com/)

NBO5 users need only to specify the LCAO transformation matrix (AOBAS matrix) for the desired orbital basis set. This set is designated for checkpointing (storage in the checkpoint file) by a command of the form "AOBAS=C" in the $NBO keylist. For example, the NBO basis (AONBO transformation matrix) can be checkpointed by the $NBO keylist of the form

```
$NBO AONBO=C $END
```

and other orbital choices can be specified analogously. By default, checkpointed NBOs or other sets are numbered as in NBO output. However, numerous options are available to reorder checkpointed orbitals according to occupancy or other specified permutation (see *NBO Manual*, Section B-12). For users of linked G09/NBO5 or GMS/NBO5 programs, the NBO checkpointing options are flexible and convenient for graphical purposes.

[Note however that these options are unavailable in NBO3 and older versions. Users of linked G09/NBO3 binaries must therefore follow an alternative path by including the "POP=SAVENBO" command on the Gaussian route card (*not* in the $NBO keylist). The

POP=SAVENBO command has been included in recent Gaussian versions to provide a simple emulation of NBO checkpointing, principally for CAS/NBO and other nongraphical applications. Although SAVENBO enables basic displays of occupied NBOs, it cannot do so for PNBOs or other visualization orbitals that provide more informative graphical displays. The SAVENBO command is, therefore, a rather inflexible and error-prone form of checkpointing that serves as a last resort for G09/NBO3 users, but is "unrecognizable" and should *not* be considered in G09/NBO5 applications.]

For programs that read native NBO plot files, such as

Jmol (http://jmol.sourceforge.net)

NBOView (http://www.chem.wisc.edu/~nbo5)

NBO5 users need only to include the PLOT keyword (together with a FILE=NAME identifier) in the $NBO keylist, namely,

$NBO FILE=MYJOB PLOT $END

This writes out the necessary plotfiles (MYJOB.31, MYJOB.32, ..., MYJOB.46) for the orbital viewer to display any chosen orbital from the broad NAO/NBO/NLMO repertoire.

[G09/NBO3 binary users must again follow a more circuitous path. As described in Sidebar 1.1, one must first obtain the ARCHIVE (.47) file, then insert the "PLOT" keyword in the $NBO keylist of the .47 file, and finally process this file with *GENNBO 5.0W* to produce valid plot files. (Note that files produced by the PLOT command in antiquated *NBO 3.1* are no longer recognized by *NBOView*.)]

For the *Spartan* program (only), the NBO program provides a "SPARTAN" keyword option, namely,

$NBO SPARTAN $END

that writes out a Spartan-style archive file.

PROBLEMS AND EXERCISES

1.1. Use the resources of the *NBO website* (www.chem.wisc.edu/~nbo5) to find the following:

(a) References to three recent applications of NBO analysis in *J. Am. Chem. Soc., J. Chem. Phys., J. Org. Chem., Inorg. Chem.*, or any other chosen journal of specialized interest.

(b) References to the original papers on NBO analysis (or STERIC analysis, or NRT resonance theory analysis, or other chosen keyword options of NBO program).

(c) Names (and links) of ESS program systems that currently provide NBO interfaces or internal linkages.

(d) Reference to a general review article describing NBO methods or applications.

(e) One or more frequently asked questions or problems that sometimes bedevil new NBO users, for which you found a helpful answer.

(f) The date of the latest posted code correction for bugs in the NBO program.

1.2. Use the Tutorials section of the *NBO website* to discover the following:

(a) What is the "natural transition state" between reactant and product species of a chemical reaction? Why is this concept applicable even in barrierless reactions, for example, of ion–molecule type?

(b) Dihaloalkenes (e.g., dichloroethylene, a common cleaning fluid) exhibit a strange preference for the *cis*-isomer, despite the obvious steric and electrostatic advantages of the *trans*-isomer which keeps the "bulky" and "polar" halide ligands further separated. What is the primary electronic effect that stabilizes the *cis*-isomer compared to the *trans*-isomer of difluoroethylene (or related dihaloalkenes)?

(c) What is the best Lewis structure formulation for phosphine oxide (H_3PO), and how would it be compared with other representations commonly found in journals or textbooks?

1.3. Prepare sample input $NBO keylists to discover (with help from Appendix C, if needed) the following:

(a) The orbital interaction integral

$$\int \varphi_i^{(NBO)} * F_{op} \varphi_j^{(NBO)} \, d\tau$$

[off-diagonal $(\mathbf{F}^{(NBO)})_{ij}$ matrix element of the NBO-based Fock matrix that represents the effective 1-electron Hamiltonian operator F_{op} of the system] between NBOs 14 and 27.

(b) The orbital energy integral

$$\int \varphi_i^{(AO)} * F_{op} \varphi_i^{(AO)} \, d\tau$$

[diagonal $(\mathbf{F}^{(AO)})_{ii}$ matrix element of the AO-based Fock matrix] for basis AO 16; and similarly the orbital energies of NAO 27, NBO 18, NLMO 23, and MOs 8, 9, and 10.

(c) The overlap integrals

$$\int \varphi_i^{(AO)} * \varphi_j^{(AO)} \, d\tau$$

[off-diagonal matrix elements of the $\mathbf{S}^{(AO)}$ overlap matrix] between basis AOs (3, 4), (3, 5), and (4, 5).

1.4. Using your favorite orbital viewer package, prepare one or more orbital images of a chosen NBO for a chosen system (such as the H-atom example of Sidebar 1.1). Explain in words what each image portrays and how different images (e.g., from different packages or different viewing options in the same package) are related, including advantages and disadvantages of each form.

Chapter 2

Electrons in Atoms

2.1 FINDING THE ELECTRONS IN ATOMIC WAVEFUNCTIONS

From a quantum mechanical perspective, electrons are described by the orbitals they occupy. Each orbital "electron container" is a three-dimensional (3D) spatial function having a positive or negative numerical value (orbital amplitude) at every point in space. Around an atomic nucleus, such electron containers are called atomic orbitals (AOs), with characteristically large amplitudes (including large amplitude swings between positive and negative values) near the nucleus, but rapidly decaying values at large distances from the nucleus. The analytical forms of such atomic orbitals are exactly known only for the hydrogen atom, but good numerical approximations are now available for all atoms of the periodic table.

In the present chapter, we examine the basic building blocks of atomic and molecular wavefunctions, the atomic spin-orbitals of individual electrons (Section 2.2), and the "configurations" of occupied spin-orbitals that characterize the chosen electronic state (Section 2.3). This leads to introduction of intrinsic "natural" orbitals that optimally describe the final wavefunction, and are often found to differ surprisingly from the assumed "basis atomic orbitals" that were used to construct the numerical wavefunction (Sidebar 2.1). We then describe how these intrinsic building blocks are "found" in natural bond orbital (NBO) output, taking advantage of the simplicity of the atomic limit to introduce general NBO terminology, output conventions, and orbital display modes that are employed throughout this book. Readers familiar with basic NBO program usage and output may prefer to skip forward to chapters dealing with systems and properties of greater chemical interest.

SIDEBAR 2.1 *WHAT ARE "NATURAL" ORBITALS?*

An "orbital" refers to a one-electron wavefunction, and more specifically to the spatial part of a one-electron "spin-orbital." Electronic orbitals are often associated with the simple Hartree–Fock (HF) approximation, a single-configuration approximation to the complex

---

Discovering Chemistry With Natural Bond Orbitals, First Edition. Frank Weinhold and Clark R. Landis.
© 2012 John Wiley & Sons, Inc. Published 2012 by John Wiley & Sons, Inc.

many-electron wavefunction Ψ, but the usefulness of the orbital concept goes beyond HF level. In HF theory, each electron is assigned to occupy a unique spin-orbital and the total wavefunction Ψ_{HF} is specified by the associated "electron configuration," a listing of its occupied spin-orbitals. For a closed-shell system with α and β spin-orbitals of identical spatial form, we usually focus on the spatial $(\mathbf{r})$ dependence of each doubly occupied orbital in the configuration.

Mathematically, the single-configuration Ψ_{HF} wavefunction is expressed as a "Slater determinant" (antisymmetrized product) of the occupied spin-orbitals. In this limit, only the chosen N occupied spin-orbitals contribute to description of the N-electron system, whereas an infinite number of remaining "virtual" spin-orbitals are ignored. This crude HF-type (or molecular orbital) description of the true many-electron $\Psi(\mathbf{r}_1, \mathbf{r}_2, \ldots, \mathbf{r}_N)$ exerts a powerful hold on chemical pedagogy, but is often seriously defective in quantitative terms.

When the errors of the single-configuration HF-type description become nonnegligible, the orbital concept seems to become problematic. More accurate "correlated" many-electron wavefunctions can still be expressed in terms of orbitals and Slater determinants, but *unlimited* numbers of determinants, each with a distinct set of N occupied spin-orbitals, are now required for precise description of Ψ. Moreover, as the list of Slater determinants increases without limit, the starting choice of orbitals becomes increasingly *un*important. Indeed, in the limit of including all possible Slater determinants (i.e., all possible ways of choosing N spin-orbitals from a complete orthonormal set), the starting choice of orbitals becomes totally immaterial, and *any* complete orthonormal set of orbitals could serve equally well to describe Ψ. Thus, one might be led to the extreme conclusion that orbitals play *no* useful conceptual role except in the uncorrelated single-configuration HF limit. In this extreme view, the familiar atomic and molecular orbitals (MOs) of freshman chemistry seem to have lost significance, and the orbital concept itself is called into question.

Fortunately, the rigorous measurement theory of many-electron quantum mechanics justifies essential *retention* of orbital-type conceptions and their applications in bonding theory. As originally formulated by J. von Neumann in his *Mathematical Foundations of Quantum Mechanics* (Princeton University Press, Princeton, NJ, 1955), the fundamental object underlying quantal measurement of a pure-state N-electron system is the "density matrix" $\Gamma^{(N)}$:

$$\Gamma^{(N)} = \Psi(\mathbf{r}_1, \mathbf{r}_2, \ldots, \mathbf{r}_N)\Psi^*(\mathbf{r}_1', \mathbf{r}_2', \ldots, \mathbf{r}_N') \qquad (2.1)$$

K. Husimi (*Proc. Phys. Math. Soc. Jpn.* **22**, 264, 1940) subsequently showed that analogous measurable properties of smaller *sub*systems of the N-electron system are expressed most rigorously in terms of corresponding pth-order "reduced" density matrices $\Gamma^{(p)}$,

$$\Gamma^{(p)} = [N!/p!(N-p)!] \int \Gamma^{(N)} d\tau_N d\tau_{N-1} \cdots d\tau_{N-p+1} \qquad (2.2)$$

in which the dependence on all but p subsystem electrons has been "averaged out" by spatial integration over electrons $p + 1, p + 2, \ldots, N$ (after equating primed and unprimed coordinates in the integrand). This reduction permits a spectacular simplification for atomic or molecular systems, because the Hamiltonian operator depends only on one-electron (kinetic energy and nuclear attraction) and two-electron (interelectron repulsion) interactions. As a result, only the *first*- and *second*-order reduced density matrices $\Gamma^{(1)}$, $\Gamma^{(2)}$ are required to evaluate *any* measurable property of a pure-state atomic or molecular species. In effect, $\Gamma^{(1)}$ and $\Gamma^{(2)}$ condense *all* the information about Ψ that is relevant to chemical questions!

Of these two objects, $\Gamma^{(1)}$ (usually referred to as "the" density matrix) is far the more important. Indeed, $\Gamma^{(1)}$ would be completely adequate for chemical questions if electron correlation effects were completely negligible rather than a $\sim 1\%$ correction to total energy. Thus, about 99% of a chemist's attention should focus on the information contained in the first-order reduced density matrix $\Gamma^{(1)}$, as is done throughout this book.

Because $\Gamma^{(1)}$ is inherently a *one*-electron operator, it is deeply connected to *orbital*-level description of the *N*-electron system. Indeed, it was recognized by P.-O. Löwdin (*Phys. Rev.* **97**, 1474, 1955) that the solutions $\{\theta_i\}$ of the characteristic "eigenvalue equation" for $\Gamma^{(1)}$

$$\Gamma^{(1)}\theta_i = n_i\theta_i, \quad i = 1, 2, \ldots, \infty \tag{2.3}$$

provide the fundamental "natural" orbitals (intrinsic eigenorbitals of $\Gamma^{(1)}$) that underlie description of an *N*-electron system of arbitrary complexity. Each natural orbital (NO) θ_i has "occupancy" n_i

$$n_i = \int \theta_i^* \Gamma^{(1)} \theta_i d\tau = \langle \theta_i | \Gamma^{(1)} | \theta_i \rangle \tag{2.4}$$

that is *maximum possible* for ordered members of a complete orthonormal set as follows from general minmax properties of eigenvalue equations such as (2.3). Accordingly, natural orbitals $\{\theta_i\}$ are intrinsically optimal for providing the most compact and rapidly convergent description of *all* one-electron properties of the *exact* many-electron Ψ. In effect, "natural orbitals" can be defined as the orbitals selected by the wavefunction itself (through its reduced $\Gamma^{(1)}$) as optimal for its own description.

Unlike HF molecular orbitals, the natural orbitals are not restricted to a low-level approximation, but are rigorously defined for *any* theoretical level, up to and including the exact Ψ. As eigenfunctions of a physical (Hermitian) operator, the NOs automatically form a *complete* orthonormal set, able to describe every nuance of the exact Ψ and associated density distribution, whereas the occupied MOs are seriously *in*complete without augmentation by virtual MOs. Furthermore, the occupancies n_i of NOs are not restricted to integer values (as are those of MOs), but can vary *continuously* within the limits imposed by the Pauli exclusion principle, namely, for closed-shell spatial orbitals,

$$0 \le n_i \le 2 \text{ (sum of } \alpha \text{ and } \beta \text{ occupancies)} \tag{2.5}$$

Nevertheless, the NOs are optimally chosen to give greatest possible condensation of electron density into the lowest few orbitals (most "HF-like" description of the exact Ψ in the maximum density sense), and they reduce back to conventional Hartree–Fock MOs in the uncorrelated limit $\Psi \to \Psi_{HF}$. Hence, familiar MO-type concepts are recovered intact when electron correlation effects are negligible, but the intrinsic NOs allow us to extend and generalize these orbital concepts for wavefunctions of *any* theoretical level.

As mentioned above, the maximum-occupancy property of NOs is a necessary and sufficient condition for satisfying the eigenvalue equation (2.3). We can therefore use the maximum-occupancy criterion to search for localized (i.e., 1- or 2-center) regions that contain high-occupancy "*local* NOs," consistent with the Pauli restriction (2.5). From elementary bonding principles, we can expect that such high-occupancy 1-center (lone pair) or 2-center (bond) local NOs are primarily located in the regions associated with electron pairs in the Lewis dot diagram. Accordingly, it turns out (J. P. Foster and F. Weinhold, *J. Am. Chem. Soc.* **102**, 7211, 1980) that density matrix and natural orbital concepts can be generalized to identify the optimal *local* bonding patterns envisioned in the chemist's Lewis structure diagram, leading to the "natural bond orbitals" that optimally correspond to such localized description of electron density. Because the forms of 1c/2c NBOs are more restrictive than those of delocalized multicenter NOs, the NBO occupancies are typically less than those of delocalized NOs, even in the single-configuration MO limit. However, as the success of Lewis structural concepts leads us to expect, the NBOs are commonly exhibit *near* double-occupancy, with slight deviations that reflect the subtle resonance effects of the molecular bonding pattern. The NBO occupancy variations also guarantee the *uniqueness* of NBO forms, whereas the exact double-occupancy of MOs (or NOs) leads to a type of unitary indeterminacy that prevents their unique definition (cf. Sidebar 11.5).

For further details, see the "What Are 'Natural' Orbitals" link on the NBO website or Chapter 1 of *Valency and Bonding* and references therein.

2.2 ATOMIC ORBITALS AND THEIR GRAPHICAL REPRESENTATION

Getting acquainted with the shapes and sizes of atomic orbitals is one of the first important skills for a chemistry student to master. Figure 2.1a–c depicts the occupied 1s atomic orbital of the ground-state hydrogen (H) atom in three different graphical representations (all obtained from PLOT output using the NBOView orbital plotter, as described in Appendix B).

Figure 2.1a shows the one-dimensional (1D) orbital amplitude profile $\varphi(r)$ as a function of distance r from the nucleus (positioned at the cross-hair symbol at the midpoint of the profile axis). As shown, the orbital amplitude "peaks" at the nucleus, but decays steeply as r increases in either direction along the chosen one-dimensional profile axis. Note that no particular "electron radius" about the nucleus is evident in such a plot. Indeed, we must begin thinking in completely nonclassical fashion when attempting to envision electrons in orbital terms, because the proper quantum mechanical description bears only remote connection to the classical-type

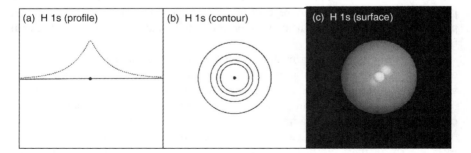

Figure 2.1   Hydrogen atom 1s orbital in (a) 1D profile, (b) 2D contour, and (c) 3D surface plot. (See the color version of this figure in Color Plates section.)

"Bohr orbit" model that students are first told about in beginning chemistry classes. (Such oversimplified models create a superficial sense of "understanding" the quantum mechanical behavior that is aptly characterized as "weirder than you think, and weirder than you *can* think!")

Figure 2.1b shows a corresponding 2D contour plot of orbital amplitude in a plane around the nucleus, centered at the middle of the contour plane. This plot shows the circular contour lines of equal "elevation" (amplitude), widely spaced over the gentle outer slopes of the orbital, but tightly bunched around the nucleus (where only the first few contours are shown approaching the cross-hair nucleus symbol). The circular contour lines clearly exhibit the spherical symmetry of the 1s hydrogenic orbital, which is not so apparent in Fig. 2.1a.

Finally, Fig. 2.1c shows the same hydrogenic 1s orbital as a 3D surface plot, which resembles a photograph of a space-filling object. The "surface" of this orbital object corresponds to the outermost contour line of Fig. 2.1b, chosen as 0.0136 a.u. to roughly match the empirical van der Waals radius of the atom. Of course, the orbital $\varphi(\mathbf{r})$ exhibits no sharp discontinuity or "surface" at any distance (as shown in Fig. 2.1a). But with a consistent cutoff amplitude, one can gain an informative visual impression of both shape and size of the orbital, i.e., the spatial region in which its amplitude contours are most highly concentrated. When the orbital is occupied, its square $|\varphi(\mathbf{r})|^2$ gives the contribution to *electron density* at point $\mathbf{r}$, which allows us to "find the electron" as nearly as that phrase makes sense in the quantum world.

With this background, we can also consider atomic orbitals of more varied shapes and sizes. Figure 2.2 shows the corresponding profile, contour, and surface plots of fluorine (F) atom 2s, 2p, and 3d orbitals. As seen in Fig. 2.2, all three orbitals now have regions of both positive and negative sign (phase), separated by *nodes* (surfaces of zero amplitude). The contour plots (middle) show contours of positive or negative phase as solid lines or dashed lines, respectively, while the corresponding phases in the surface plots (right) are shown as blue or yellow, respectively. The contour plots clearly show both the angular shape (e.g., the "dumbbell" shape of the 2p orbital or "four-leaf clover" shape of the 3d orbital) and the radial "strength" of each orbital. Because the orbital phase patterns play an important role in understanding chemical behavior, the 2D contour or 3D surface plots usually provide the more useful orbital visualization, but the 1D radial profiles may also be useful in showing orbital details that are relevant to chemical behavior.

The qualitative orbital forms shown in Fig. 2.2 are *modulated* by subtle variations of overall electronic configuration and charge state. These variations are shown in greater detail for the $2p_z$ orbital in Fig. 2.3. Slight variations in $2p_z$ spin-orbital size (diffuseness) are seen to distinguish the occupied $2p_z^\uparrow$ (a) from the vacant $2p_z^\downarrow$ spin-orbital (b) of neutral fluorine atom, or from the corresponding doubly occupied orbital of the F^- anion (c). In the 1D amplitude profiles, the subtle differences can be seen most clearly in the slightly lower "peak" heights and correspondingly expanded "wing" spans of anionic fluoride (c) compared to that of neutral fluorine (a and b). Such anionic orbital expansion might be expected from the weakened attractive forces

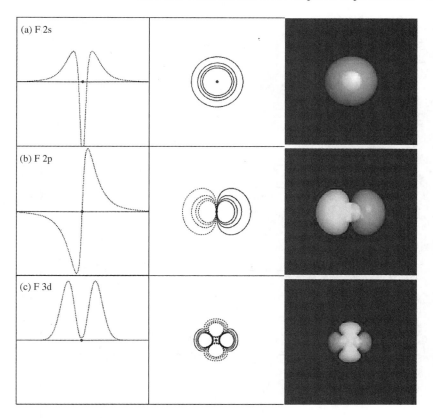

Figure 2.2   Fluorine atom (a) 2s, (b) 2p, and (c) 3d orbitals in 1D profile (left), 2D contour (middle), and 3D surface plot (right). The depicted orbitals have respective occupancies of 2, 1, and 0 in the F atom ground state. (Note that the four outermost contour lines of default *NBOView* contour output do not include the negative 2s "inner spike" near the nucleus, which is better seen in the 1D profile plot.) (See the color version of this figure in Color Plates section.)

when the nuclear charge is further "screened" by the added electron. (Corresponding orbital contraction is found when electrons are removed from other orbitals to form cations.) Such "breathing" size changes that accompany gain or loss of electrons are among the most important physical effects to be captured in accurate orbital visualizations.

A similar, but weaker, form of orbital-breathing variation can be found even in neutral atoms. The spin-orbital plotted in Fig. 2.3a is the *singly* occupied $2p_z$ orbital ($2p_z{}^{\uparrow}$, of "up spin") of the atomic fluorine radical, whereas that in (b) is the corresponding $2p_z{}^{\downarrow}$ β ("down") spin-orbital that is vacant in this formal configurational assignment and slightly less tightly attracted to the nucleus. Comparison of the singly occupied $2p_z{}^{\uparrow}$ with the doubly occupied $2p_{x,y}{}^{\uparrow}$ or $2p_{x,y}{}^{\downarrow}$ spin-orbitals would reveal still more subtle size variations, corresponding to the differing

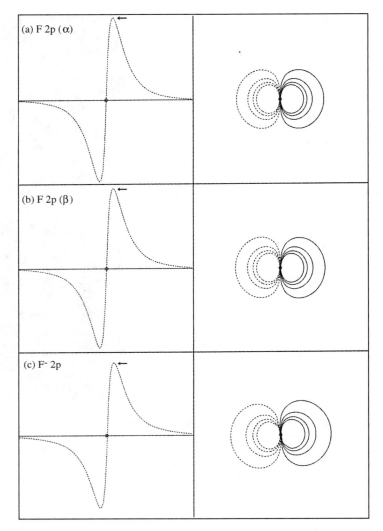

Figure 2.3   Fluorine valence $2p_z$ natural spin-orbital in profile (left) and contour (right) plots for F atom: (a) α spin, (b) β spin, and (c) F^- anion, showing subtle variations of diffuseness with differences in electron configuration and overall charge. (The small arrows in the left panels call attention to variations in maximum lobe amplitude, which are among the most "obvious" of the virtually imperceptible graphical differences.)

electronic environments in doubly occupied versus singly occupied orbital regions. Thus, we should anticipate that accurate representations of atomic orbitals (in contrast to cartoon-like textbook representations) should depict the subtle size variations resulting from altered Coulomb and exchange forces in electronic states of differing charge or spin multiplicity.

Table 2.1 Orbital energies (a.u.) for F and F^-, showing differences between singly and doubly occupied orbitals and between α and β spin-orbitals in the open-shell neutral species.

Orbital	Orbital energy (a.u.)	
	$\varepsilon(\alpha)$	$\varepsilon(\beta)$
F atom		
$2p_{x,y}$	−0.48378	−0.45163
$2p_z$	−0.57138	−0.27528
F^- anion		
$2p_{x,y,z}$	+0.01314	+0.01314

The subtle variations in fluorine $2p_z$ atomic orbitals are more conspicuously exhibited in the numerical values of atomic orbital energies. Table 2.1 shows the calculated orbital energies (in atomic units; Appendix E) for 2p-type spin-orbitals in open-shell F and closed-shell F^-. From the energy conversion factor to common thermochemical units (1 a.u. = 627.51 kcal/mol; Appendix E), one can see that the fluorine atom $2p_z{}^\uparrow$ spin-orbital is about 55 kcal/mol lower in energy than the degenerate $2p_x{}^\uparrow$, $2p_y{}^\uparrow$ levels. Furthermore, each of the "doubly occupied" 2p orbitals consist of split spin-orbital levels, with that of α (majority) spin lying about 20 kcal/mol below the β level, and the orbital energies of the neutral atom are in all cases significantly different from those of the anion. Such energy differences indeed reflect chemically significant variations that a chemically useful orbital description must correctly represent.

Figures 2.1–2.3 and Table 2.1 emphasize the important *differences* between spin-orbitals that are considered "equivalent" in elementary treatments. This is particularly true for open-shell species, where the notion of "pairing" electrons of opposite spin in the *same* spatial orbital is generally unrealistic. Instead, one should visualize open-shell electronic distributions in terms of *different orbitals for different spins* (DODS), recognizing that distinct Coulomb and exchange forces will generally *split* "paired" electrons into spatially distinct spin-orbitals (see Sidebar 2.2). The DODS concept is automatically incorporated into open-shell NBO analysis, where analysis of α and β spin sets proceeds independently in separate output sections, with no presumed relationship between natural orbitals of the two spin sets. Only in rather exceptional cases (e.g., uncorrelated closed-shell singlet species in near-equilibrium geometry) will electrons be found to "pair up" in the restrictive manner envisioned in elementary textbooks. The DODS concept is generally a more satisfactory conceptual foundation on which to build an accurate and robust picture of closed- and open-shell electronic phenomena.

As discussed in greater detail in the following section, the conceptual confusion over DODS-type splitting and orbital breathing effects often stems from mistaken attribution of physical significance to the numerical *basis functions* that are employed

SIDEBAR 2.2   *DIFFERENT ORBITALS FOR DIFFERENT SPINS CONCEPT FOR OPEN-SHELL SPECIES*

The concept that electrons of different spin must generally be associated with *spin*-orbitals of distinct spatial form (different orbitals for different spins) is fundamental to accurate description of open-shell species. Even if the original wavefunction was formulated under the restriction of identical spatial forms for "paired" electrons of α and β spin as in the restricted open-shell Hartree–Fock (ROHF) approximation and its variants, the optimal natural orbitals for describing the final wavefunction are generally found to be of DODS form, with presumed "pairs" split into spin-orbitals of spatially and energetically *distinct* form. In highly correlated wavefunctions, such DODS-type splittings may even become important in closed-shell singlet systems. The DODS-type (unrestricted) description is the starting point for "unrestricted Hartree–Fock" (UHF) wavefunctions, which are generally superior to ROHF-type wavefunctions constructed from the same basis functions.

NBO analysis is strongly oriented toward such UHF-type DODS description, with completely independent analyses (and separate output sections) for α and β spin sets. In nonrelativistic quantum theory, the potential energy contains no explicit spin dependence, so electrons of opposite spin "live" in opposite worlds, coupled to opposite spin electrons only by Coulomb interactions (which actually tend to disfavor electrons "pairing" into the same spatial region), but strongly coupled to same spin electrons by Pauli-type exchange forces. The beginning student is, therefore, encouraged to think of α and β spin sets as having *different* orbital forms, spatial distributions, and bonding propensities, in accordance with the general DODS viewpoint that is stressed throughout this book.

in constructing the wavefunction. Indeed, popular basis functions are commonly referred to as "atomic orbitals", but this designation is quite inappropriate and misleading. For example, the basis AOs would typically be chosen *identically* for the F and F$^-$ species of Table 2.1, whereas the actual natural orbitals of these species are found to differ significantly. We shall continue to focus on the intrinsic natural orbitals that are finally found to be optimal for describing the atomic wavefunction rather than for preselected "basis AOs" (primarily chosen for numerical convenience) that were initially employed in its construction.

2.3   ATOMIC ELECTRON CONFIGURATIONS

The concept of an atomic electron configuration refers to the assignment of each electron to a specific spin-orbital, consistent with the restrictions of the Pauli exclusion principle (no more than one electron per spin-orbital). In actuality, a more accurate quantum mechanical description usually involves weighted contributions from multiple configurations, so that the overall occupancy of each atomic spin-orbital becomes a fractional number (still not exceeding unity). However, for most atoms in their low-lying states, the dominance of one particular configuration is so strong that we can reasonably describe the state in terms of a single electron configuration, assigning to each available spin-orbital an occupancy of one (occupied) or zero (vacant) (cf. Sidebar 2.3). Such single-configuration assignments underlie the

SIDEBAR 2.3 *INTRINSIC (NATURAL) ORBITALS OF A MULTICONFIGURATIONAL STATE*

As mentioned in the main text, the precise details of the atomic orbitals depend on the chosen electron configuration. Although atomic orbitals of a secondary configuration may closely resemble those of a primary configuration, they are not identical, nor are *either* set of orbitals optimal for describing the true multiconfigurational state. However, following the concept first introduced by P.-O. Löwdin (see *V&B*, Section 1.5), one can obtain a unique set of intrinsic "natural" orbitals that give the most compact and efficient "single configuration"-like description of overall electron density, with *fractional* occupancies replacing the integers of idealized single configuration description. Such natural orbitals intrinsically incorporate the multiconfigurational averaging effects, and they become identical to simple "Hartree–Fock" orbitals (*V&B*, Section 1.3) in the single configuration limit.

Many types of "atomic orbitals" might be considered as candidates for building chemical valency and bonding concepts. (Even the numerical "basis atomic orbitals" that underlie ESS calculations have been employed by some authors for this purpose; cf. Sidebar 2.4) However, in the present book, we are always implicitly envisioning the "natural" choice of these orbitals, because that is the set of orbitals that the wavefunction itself selects as optimal for its own description. Experience shows that even wavefunctions of widely varying mathematical form (if sufficiently accurate) tend to yield remarkably similar natural orbitals. These orbitals, therefore, provide a convenient *lingua franca* for expressing and comparing the content of the many possible forms of wavefunction in current usage. For further details, visit the *NBO website* (www.chem.wisc.edu/~nbo5/web_nbo.htm).

Aufbau principle of periodic table structure and associated elementary theories of chemical valency and bonding. Accordingly, to "find the electrons," we need to determine the *occupancy* of each atomic orbital, as well as its size and shape.

As an example, the expected electron configuration of a fluorine atom ground state can be expressed as

$$(1s)^2(2s)^2(2p_x)^2(2p_y)^2(2p_z)^1 \tag{2.6}$$

which is shorthand for the more complete spin-orbital description:

$$[(1s\uparrow)^1(2s\uparrow)^1(2p_x\uparrow)^1(2p_y\uparrow)^1(2p_z\uparrow)^1][(1s\downarrow)^1(2s\downarrow)^1(2p_x\downarrow)^1(2p_y\downarrow)^1] \tag{2.7}$$

$$\underset{\alpha \text{ spin}}{\qquad\qquad} \underset{\beta \text{ spin}}{\qquad\qquad}$$

(Each list might be considered to include *all* possible spin-orbitals of a complete orthonormal set, with most having zero occupancy.)

Among the occupied fluorine orbitals of ground configuration (2.6), (2.7), we may first consider some details of the s-type orbitals of "core" 1s and "valence" 2s type, which lie below the vacant "Rydberg" (extravalence) 3s, 4s, ... orbitals of higher principal quantum number. As seen in Table 2.1, the core 1s electrons lie much deeper in energy than 2s and other valence shell electrons, and are often considered "inert" for chemical purposes. However, the occupied core orbitals exert significant influence on higher lying valence orbitals of the same symmetry through the strong exchange-type repulsions associated with the Pauli exclusion principle.

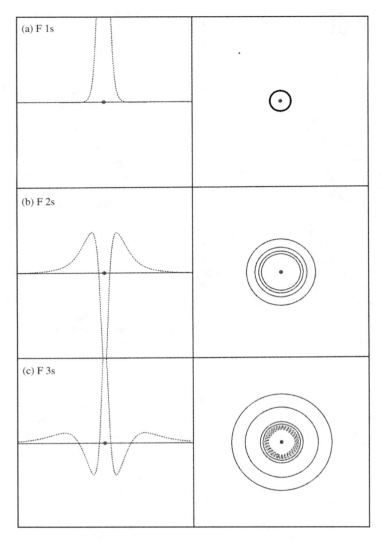

Figure 2.4   Fluorine (a) core 1s, (b) valence 2s, and (c) Rydberg 3s orbitals in profile (left) and contour (right) plots, showing radial oscillations and nodal patterns that preserve orthogonality to lower orbitals of the same symmetry.

Figure 2.4 compares low-lying ns-type orbitals ($n = 1-3$) of the neutral F atom, showing that all have similar spherical surface plots but quite different radial dependence, most notable in the increasing number of radial oscillations and nodes as principal quantum number n increases. Among the vacant orbitals in configuration (2.6), (2.7), we would also find orbitals of higher angular momentum (3d, 4f, . . .) that exhibit increasing numbers of angular oscillations and nodal planes. The higher number of radial and angular oscillations can generally be associated with higher kinetic energy, which is avoided as long as possible in the *Aufbau* sequence that leads to the ground-state configuration. Such ripple-type oscillations are required by deep principles of quantum mechanics to ensure that orbitals remain properly *orthogonal* ("perpendicular" in wave-like sense) to one another. The details of how these orbital ripple patterns coadjust are of great importance in many chemical phenomena, as will be described in examples throughout this book.

As mentioned in Section 2.2, one must be careful not to confuse "basis AOs" (as employed by the host ESS in numerical calculation of the wavefunction) with the intrinsic (natural) orbitals that underlie an optimal configurational description of the physical atomic system. Sidebar 2.4 illustrates this important distinction for the *n*s-type orbitals of Fig. 2.4. The student may safely ignore the usual pages of computer output devoted to "Mulliken population analysis" and similar descriptors of basis AOs compared to analyses of natural orbitals.

SIDEBAR 2.4   *NUMERICAL BASIS AOs VERSUS PHYSICAL ATOMIC ORBITALS*

As remarked in the main text, "basis AOs" employed by a host ESS seldom have realistic resemblance to the physical orbitals of an atomic system. We may illustrate the differences by considering the fluorine 1s, 2s, and 3s atomic orbitals previously exhibited in Fig. 2.4. For the standard basis set (6-311++G**) of contracted Gaussian-type functions that were employed to construct the atomic wavefunction, we consider the corresponding lowest three s-type basis AOs (often labeled "1s," "2s," and "3s" in ESS program output).

The qualitative differences between basis AOs and physical orbitals of the atom are exhibited in the gross disparities between corresponding orbital energies diagonal expectation values of the effective one-electron Hamiltonian (Fock or Kohn–Sham) operator, as tabulated below:

AO	Orbital energy (a.u.)	
	Natural	Basis
1s	−24.772	−21.219
2s	−1.258	−20.035
3s	0.658	−4.647

As seen in the table, the "basis AO" energetics are erroneous by multiple atomic units (*thousands* of kcal/mol) with respect to actual valence shell energy levels.

We can also see the disparities in the graphical forms of the basis AOs compared to realistic natural orbitals. The leading three s-type basis AOs are displayed in Fig. 2.5 in profile and contour plots that can be directly compared with the physical 1s, 2s, and 3s orbitals of Fig. 2.4.

Except for the 1s orbital, the basis AOs are seen to have practically *no* resemblance to physical s-type orbitals. Most conspicuous is the absence of any internal nodal structure near the nucleus, corresponding to unphysical "overlap" (nonorthogonality) between core and valence functions. Within the ESS computer program, proper core–valence orthogonality is maintained so that the final wavefunction satisfies the Pauli principle; however, a naive user who simply takes printed "atomic orbital" labels at face value will have little warning of the associated conceptual errors.

Although the present numerical results refer specifically to 6-311++G** basis AOs, similar defects would be seen in other popular basis types, including those (e.g., STO-3G) based on Slater-type orbitals.

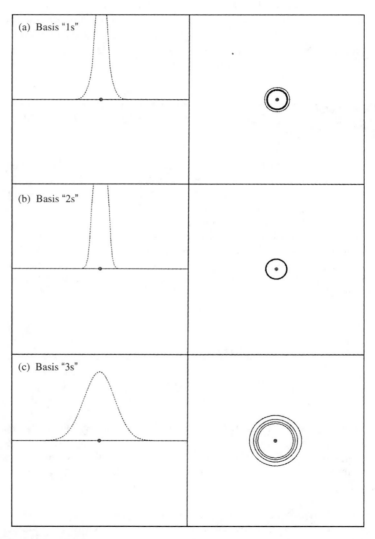

Figure 2.5   Fluorine "basis AOs" (6-311++G** basis set) for the lowest three s-type functions, showing the unphysical (nodeless) character near the nucleus (cf. Fig. 2.4).

2.4   HOW TO FIND ELECTRONIC ORBITALS AND CONFIGURATIONS IN NBO OUTPUT

Let us assume you have successfully obtained a wavefunction and NBO analysis for a fluorine atom, for example, with a Gaussian/NBO5 input file of the form

```
%mem=10000000
#B3LYP/6-311++G** POP=NBOREAD

F atom

0 2
F

$NBO FILE=fluorine $END
```

You should first look in the output file for the starting NBO banner as shown in I/O-2.1.

```
 ─────────────────────── I/O-2.1 ───────────────────────
********************************* NBO 5.G *********************************
          N A T U R A L   A T O M I C   O R B I T A L   A N D
          N A T U R A L   B O N D   O R B I T A L   A N A L Y S I S
**************************************************************************
   (c) Copyright 1996-2008 Board of Regents of the University of Wisconsin System
       on behalf of the Theoretical Chemistry Institute.  All Rights Reserved.


       Cite this program as:


       NBO 5.G.  E. D. Glendening, J. K. Badenhoop, A. E. Reed,
       J. E. Carpenter, J. A. Bohmann, C. M. Morales, and F. Weinhold
       (Theoretical Chemistry Institute, University of Wisconsin,
       Madison, WI, 2001); http://www.chem.wisc.edu/~nbo5


       /FILE   / : Set to fluorine

 Analyzing the SCF density

 Job title: F atom

 Storage needed:      2628 in NPA,      2642 in NBO (   10000000 available)
```

Immediately below is the "NATURAL POPULATIONS" table of Natural Atomic Orbital (NAO) labels and occupancies (I/O-2.2).

Looking under the "Type(AO)" and "Occupancy" columns, one can see that NAO 1 is a core 1s of occupancy 2.0000 (i.e., "doubly occupied"). Similarly, NAOs 2, 6, and 10 are the doubly occupied valence 2s, $2p_x$, and $2p_y$ orbitals, and NAO 14 is the singly occupied $2p_z$ orbital. (One could readily confirm these identifications from PLOT files

---

I/O-2.2

NATURAL POPULATIONS: Natural atomic orbital occupancies

NAO	Atom	No	lang	Type(AO)	Occupancy	Spin
1	F	1	s	Cor(1s)	2.00000	0.00000
2	F	1	s	Val(2s)	1.99961	-0.00017
3	F	1	s	Ryd(3s)	0.00014	0.00000
4	F	1	s	Ryd(4s)	0.00000	0.00000
5	F	1	s	Ryd(5s)	0.00000	0.00000
6	F	1	px	Val(2p)	1.99970	0.00030
7	F	1	px	Ryd(3p)	0.00030	-0.00030
8	F	1	px	Ryd(4p)	0.00000	0.00000
9	F	1	px	Ryd(5p)	0.00000	0.00000
10	F	1	py	Val(2p)	1.99970	0.00030
11	F	1	py	Ryd(3p)	0.00030	-0.00030
12	F	1	py	Ryd(4p)	0.00000	0.00000
13	F	1	py	Ryd(5p)	0.00000	0.00000
14	F	1	pz	Val(2p)	0.99885	0.99885
15	F	1	pz	Ryd(3p)	0.00115	0.00115
16	F	1	pz	Ryd(4p)	0.00000	0.00000
17	F	1	pz	Ryd(5p)	0.00000	0.00000
18	F	1	dxy	Ryd(3d)	0.00000	0.00000
19	F	1	dxz	Ryd(3d)	0.00000	0.00000
20	F	1	dyz	Ryd(3d)	0.00000	0.00000
21	F	1	dx2y2	Ryd(3d)	0.00000	0.00000
22	F	1	dz2	Ryd(3d)	0.00024	0.00017

similar to Figs. 2.1–2.5.) The remaining 17 NAOs correspond to vacant 3s, 3p, 3d, . . . Rydberg-type orbitals lying outside the valence shell, usually ignorable for chemical purposes.

However, the NAOs are not yet the "best possible" orbitals for the atomic configuration (as can be seen by the slight occupancies in other than valence 2s, 2p orbitals). NAOs are idealized atom-like orbitals that always have perfect rotational symmetry in both coordinate and spin space, even if the actual electron configuration (an open-shell doublet radical in this case) breaks symmetry. For the closed-shell F^- species, NAOs are already "best possible." Some properties of NAOs and other spin-orbitals with respect to description of "spin density" and magnetic behavior are summarized in Sidebar 2.5.

SPIN-ORBITALS, SPIN CHARGE, SPIN DENSITY, AND MOLECULAR MAGNETISM

The principal properties of an electron are its charge (e) and "spin" angular momentum orientation (α "up" or β "down"). As usual, "spinning charge" leads to a magnetic field, so each electron can be pictured as a tiny bar magnet that is oriented "↑" or "↓" with respect to any chosen external magnetic field direction. (Note that Gaussian and other ESS programs commonly identify "↑" with "majority spin" and "↓" with "minority spin" in radical species.) Although transport of total electron charge leads to well-known *electronic* properties of materials, the analogous transport of spin-up or spin-down electrons leads to magnetic *spintronic* properties. Molecular level spintronic properties are important for current magnetic storage devices as well as quantum computing technologies of the future.

In nonrelativistic MO theory, electrons of opposite spin occupy distinct α or β spin-orbitals. The total charge (q_A, atomic units) on atomic site A is evaluated by summing the occupancies of α ($^{\uparrow}n_{iA}$) and β ($^{\downarrow}n_{iA}$) spin-orbitals on the site:

$$q_A = Z_A - \sum_i (^{\uparrow}n_{iA} + {}^{\downarrow}n_{iA}) \tag{2.8}$$

In ordinary closed-shell species, the α and β spin-orbitals are equally populated, leading to overall "diamagnetism" (weak repulsion to a magnetic field). However, for radicals and other open-shell species, the total charge q_A is the sum of *distinct* $^{\uparrow}q_A$, $^{\downarrow}q_A$ "spin charges":

$$^{\uparrow}q_A = Z_A/2 - \sum_i {}^{\uparrow}n_{iA} \tag{2.9}$$

$$^{\downarrow}q_A = Z_A/2 - \sum_i {}^{\downarrow}n_{iA} \tag{2.10}$$

$$q_A = {}^{\uparrow}q_A + {}^{\downarrow}q_A \tag{2.11}$$

leading to net up or down magnetism at local site A and overall "paramagnetism" (strong attraction to a magnetic field) that is the signature of a molecular magnet. It is therefore important that such spin charge distributions be accurately characterized for open-shell species.

For many purposes, the key magnetic property at each atomic site is the net "spin density" ($^{\delta}\rho_A$), the *difference* of α and β spin charges,

$$^{\delta}\rho_A = {}^{\uparrow}q_A - {}^{\downarrow}q_A \tag{2.12}$$

In natural population analysis (NPA) of open-shell species (see, for example, I/O-3.10), the "natural spin density" is evaluated for each NAO, then summed over NAOs on each atom to give $^{\delta}\rho_A$, and finally over all atoms to give net overall spin density $^{\delta}\rho$ of the species (as measured by ESR spectroscopy; see Chapter 7). This provides a very detailed picture of spin charge and spin polarization distributions throughout the molecule, allowing one to quantify (or rationally design) specific magnetic properties of interest.

Note that NAOs for α and β spin have *identical* spatial forms (as required to ensure rotational invariance against different coordinate choices in spin space). The populations $^{\uparrow}n_{iA}$, $^{\downarrow}n_{iA}$ of spin-up and spin-down electrons "in NAO i of atom A," therefore, have well-defined meaning in the NPA framework.

The optimal atomic orbitals we are seeking are obtained in the Natural Bond Orbital search, which can only find nonbonding 1-center "lone particle" (LP) spin-orbitals in this atomic case. The optimal atomic spin-orbitals are found separately in the sections for α spin search, labeled by

```
*****************************************************
*******          Alpha spin orbitals      ******
*****************************************************
 :
NATURAL BOND ORBITAL ANALYSIS, alpha spin orbitals:
```

and later in the analogous section for β spin. (For closed-shell systems such as F^-, separate output sections for α and β spin are not required, because the spatial orbitals are identical in the two spin sets.) Partial output for the α spin-orbital set is shown in I/O-2.3.

I/O-2.3

```
      (Occupancy)   Bond orbital/ Coefficients/ Hybrids
    ----------------------------------------------------------------
   1. (1.00000) CR ( 1) F  1        s(100.00%)
                                     1.0000   0.0000   0.0000   0.0000   0.0000
                                     0.0000   0.0000   0.0000   0.0000   0.0000
                                     0.0000   0.0000   0.0000   0.0000   0.0000
                                     0.0000   0.0000   0.0000   0.0000   0.0000
                                     0.0000   0.0000

   2. (1.00000) LP ( 1) F  1        s(  0.00%)p 1.00(100.00%)
                                     0.0000   0.0000   0.0000   0.0000   0.0000
                                     1.0000  -0.0003   0.0003   0.0000   0.0000
                                     0.0000   0.0000   0.0000   0.0000   0.0000
                                     0.0000   0.0000   0.0000   0.0000   0.0000
                                     0.0000   0.0000

   3. (1.00000) LP ( 2) F  1        s(  0.00%)p 1.00(100.00%)
                                     0.0000   0.0000   0.0000   0.0000   0.0000
                                     0.0000   0.0000   0.0000   0.0000   1.0000
                                    -0.0003   0.0003   0.0000   0.0000   0.0000
                                     0.0000   0.0000   0.0000   0.0000   0.0000
                                     0.0000   0.0000

   4. (1.00000) LP ( 3) F  1        s(  0.00%)p 1.00(100.00%)
                                     0.0000   0.0000   0.0000   0.0000   0.0000
                                     0.0000   0.0000   0.0000   0.0000   0.0000
                                     0.0000   0.0000   0.0000   0.9994  -0.0338
                                    -0.0002   0.0000   0.0000   0.0000   0.0000
                                     0.0000   0.0000

   5. (1.00000) LP ( 4) F  1        s( 99.98%)p 0.00(  0.00%)d 0.00(  0.02%)
                                     0.0000   0.9999  -0.0085   0.0000   0.0000
                                     0.0000   0.0000   0.0000   0.0000   0.0000
                                     0.0000   0.0000   0.0000   0.0000   0.0000
                                     0.0000   0.0000   0.0000   0.0000   0.0000
                                     0.0000   0.0144

   6. (0.00000) RY*( 1) F  1        s(100.00%)p 0.00(  0.00%)d 0.00(  0.00%)
    :
```

In this output, NBO 1 is identified as a CR (core) orbital of unit occupancy and 100% s-character. The 17 numbers listed below tell how to compose this NBO from the 17 NAOs of I/O-2.1. In this case, only the first coefficient is nonvanishing, corresponding to the "Cor(1s)" NAO. Thus, NAO 1 and NBO 1 each represent the core 1s atomic orbital and are essentially and unsurprisingly identical.

In a similar manner, NBO 2 is identified as a singly occupied valence LP (1-center, nonbonding) orbital of 100% p-character, composed almost entirely of NAO 6 (the valence $2p_x$ NAO), but with tiny admixtures of NAOs 7 and 8 (Rydberg-type $4p_x$, $5p_x$), corresponding to the $2p_x{}^\uparrow$ spin-orbital discussed above. NBO 3 is similarly the $2p_y{}^\uparrow$ (spatially equivalent to NBO 2), while NBO 4 is the slightly inequivalent $2p_z{}^\uparrow$ spin-orbital (in the "singly occupied" direction) that was plotted in Fig. 2.2a–c. Finally, NBO 5 is the singly occupied valence $2s^\uparrow$ spin-orbital that completes the formal valence shell. (Why it is shown as only "99.98%" s-character is a long story involving Gaussian cartesian d-functions that need not concern us here). NBOs 6–22 are a long list of "leftover" RY* (Rydberg-type) spin-orbitals of zero occupancy, which therefore, making zero contribution to any measurable property of this atomic state. The NAO composition coefficients of such negligibly occupied RY* orbitals are not included in the default printed output, but zealots can consult Appendix C to obtain such numerical details or use the orbital plotting methods described in Appendix B to visualize the orbitals.

The formal electronic configuration is also displayed in "Natural Electron Configuration" output, initially for the total atom

```
 Atom No          Natural Electron Configuration
-----------------------------------------------------
   F   1      [core]2s( 2.00)2p( 5.00)
```

and subsequently in the separate sections for α spin

```
 Atom No          Natural Electron Configuration
-----------------------------------------------------
   F   1      [core]2s( 1.00)2p( 3.00)
```

and β spin

```
 Atom No          Natural Electron Configuration
-----------------------------------------------------
   F   1      [core]2s( 1.00)2p( 2.00)
```

Thus, the NBO description of the F atom corresponds in all qualitative respects to the simple picture presented in freshman-level introduction to periodic table regularities.

Finally, to obtain the energies associated with these orbitals (as presented in Table 2.1), we can look down to the NBO Summary near the end of each spin section of output. The α-spin summary is shown in I/O-2.4. The spin-orbital energies are those quoted in Table 2.1, and the unit occupancies of NBOs 1–5 correspond to exact

-- *I/O-2.4* --

```
NATURAL BOND ORBITALS (Summary):


                                                        Principal Delocalizations
              NBO              Occupancy    Energy      (geminal,vicinal,remote)
===================================================================================

Molecular unit  1   (F)
   1. CR ( 1) F  1             1.00000    -24.77226
   2. LP ( 1) F  1             1.00000     -0.48378
   3. LP ( 2) F  1             1.00000     -0.48378
   4. LP ( 3) F  1             1.00000     -0.57138
   5. LP ( 4) F  1             1.00000     -1.25763
   6. RY*( 1) F  1             0.00000      0.65837
      :
  22. RY*(17) F  1             0.00000      3.62988

            ------------------------------
               Total Lewis     5.00000   (100.0000%)
          Valence non-Lewis    0.00000   (  0.0000%)
          Rydberg non-Lewis    0.00000   (  0.0000%)
            ------------------------------
              Total unit  1    5.00000   (100.0000%)
              Charge unit 1   -0.50000
```

-- *I/O-2.5* --

```
NATURAL BOND ORBITALS (Summary):


                                                        Principal Delocalizations
              NBO              Occupancy    Energy      (geminal,vicinal,remote)
===================================================================================

Molecular unit  1   (F)
   1. CR ( 1) F  1             1.00000    -24.74144
   2. LP ( 1) F  1             1.00000     -0.45163
   3. LP ( 2) F  1             1.00000     -0.45163
   4. LP ( 3) F  1             1.00000     -1.16523
   5. LP*( 4) F  1             0.00000      0.69190
   6. RY*( 1) F  1             0.00000     25.28099
   7. RY*( 2) F  1             0.00000     42.25181
      :
  22. RY*(17) F  1             0.00000      3.77100

            ------------------------------
               Total Lewis     4.00000   (100.0000%)
          Valence non-Lewis    0.00000   (  0.0000%)
          Rydberg non-Lewis    0.00000   (  0.0000%)
            ------------------------------
              Total unit  1    4.00000   (100.0000%)
              Charge unit 1    0.50000
```

representation (100% "Total Lewis" accuracy) of the "electron-dot" configurational description for α spin. The corresponding β-spin summary is shown in I/O-2.5. In this case, NBO 5 is the empty "LP*" (vacant valence shell nonbonding) $2p_z^{\downarrow}$ spin-orbital that was plotted in Fig. 2.5.

Note that each spin set in I/O-2.4 and I/O-2.5 is associated with a "charge" of -0.5 for α spin and $+0.5$ for β spin. Such formal "spin charge" (see Sidebar 2.5) cancels overall, but reminds us that the α configuration is "anion-like" (i.e., like that of F^-), while the β configuration is "cation-like" (i.e., like F^+). Note also that the "Principal Delocalizations" column is blank in both spin sets, because such electronic mischief can only occur in polyatomic species, as discussed in later chapters.

For further practice in reading tabular NBO output accurately, refer to the "self-explaining" examples of the NBO website (http://www.chem.wisc.edu/~nbo5/mainprogopts.htm) or the explanations of sample output given in the *NBO Manual*.

2.5 NATURAL ATOMIC ORBITALS AND THE NATURAL MINIMAL BASIS

As emphasized in Fig. 2.3, the final natural orbitals of an atomic wavefunction will reflect subtle asymmetries of an open-shell configuration, such as slight differences between p_x, p_y, and p_z spatial orbitals or between $p_z^{\uparrow}$, and $p_z^{\downarrow}$ spin-orbitals. However, for many purposes, it is preferable to consider slightly-modified forms of these orbitals that exhibit the expected free-atom rotational symmetries of both position and spin space. Such "natural atomic orbitals" have the advantage of complete rotational invariance with respect to arbitrary choices of coordinate axes in either position or spin space, a highly desirable property for analysis purposes.

For closed-shell singlet species of overall ^1S symmetry (such as F^-), the NAOs are perfectly equivalent to final natural orbitals of the atomic wavefunction. However, for open-shell atoms, the NAOs (I/O-2.2) are very slightly different from the final atomic natural orbitals (I/O-2.3 and I/O-2.4). For qualitative conceptual purposes, however, the differences are immaterial. We shall henceforth consider NAOs to be effectively equivalent to the physical "natural orbitals of the atomic wavefunction.

The NAOs form a complete orthonormal set (identical for α and β spin sets) that can be used to *exactly* represent any aspect of the numerical wavefunction. The NAOs can, therefore, replace ESS basis AOs as numerical building blocks for reconstructing the wavefunction (exactly!) in much more compact and transparent form. (The exact transformation between AOs and NAOs is obtained from the AONAO keyword, as described in Appendix C.)

As shown in I/O-2.2, the NAOs effect a strong separation between highly occupied core (Cor) and valence (Val) orbitals and the negligibly occupied Rydberg (Ryd) orbitals beyond the formal valence shell. The former set is identified as the "natural minimum basis" (NMB) and the latter as the "natural Rydberg basis" (NRB). Although a "minimum basis" (MB) calculation is usually considered of unacceptable accuracy with the common basis AOs of ESS calculations, the corresponding NMB calculation in the basis of NAOs (i.e., using only 5 of the 22 NAOs in I/O-2.2) gives superb numerical accuracy, practically equivalent to that of all 22 AOs in the original ESS basis! As will be shown, the astonishing accuracy of NMB-type representation also extends to the domain of molecular calculations.

The NAO-based NMB concept differs in subtle ways from the AO-based minimum basis concept as commonly implemented in ESS packages. In each case, basis functions are added in (n, l)-subsets of increasing principal (n) and angular (l) quantum numbers, and atoms of a given period (row) and angular block of the periodic table are treated comparably. In the NAO case, however, the NMB is incremented by a new (n, l)-subset if, and only if, at least *one* atom in the same row of the angular block actually contains an occupied (n, l)-type orbital in its ground-state configuration. (In standard ESS treatments, s- and p-type basis groups are added simultaneously, even if s-block atoms never contain an occupied valence p-type orbital in their ground-state configuration.) The NMB set, therefore, often contains *fewer* basis functions than the corresponding ESS-based MB set, but even so is *far* more accurate for representing even the "complete basis set" wavefunction for the system.

Because the Rydberg-type NAOs tend to have negligible occupancies (typically, 0.0001e or less), it is generally safe to completely *ignore* NRB contributions (i.e., 17 of the 22 orbitals in the 6-311++G** description of F) for general analysis purposes. In effect, valence NAOs play the role of the "effective minimal basis AOs" envisioned in the simplest semiempirical treatments. Throughout this book, we shall therefore often truncate I/O displays (as in I/O-2.4) to focus on dominant NMB contributions, ignoring the increasingly large number of "leftover" Rydberg-type orbitals that are typical of large AO basis sets.

As emphasized in the comparisons of Sidebar 2.4, much of the numerical efficiency of NAOs can be traced to their accurate maintenance of mutual orthogonality. Particularly important are the radial oscillations and nodal patterns near the nucleus that prevent unphysical (Pauli violating) collapse of valence electrons into the atomic core region. However, this characteristic nodal structure will be absent in NAOs calculated in the framework of "effective core potential" (ECP) theory, where inner shell electrons are replaced by an effective repulsive (pseudo-) potential that prevents valence-level orbitals from penetrating the core region. If an ECP basis set is employed (e.g., LANL2DZ), the core-type NAOs will be absent and valence NAOs will lack the characteristic nodal patterns near the nucleus. But the high overall accuracy of NMB-level description will be maintained with "missing" core electrons assigned to the ECP for formal electron budgeting in NBO output. However, in the present study, we restrict attention to all-electron calculations in which core–valence exchange repulsion is treated explicitly and the NAOs have the general forms shown in Figs. 2.2 and 2.4.

Note finally that in open-shell systems, NAOs may still exhibit different orbital energies in α and β spin sets, due to the way in which "orbital energy" depends on specific configurational exchange forces. For example, the energies of carbon 2s, 2p orbitals in the free atom $C(2s^2 2p^2)$ configuration need not be identical to those of the "promoted" $C(2s^1 2p^3)$ configuration for chemical bonding, due to the distinct shielding effects in the two configurations. Similarly, the energies of p_x, p_y, p_z NAOs need not be degenerate in an open-shell species, despite the fact that their spatial forms are rotationally equivalent. Such configurational dependencies are seldom mentioned in elementary presentations of orbital theory, but they are essential in modern self-consistent field computational implementations of orbital-based concepts.

PROBLEMS AND EXERCISES

2.1. In SCF theory, the energy operator (F_{op}) has contributions from kinetic energy (K_{op}), nuclear–electron attractions (V_{op}), and electron–electron repulsions (R_{op}), expressed in operator form by the equation

$$F_{op} = K_{op} + V_{op} + R_{op}$$

or in matrix form (in any chosen basis "representation") by

$$\mathbf{F} = \mathbf{K} + \mathbf{V} + \mathbf{R}$$

In a chosen basis set of orbitals $\{\varphi_i\}$, the matrix elements are related to the operator F_{op} by

$$(\mathbf{F})_{ij} = \int \varphi_i^* F_{op} \varphi_j d\tau = \langle \varphi_i | F_{op} | \varphi_j \rangle$$

The "orbital energy" ε_i of orbital φ_i is the "diagonal" ($i=j$) matrix element,

$$\varepsilon_i = (\mathbf{F})_{ii} = \int \varphi_i^* F_{op} \varphi_i d\tau = \langle \varphi_i | F_{op} | \varphi_j \rangle$$

These equations allow orbital interactions and energies (as well as their K_{op}, V_{op}, R_{op} components) to be evaluated for any chosen basis set of interest (AOs, NAOs, NBOs, . . ., MOs) from matrix elements that are easily obtainable from the NBO program (see Appendix C and problems from Chapters 1 and 2).

(a) For the Ne atom, find the orbital energy of the 2s NAO and its contributions from kinetic energy (K_{op}), nuclear–electron attractions (V_{op}) and electron–electron repulsions (R_{op}).

(b) Similarly, find the orbital energy and K_{op}, V_{op}, and R_{op} contributions for basis orbitals AO 1, AO 2, and AO 3 (or any other AOs you think interesting) in the same Ne atom calculation. Do any of these basis AOs resemble the physical 2s NAO in energetic characteristics? Discuss the nature and magnitude of discrepancies for each AO.

2.2. Hund's rule states that electrons in degenerate singly occupied atomic orbitals (e.g., $2p_x$ and $2p_y$) prefer to be in triplet ($2p_x{}^\uparrow 2p_y{}^\uparrow$) rather than singlet ($2p_x{}^\uparrow 2p_y{}^\downarrow$) spin configuration. The singlet–triplet energy difference

$$^{1,3}\Delta E_{Hund} = {}^1E(2p_x{}^\uparrow 2p_y{}^\downarrow) - {}^3E(2p_x{}^\uparrow 2p_y{}^\uparrow)$$

can be readily evaluated for atoms of various atomic numbers (Z) and net charge (q) with a chosen ESS program.

(a) Evaluate $^{1,3}\Delta E_{Hund}$ for $^{1,3}C$ ($Z = 6$, $q = 0$), $^{1,3}O$ ($Z = 8$, $q = 0$), $^{1,3}F^+$ ($Z = 9$, $q = 1$), and $^{1,3}Ne^{2+}$ ($Z = 10$, $q = 2$) and find the singly occupied NAOs i, j for each species. Describe how $^{1,3}\Delta E_{Hund}$ varies with changes in Z or q (if necessary, examining similar species with other Z, q combinations).

(b) Hund's rule is usually attributed to differences in electron–electron repulsions (R_{op}) between singly occupied orbitals in the two configurations. Evaluate the K_{op}, V_{op}, and R_{op}-type interactions between singly occupied NAOs i, j for each species in (a). Which (if any) energy component seems to best account for $^{1,3}\Delta E_{Hund}$ and its Z, q variations? (see *WIRESs Comp. Mol. Sci.* **2**, 1, 2012 for discussion.)

2.3. Koopmans' approximation (see *V&B*, p. 119ff) states that ionization energy (IE) of an atom

$$A \xrightarrow{\text{IE}} A^+ + e^-$$

is approximately the (negative of) orbital energy ε_i of the parent orbital from which the electron was removed

$$IE \cong -\varepsilon_i$$

typically the highest occupied orbital of the parent species.

(a) Evaluate the ionization energies IE_1, IE_2, …, IE_8 for successive removal of valence electrons from a Ne atom:

$$Ne \xrightarrow{IE_1} Ne^+ \xrightarrow{IE_2} Ne^{2+} \xrightarrow{IE_3} Ne^{3+} \xrightarrow{\quad} \cdots \xrightarrow{IE_8} Ne^{8+}$$

and identify the "originating" NAO i and orbital energy ε_i in each parent species. Plot IP versus ε_i for each ionization step and comment on the observed success or failure of Koopmans' approximation. (Use the successive multiplicities 1, 2, 3, 4, 3, 2, 1, 2 for Ne, Ne^+, …, Ne^{8+}.)

(b) Similarly, evaluate the first ionization energy (IP_1) for each of the neutral species ^1Ne, ^2F, ^3O, ^4N, ^3C, ^2B, ^1Be, ^2Li, and compare with $-\varepsilon_i$ for the originating NAO i of the parent neutral. As above, comment on the accuracy of Koopmans' approximation.

2.4. In the formal *Aufbau* procedure for atomic electron configurations, an electron is added to the "lowest unfilled orbital" with each increment in nuclear charge Z.

(a) Consider the *Aufbau* from Ca to Sc. What NAO of Ca is the "lowest unfilled" orbital? Which NAO of Sc is the "newly occupied" orbital of the configuration? Is the "newly occupied" orbital the same as the "highest occupied" orbital of Sc?

(b) When an electron is removed from Sc by ionization

$$Sc \rightarrow Sc^+ + e^-$$

which NAO of Sc (the "newly occupied" or "highest occupied") loses the electron? Explain why this curious reversal does not contradict Bohr's *Aufbau* concept.

2.5. Evaluate ground-state wavefunctions for atoms of the first long period (Li, Be, B, C, N, O, F, Ne) by both "standard" (B3LYP/6-311++G**) DFT methodology and alternative methods:

(a) CAS(2,2)/6-311++G**

(b) CISD/6-311++G**

(c) MP2/6-311++G**

Can you can find any significant NAO occupancy differences for any of these species. [*Hint:* Beryllium exhibits unusually strong two-configurational character (static correlation) that gives significant occupancy to the 2p NAO as well as to the expected 2s NAO.] If using the Gaussian program, your Li input deck may be prepared as shown below (e.g., for the MP2 job):

```
#UMP2/6-311++G** POP=NBOREAD DENSITY=CURRENT NOSYMM GUESS=MIX


Li atom


0 2
Li


$NBO FILE=Li PLOT FIXDM $END
```

Chapter 3

Atoms in Molecules

The concept that substances are composed of molecules, and molecules are composed of atoms, can be traced back to chemical antiquity. Nevertheless, in modern molecular electronic structure theory, the "atomic" constituents differ appreciably from the immutable, indivisible particles envisioned by the ancients. Of course, the signature properties of an atom are only indirectly linked to the positively charged nucleus, which carries virtually the entire atomic mass but occupies only an infinitesimally small portion of the apparent atomic volume. We now understand "the atom" to be composed of the surrounding quantum mechanical distribution of *electrons* that occupy the characteristic set of *orbitals* associated with the nucleus in question. "Finding the atom" in a molecular wavefunction therefore reduces (as in Chapter 2) to the problem of finding the *atomic orbitals* and the associated *electronic configuration* (number of electrons occupying each available atomic orbital) around each nuclear center.

Of course, in a molecular species we can no longer expect integer (or near-integer) numbers of electrons in each atom-like orbital; for example, the simplest imaginable diatomic species, H_2^+, could have no more than 0.5e associated with each nuclear center. In general, we must expect to deal with *fractional* occupation numbers reflecting the characteristic "electron sharing" that underlies the chemical bonding phenomenon. Such equal or unequal sharing of electrons between nuclear centers also leads to noninteger *atomic charges*, which express the net gain or loss of electrons at each nuclear center (relative to the isolated neutral atom) that accompanies molecular binding. Except for these characteristic noninteger electronic occupancy values, the problem of "finding the atom in the molecule" is closely analogous to the problem considered in Chapter 2 of characterizing the atomic orbitals and associated configurational "populations" for the wavefunction of an isolated atom or ion.

As before, the Natural Atomic Orbitals (NAOs) serve as the optimal "effective" atom-like orbitals for describing the overall electron density distribution of the molecular wavefunction, so that finding the atomic electrons in NBO output is not more difficult than in Chapter 2. We shall first examine how the NAOs within the molecular environment *differ* from the free-space forms encountered in Chapter 2. We use the experience gained there to anticipate the "breathing"

Discovering Chemistry With Natural Bond Orbitals, First Edition. Frank Weinhold and Clark R. Landis.
© 2012 John Wiley & Sons, Inc. Published 2012 by John Wiley & Sons, Inc.

changes associated with net gain or loss of electronic charge in molecule formation, as well as proper maintenance of orthogonality (and Pauli-compliance) with
respect to filled orbitals of other regions.

3.1 ATOMIC ORBITALS IN MOLECULES

In a molecular environment, the effective atom-like orbitals are expected to be
modified by two principal physical effects:

(1) As a result of the electronic give and take of chemical bonding, the net
population of electrons around each nuclear center may increase or decrease,
resulting in partial anionic or cationic character. In accordance with the
breathing variations noted in Chapter 2, this leads to expansions or contractions of orbital size, relative to the free-atom neutral species.

(2) As valence orbitals of one atom come into "contact" (overlap) with those of
another center, each must develop outer "ripple patterns" (analogous to the
much stronger such features that each maintains with respect to its own core
orbitals) to preserve mutual orthogonality. Such ripple patterns inherently
correspond to increased kinetic energy (increased "curvature" as seen by the
quantum mechanical Laplacian operator for kinetic energy) and consequent
"repulsive" contribution to total energy, if both orbitals are fully occupied.
(This is the essential electronic origin of the "steric repulsion" mandated by
the Pauli exclusion principle; see Chapter 6.) Such asymmetric outer rippling
toward another atom can be partially avoided by contracting the orbital
toward its own center, but this in turn requires adverse readjustment of the
inner oscillations with respect to its own core orbital(s).

As a result of these effects, the original free-atom orbital symmetries are expected to
be lowered (particularly by the outer rippling patterns) and average orbital size is
expected to decrease due to such asymmetric "confinement" by the molecular
environment. However, the average size change due to molecular confinement, effect
(2), will be modulated by the net gain or loss of electrons in chemical bonding, effect
(1). Thus, the optimal atom-like orbitals (the NAOs) of the molecular wavefunction
are expected to differ appreciably from their free-atom counterparts, reflecting the
competition between the two effects.

For qualitative visualization purposes, it is often desirable to employ idealized
NAO-like orbitals of idealized free-atom symmetry, neglecting the "rippling" effects
of interatomic orthogonality. Such *preorthogonal NAOs* (PNAOs) are provided by the
NBO program to enable direct visualization of "orbital overlap." Because PNAO
overlaps convey a powerful visual impression of the actual quantum mechanical
interaction integrals (as expressed by the Mulliken approximation; cf. *V&B*, p. 31),
the PNAOs will be used extensively for orbital illustrations throughout this book. In
this section, we wish to illustrate the atomic orbitals in molecules for a simple example
and show how PNAOs are used to visually assess the strength of NAO interactions in
the molecular framework.

3.1.1 Atomic Orbital Interactions in Hydrogen Fluoride

As a simple example, let us first consider the hydrogen fluoride (HF) molecule, using the output from a wavefunction calculation that might be obtained as follows (e.g., for Gaussian input):

```
#B3LYP/6-311++G** POP=NBOREAD

HF molecule test job

0 1
F
H   1   0.925

$NBO ARCHIVE FILE=HF PLOT $END
```

For this closed-shell example (as in the F^- case of Chapter 2), only one main section of NBO output is produced (rather than separate sections for α and β spin), and each "orbital" refers to a spatial orbital of occupancy 0-2 (rather than a spin-orbital of occupancy 0-1).

Let us first examine the "natural populations" output shown in I/O-3.1. As expected, the NAO listing now includes entries for F (NAOs 1–22) and H (NAOs 23–29), similar to those for the individual atoms except for the evident changes in electronic occupancy and orbital energy. For example, the valence 2s of F (NAO 2) now contains 1.9093e instead of 1.9996e, and the $2p_z$ (NAO 14) contains 1.6390e instead of 0.9989e (corresponding to an overall net gain of about 0.55e), whereas the occupancies of other NAOs do not differ appreciably from their values in the free F atom (cf. I/O-2.2). At the other end of the molecule, we can see that the H 1s occupancy has dropped to 0.4459e, corresponding to net ~0.55e loss to the more electronegative F atom in molecule formation (as required by the strict electronic bookkeeping). Note that although more precise numerical values are printed in the output, we shall generally quote occupancies only to the nearest 0.0001e, consistent with the maximum number of decimals expected to contain chemically interesting detail, according to mathematical relationships to be discussed in Chapter 5.

[The I/O-3.1 output includes messages about "population inversions" on F and H that are for informational purposes only. Such a message appears whenever the NBO program notices an unusual "inversion" in energy ordering compared to occupancy ordering. On the H atom, for example, NAO 24 is seen to have slightly higher occupancy than NAO 25 (0.00058e versus 0.00003e) despite the fact that its orbital energy is higher (1.74904 versus 0.88288 a.u.). Because both occupancies are near zero on the "chemically interesting" scale, this incidental detail of Rydberg-type "left-over" basis function numerics has no chemical significance, and hence can be safely ignored in this and most other cases.]

3.1.2 Visualizing Atomic Orbital Interactions with PNAOs

As anticipated in the discussion of Section 2.2, the NAOs of the hydrogen fluoride molecule differ perceptibly from those of isolated H and F atoms. Figure 3.1 shows

```
─────────────────────── I/O-3.1 ───────────────────────
NATURAL POPULATIONS:  Natural atomic orbital occupancies

NAO Atom No lang   Type(AO)    Occupancy      Energy
--------------------------------------------------------
  1   F   1  s     Cor( 1s)     1.99995      -24.42473
  2   F   1  s     Val( 2s)     1.90928       -1.33535
  3   F   1  s     Ryd( 4s)     0.00153        1.59043
  4   F   1  s     Ryd( 3s)     0.00000        1.16611
  5   F   1  s     Ryd( 5s)     0.00000       65.88014
  6   F   1  px    Val( 2p)     1.99724       -0.42088
  7   F   1  px    Ryd( 3p)     0.00025        1.26868
  8   F   1  px    Ryd( 4p)     0.00000        1.86557
  9   F   1  px    Ryd( 5p)     0.00000        4.54526
 10   F   1  py    Val( 2p)     1.99724       -0.42088
 11   F   1  py    Ryd( 3p)     0.00025        1.26868
 12   F   1  py    Ryd( 4p)     0.00000        1.86557
 13   F   1  py    Ryd( 5p)     0.00000        4.54526
 14   F   1  pz    Val( 2p)     1.63899       -0.40833
 15   F   1  pz    Ryd( 3p)     0.00126        0.98515
 16   F   1  pz    Ryd( 4p)     0.00005        2.76693
 17   F   1  pz    Ryd( 5p)     0.00000        3.55886
 18   F   1  dxy   Ryd( 3d)     0.00000        3.78278
 19   F   1  dxz   Ryd( 3d)     0.00048        3.91492
 20   F   1  dyz   Ryd( 3d)     0.00048        3.91492
 21   F   1  dx2y2 Ryd( 3d)     0.00000        3.78278
 22   F   1  dz2   Ryd( 3d)     0.00184        4.30043

 23   H   2  s     Val( 1s)     0.44591        0.10358
 24   H   2  s     Ryd( 4s)     0.00058        1.74904
 25   H   2  s     Ryd( 2s)     0.00003        0.88288
 26   H   2  s     Ryd( 3s)     0.00000        0.89478
 27   H   2  px    Ryd( 2p)     0.00204        1.44062
 28   H   2  py    Ryd( 2p)     0.00204        1.44062
 29   H   2  pz    Ryd( 2p)     0.00059        2.38160

     Population inversion found on atom  F  1
     Population inversion found on atom  H  2
```

orbital amplitude profiles for the fluorine atom 1s, 2s, and $2p_z$ NAOs and overlapping pre-NAO counterparts in the HF molecule. As seen in the figure, the molecular PNAOs (on the right) are slightly more diffuse, as expected from the partial anionic "fluoride" character in HF [effect (1)]. However, in other respects the PNAOs are virtually indistinguishable from the free-atom NAOs of Figs. 2.2–2.4 (where NAOs and PNAOs are equivalent), thus confirming the strong persistence of such atom-like features into the molecular environment. The final NAOs (on the left) exhibit the asymmetric "ripple patterns" toward the adjacent H atom [effect (2)], tending to contract the orbital profile and introduce an additional node near the H nucleus. Although the NAOs no longer have ideal free-atom symmetries (due to the asymmetric perturbations of the molecular environment), they clearly retain their recognizable atom-like character in the final molecule.

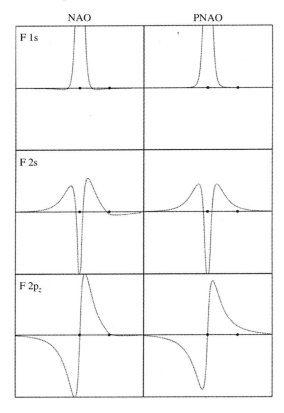

Figure 3.1   Orbital profiles for F 1s (upper), 2s (middle), and 2p$_z$ (lower) NAOs (left) and PNAOs (right) in the HF molecule. (The off-center cross-hair symbol marks the position of the H nucleus.)

Because of its small size and lack of repulsive inner core, the H 1s orbital can approach and overlap the orbitals on other atoms to an unusual extent. The PNAO versus NAO differences are therefore particularly large for this orbital. Figure 3.2 depicts the 2p$_z$(F)-1s(H) orbital interaction in a variety of graphical displays (NAO versus PNAO, profile versus contour) to illustrate the strongly perturbed form of the 1s(H) NAO. Whereas the plotted NAOs on the left become more confusingly distorted as the strength of interaction increases, the corresponding PNAO diagrams on the right show how the two atoms align their orbital lobes for "maximum overlap," consistent with freshman-level description. Indeed, one can see (particularly in the PNAO contour diagram at the lower right) that the 1s(H) orbital positions itself for maximum possible overlap with the positive lobe of the 2p$_z$(F) orbital, while avoiding the destructive overlap with the negative backside lobe that would be incurred if the two atoms moved closer together. These comparisons illustrate how the graphical depiction of PNAO overlap becomes an increasingly valuable visual cue as the strength of orbital interaction increases, whereas the corresponding NAO plots (in which orbital overlap is always *zero*, by construction) become increasingly distorted and difficult to recognize. Use of PNAO overlap diagrams to suggest the strength of NAO orbital interactions (the essence of Mulliken's approximation) is thus

NAO   PNAO

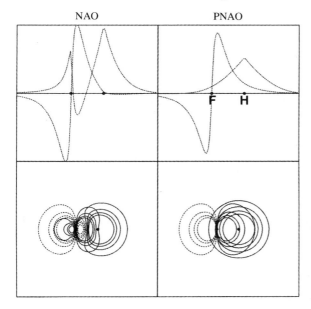

Figure 3.2 $2p_z(F)-1s(H)$ orbital interaction in HF, shown in profile (upper) and contour (lower) plots for NAOs (left) and PNAOs (right).

a powerful visual aid to gain qualitative conceptual insights into the electronic logic of chemical bonding interactions. Even though the orbital interaction integrals are properly calculated with NAOs (for reasons described in Chapter 6), the corresponding PNAOs are strongly preferred for graphical visualization purposes and (unless otherwise indicated) are always employed in the "NAO illustrations" of this book.

3.2   ATOMIC CONFIGURATIONS AND ATOMIC CHARGES IN MOLECULES

The NAO populations, as tabulated in I/O-3.1, make it easy to sum up the total number of electrons in each (n,l) subshell to obtain the "natural electron configuration." Similarly, the total electronic population on the atom can be combined with the nuclear charge to obtain the net "natural charge" for each atom, and other aspects of the electronic bookkeeping can be summarized for informational purposes. Such summaries follow the table of detailed NAO orbital occupancies (I/O-3.1) in the "natural population analysis" (NPA) section of NBO output.

I/O-3.2 displays the remaining portion of NPA output for the HF molecule. The first part of this output is the "summary of natural population analysis," which gives the atomic natural charge on each atom (corresponding simply to the orbital gains and losses noted in the preceding section) and the partitioning of total atomic populations into core, valence, and Rydberg contributions. The second portion of output shows a further partitioning of total electronic occupancy into NMB and NRB components, documenting the overwhelming dominance of the contributions from natural minimal basis orbitals (which comprise only 6 of the 29 total orbitals in this basis set, but describe about 99.89% of the total electron density). The final portion of this output

```
────────────────────────────── I/O-3.2 ──────────────────────────────
 Summary of Natural Population Analysis:

                                  Natural Population
                   Natural    ----------------------------------------------
    Atom No        Charge        Core     Valence    Rydberg      Total
 ------------------------------------------------------------------------
     F   1        -0.54882     1.99995    7.54275    0.00612     9.54882
     H   2         0.54882     0.00000    0.44591    0.00527     0.45118

 =======================================================================
 * Total *         0.00000     1.99995    7.98866    0.01139    10.00000

                                  Natural Population
              ---------------------------------------------------
    Core                          1.99995 ( 99.9974% of    2)
    Valence                       7.98866 ( 99.8583% of    8)
    Natural Minimal Basis         9.98861 ( 99.8861% of   10)
    Natural Rydberg Basis         0.01139 (  0.1139% of   10)
              ---------------------------------------------------


    Atom No          Natural Electron Configuration
 ------------------------------------------------------------------------
     F   1        [core]2s( 1.91)2p( 5.63)
     H   2              1s( 0.45)
```

summarizes the "natural electron configuration" on each atom, corresponding, for example, to $(2s)^{1.91}(2p)^{5.63}$ valence configuration on F and $(1s)^{0.45}$ on H. Strict mutual consistency of "electronic bookkeeping" entries and rigorous compliance with the Pauli exclusion principle are assured by the mathematical theorems that underlie the NPA algorithm, which must be strongly distinguished from alternative "population analysis" algorithms still in common usage (Sidebar 3.1).

SIDEBAR 3.1   *THE MANY VARIETIES OF "ATOMIC CHARGE"*

At the beginning, users are often bewildered by the confusing assortment of "atomic charge" values that may be offered to describe the electronic charge distribution (sometimes by the same ESS program). Whereas energy, angular momentum, dipole moment, and other properties of the many-electron wavefunction are unambiguously determined as expectation values of well-defined Hermitian operators, the concept of partial "charge on each atom" lacks such clear-cut definition. The chemical importance of the atomic charge concept naturally leads to spirited discussions as to which of the many possible definitions is to be considered "correct." Here we wish to briefly describe a number of alternative proposals that have been put forward, comparing and contrasting their definitions and numerical values with the "natural atomic charges" (NPA charges) adopted throughout this book.

A key distinguishing characteristic of atomic charge definitions is their intended usage in describing some aspect of the electronic charge distribution, i.e., the spatial variation of electron density $\rho(\mathbf{r})$. In the most superficial usages, this *distribution* is replaced by supposed *point* charges (Dirac delta functions) at each nucleus, namely,

$$\rho(\mathbf{r}) \cong \sum_A q_A \delta(\mathbf{r} - \mathbf{r}_A) \tag{3.1}$$

Of course, this equation is grossly incorrect at *every* spatial $\mathbf{r}$. However, one may consider various *moments* of the electron density distribution by multiplying both sides of (3.1) by $\mathbf{r}^n$ (for chosen n) and integrating over all space to obtain

$$\langle \mathbf{r}^n \rangle = \int \mathbf{r}^n \rho(\mathbf{r})d\tau \cong \sum_{A} q_A \mathbf{r}_A^{\ n} \tag{3.2}$$

The q_A's may then be defined as the numerical parameters that make Equation (3.2) "correct" for the chosen n value. For the *dipole* moment ($n = 1$), for example, the point-like q_A's on the right are chosen (together with known nuclear positions $\mathbf{r}_A$) to match the dipole integral on the left as closely as possible in each spatial direction. Similar fitting procedures (leading, of course, to different q_A values) can be employed for other chosen n values (e.g., $n = -3$ for "atomic polar tensor" charges), with each such set of "charges" having value within its intended framework of usage.

A somewhat related definition can be based on defining a classical-type "electrostatic potential" V_{ELP}:

$$V_{ELP} \equiv \sum_{A<B} q_A q_B / |\mathbf{r}_A - \mathbf{r}_B| \tag{3.3}$$

which is equated to a properly evaluated quantum mechanical interaction energy (e.g., with an external test charge or other probe species at chosen separation) and fitted, as best possible, to chosen q_A, q_B values. Such a V_{ELP}-based definition of atomic charges is based on the assumption that intermolecular interactions are of classical electrostatic nature, but as discussed in Chapter 9, this assumption is often unjustified (even if successful fitting to V_{ELP} at some chosen geometry seems to make it tautologically "correct"). Although such numerical charges may have value in the framework of their intended usage (i.e., to replace difficult quantal interaction integrals by classical point charge formulas in a molecular dynamics simulation), they are unlikely to correspond to common understanding and usage of the atomic charge concept for more general chemical phenomena.

Still another orbital-free philosophy of atomic charge evaluation is based on the "atoms in molecules" topological formalism of R. W. F. Bader. The Bader charges $q_A^{(B)}$ are evaluated by integrating electron density within nonoverlapping "cell" boundaries (C_A) that serve to define "the atom" in Bader's topological partitioning of three-dimensional coordinate space:

$$q_A^{(B)} \equiv \int_{C_A} \rho(\mathbf{r})d\tau \tag{3.4}$$

The shapes and volumes of such atomic cells differ widely from molecule to molecule, and their boundaries exhibit sharp discontinuities that are quite surprising compared to the solutions of Schrödinger's equation for atoms. Furthermore, as pointed out by C. Perrin (*J. Am. Chem. Soc.* **113**, 2865, 1991), in model LCAO-MO diatomic wavefunctions prepared from known atomic orbital contributions, the Bader topological boundary systematically exaggerates the apparent ionicity of the bond (as compared to the LCAO coefficients of the orbitals that produced the original density). This "Perrin effect" is due to the manner in which diffuse orbitals (and associated density) "cross over" the Bader boundary more extensively than do contracted orbitals, leading to skewed charge assignment in the bonding region. Despite these differences in philosophy and numerical detail, the Bader charges are usually *closer* to NPA charges than are those calculated by other methods mentioned above, showing that quite different theoretical assumptions can lead to similar descriptors of charge distribution.

Prior to introduction of NPA concepts (A. E. Reed, R. B. Weinstock, and F. Weinhold, *J. Chem. Phys.* **83**, 735, 1985), the most widely used atomic charges were those based on "Mulliken population analysis" (MPA). Like their NPA counterparts $q_A^{(N)}$, the "Mulliken charges" $q_A^{(M)}$ are obtained by simply summing the populations of all "orbitals on atom A" $\{\varphi_i^{(A)}\}$:

$$q_A^{(M)} = \sum_i q_i^{(M)} \tag{3.5}$$

However, the MPA quantities are identified with basis AOs (Sidebar 2.3) that are generally nonorthogonal,

$$\int \varphi_i^{(A)}{}^* \varphi_j^{(B)} d\tau = S_{ij} \neq 0 \tag{3.6}$$

so it is inherently ambiguous whether overlapping portions of $\varphi_i^{(A)}$, $\varphi_j^{(B)}$ are "on A" or "on B." The Mulliken algorithm rather arbitrarily awards *half* the overlap to each atom. This allocation is admirably democratic and defensible for orbitals of similar diffuseness and shape. However, it rapidly becomes unreasonable for atoms of different electronegativity or hybrids of different composition (similar in origin, but opposite in direction, to the Perrin effect noted above for Bader charges).

Due to basis AO overlap (3.6), two types of Mulliken AO populations must be considered:

(1) "Gross" Mulliken AO populations $q_i^{(M)}$ give atomic charges that sum to the proper overall species charge, but are often found to have *un*physical negative or Pauli violating values:

$$q_i^{(M)} < 0 \quad \text{or} \quad q_i^{(M)} > 2 \tag{3.7}$$

(2) "Net" Mulliken AO populations $q_i'^{(M)}$, on the other hand, satisfy proper physical constraints of nonnegativity and Pauli-compliance:

$$0 \leq q_i'^{(M)} \leq 2 \tag{3.8}$$

but the associated atomic charges $q_A'^{(M)}$, defined as

$$q_A'^{(M)} = Z_A - \sum_i q_i'^{(M)} \tag{3.9}$$

fail to properly sum to overall species charge:

$$q_{\text{total}} \neq \sum_A q_A'^{(M)} \tag{3.10}$$

Because (3.10) is a more conspicuous and frequent failure than (3.7), the gross Mulliken populations and charges are usually quoted.

The pathologically unphysical behavior (3.7) of Mulliken AO populations actually becomes *worse* as the basis set is improved. As noted by Mulliken himself (see R.S. Mulliken and W.C. Ermler, *Diatomic Molecules: Results of Ab Initio Calculations*, Academic, New York, 1977, pp. 33–38), each $q_i^{(M)}$ can have *any* value in the range

$$-\infty \leq q_i^{(M)} \leq +\infty \tag{3.11}$$

as the basis is extended to completeness. Many workers have called attention to the severe artifacts and convergence failures of Mulliken populations and charges, and their usage has not been recommended by any recent authority.

All such unphysical artifacts and convergence failures are avoided by the natural populations $q_i^{(N)}$ of overlap-free NAOs, which automatically satisfy the physical constraints:

$$0 \leq q_i^{(N)} \leq 2 \qquad (3.12)$$

(or $0 \leq q_i^{(N)} \leq 1$ for spin NAOs) and sum strictly to the correct overall charge:

$$q_A^{(N)} = \sum_i q_i^{(N)} \qquad (3.13)$$

$$q_{\text{total}} = \sum_A q_A^{(N)} \qquad (3.14)$$

Numerous theoretical comparisons and practical applications (see, e.g., the *NBO website* bibliography links) testify to the superiority of NPA populations and charges for general chemical usage (see, e.g., K. C. Gross and P. G. Seybold, *Int. J. Quantum Chem.* **80**, 1107–1115, 2000; **85**, 569–579, 2001).

For completeness, we mention finally the "Löwdin population analysis," which is displayed by some ESS programs (although apparently never advocated by P.-O. Löwdin himself) and is based on modified Mulliken-type formulas for AOs that have been symmetrically orthogonalized. While this algorithm avoids the worst artifacts of orbital nonorthogonality, it shares with MPA the unphysical failure to converge as the AO basis is extended toward completeness, contrary to the excellent convergence characteristics that are a signature of NPA populations and charges.

As a slightly more complex example, consider the model methanol molecule (CH_3OH) in idealized geometry (tetrahedral bond angles, equal CH bond lengths), as specified by Gaussian input with the atom numbers as shown in the following diagram:

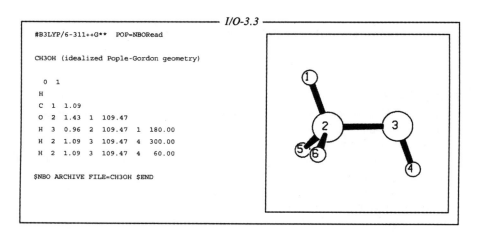

I/O-3.3

```
#B3LYP/6-311++G**  POP=NBORead

CH3OH (idealized Pople-Gordon geometry)

  0  1
  H
  C  1  1.09
  O  2  1.43  1  109.47
  H  3  0.96  2  109.47  1  180.00
  H  2  1.09  3  109.47  4  300.00
  H  2  1.09  3  109.47  4   60.00

$NBO ARCHIVE FILE=CH3OH $END
```

```
───────────────────────── I/O-3.4 ─────────────────────────
Summary of Natural Population Analysis:

                                  Natural Population
                 Natural   -------------------------------------------
   Atom No       Charge      Core     Valence    Rydberg      Total
   --------------------------------------------------------------------
     H   1      0.17758    0.00000    0.82046    0.00196     0.82242
     C   2     -0.19638    1.99930    4.17841    0.01867     6.19638
     O   3     -0.73839    1.99980    6.72368    0.01491     8.73839
     H   4      0.45281    0.00000    0.54263    0.00456     0.54719
     H   5      0.15219    0.00000    0.84533    0.00248     0.84781
     H   6      0.15219    0.00000    0.84533    0.00248     0.84781
   ====================================================================
   * Total *   0.00000    3.99911   13.95584    0.04505    18.00000
```

The calculated table of NPA charges is shown above.

An interesting feature of the atomic charge distribution is the evident *in*equivalency of methyl H atoms, with in-plane H(1) being distinctly more positive (by 0.0254e) than out-of-plane H(4), H(5), despite the fact that the methyl group was constructed with exact threefold geometrical symmetry. The asymmetric methyl charges reveal subtle "stereoelectronic" influences of the neighboring OH group (to be discussed in Chapter 5) that are expected to slightly *distort* the idealized threefold methyl geometry under full geometry optimization. The electronic origins of such subtle structural distortions exemplify the type of question that can be readily answered with the tools of NBO analysis.

Note that calculation of NPA populations and atomic charges is completely independent of subsequent analysis of hybridization, bonding, or resonance in the species. Nevertheless, final details of hybrid composition, bond occupancies, and resonance weightings will all be found to be strictly consistent with the NAO occupancies and atomic charge distributions obtained in this initial step of full NAO/NBO/NRT analysis.

3.3   ATOMS IN OPEN-SHELL MOLECULES

3.3.1   HF$^+$ Radical Cation

Finding the NAO populations and atomic charges in an open-shell species is similar to the closed-shell case, except that a composite NPA overview precedes the separate sections for α- and β-spin output.

As a simple example, let us consider the open-shell HF$^+$ cation produced by vertical (fixed bond length) ionization of HF, using the same input file as in Section 3.1.1 except for replacement of the net charge and multiplicity for the neutral singlet ("0 1") by the corresponding cation doublet values ("1 2") in the fifth line of input.

The initial composite ($\alpha + \beta$) NPA output closely resembles I/O-3.1 and 3.2, but with an additional "spin" column (spin density difference: α-NAO minus β-NAO

─────────── *I/O-3.5* ───────────

```
NATURAL POPULATIONS:  Natural atomic orbital occupancies


NAO Atom No lang   Type(AO)    Occupancy     Spin
-----------------------------------------------------
   1    F  1  s    Cor( 1s)     1.99995     0.00000
   2    F  1  s    Val( 2s)     1.92657     0.00571
   6    F  1  px   Val( 2p)     0.99886     0.99886
  10    F  1  py   Val( 2p)     1.99706     0.00055
  14    F  1  pz   Val( 2p)     1.70195     0.01219


  23    H  2  s    Val( 1s)     0.36643    -0.01793
```

occupancy), as shown above (I/O-3.5) for the NMB set (with Rydberg-type NAOs omitted for simplicity). The large spin density (0.9989) at the F $2p_x$ (NAO 6) shows that the ionized electron was essentially removed from this orbital (an off-axis fluorine lone pair).

However, slight nonzero spin density is also seen in other NAOs, including surprising *negative* spin density (-0.0179) in NAO 23, the H 1s orbital. The negative value means that there is slightly *more* spin-down density in H 1s than before a spin-down electron was removed from the system. Such counterintuitive spin redistribution cannot be described by a wavefunction of ROHF form, but both experiment and higher-level theory confirm that it is a real physical effect in many similar systems, and its successful calculation by UHF-based methods argues strongly for the superiority of this type of open-shell description (see Sidebar 2.1).

The composite NPA summary table (I/O-3.6) also includes the additional "natural spin density" column that gives the net $\alpha-\beta$ occupancy difference at each atom. This shows (as above) the curious negative spin density at H and corresponding "overshoot" of positive spin density at F. However, the overall picture corresponds closely to simple removal of a spin-down electron from a nonbonding $2p_x$ NAO on F, thereby increasing the natural charge on F by about one unit (cf. I/O-3.2) and leaving the resulting net spin-up density concentrated predominantly at this center.

─────────── *I/O-3.6* ───────────

```
Summary of Natural Population Analysis:


                         Natural Population            Natural
              Natural  ----------------------------    Spin
 Atom No      Charge    Core    Valence   Rydberg   Total    Density
----------------------------------------------------------------------
   F   1      0.37023  1.99995  6.62443   0.00539  8.62977   1.01761
   H   2      0.62977  0.00000  0.36643   0.00380  0.37023  -0.01761
======================================================================
* Total *     1.00000  1.99995  6.99086   0.00919  9.00000   1.00000
```

─────────────────────── *I/O-3.7* ───────────────────────

```
*****************************************************
*******        Alpha spin orbitals        *******
*****************************************************


NATURAL POPULATIONS:  Natural atomic orbital occupancies

NAO Atom No lang   Type(AO)    Occupancy      Energy
------------------------------------------------------
 1    F  1   s     Cor( 1s)    0.99997     -25.08640
 2    F  1   s     Val( 2s)    0.96614      -1.97960
 6    F  1   px    Val( 2p)    0.99886      -1.08857
10    F  1   py    Val( 2p)    0.99880      -0.99552
14    F  1   pz    Val( 2p)    0.85707      -0.99165


23    H  2   s     Val( 1s)    0.17425      -0.33615
```

Details of the spin NAO occupancies in the separate spin sets then follow, first for α spin, as shown in abridged NMB form in I/O-3.7 above. The corresponding results for β spin follow in the lower half of the output.

Comparison of these tables immediately shows (as inferred above) that ionization has occurred out of the fluorine $2p_x^{\downarrow}$ (occupancy "0.00000" in NAO 6 of β output). The same could be inferred from the "natural electron configuration" output for each spin set (not shown), which corresponds to an assigned fluorine configuration of $[\text{core}](2s^{\uparrow})^{0.97}(2p^{\uparrow})^{2.85}$ for α spin and $[\text{core}](2s^{\downarrow})^{0.96}(2p^{\downarrow})^{1.84}$ for β spin.

Note that Rydberg-type contributions are found to be essentially negligible in each spin set, confirming the high accuracy ($\sim$99.9%) of "freshman-level" NMB description of the open-shell radical cation species. [In fact, the percentage accuracy of NMB-level description is marginally *higher* for each spin set of

─────────────────────── *I/O-3.8* ───────────────────────

```
*****************************************************
*******        Beta  spin orbitals        *******
*****************************************************


NATURAL POPULATIONS:  Natural atomic orbital occupancies

NAO Atom No lang   Type(AO)    Occupancy      Energy
------------------------------------------------------
 1    F  1   s     Cor( 1s)    0.99997     -25.05635
 2    F  1   s     Val( 2s)    0.96043      -1.87568
 6    F  1   px    Val( 2p)    0.00000      -0.77193
10    F  1   py    Val( 2p)    0.99826      -0.96084
14    F  1   pz    Val( 2p)    0.84488      -0.95114


23    H  2   s     Val( 1s)    0.19218      -0.34226
```

open-shell HF^+ than for closed-shell HF, testimony to the high accuracy of the DODS-type description of open-shell species.] Although smaller details of the NAO spin distributions and orbital energies are also of some interest, the overall picture in this simple system conforms closely to idealized removal of an electron from the F $2p_x$ NAO, with near-perfect "pairing" persisting in other orbitals of the radical cation.

Where would the electron go if we *added* one electron to HF to form the HF^- radical anion? We leave this as an exercise to the student explorer. [Answer: 92% on H, 8% on F, mostly in H 1s and F 2s, $2p_z$ NAOs along the bonding axis, but involving significant contributions from Rydberg-type NAOs as well.]

3.3.2 Ozone

As a somewhat more complex example, let us now consider the case of ozone (O_3), which has an open-shell singlet ground state (Sidebar 3.2). The Gaussian input file to obtain the open-shell wavefunction and default NBO analysis for experimental equilibrium geometry ($R_{OO} = 1.272$, $\theta = 116.8°$) is shown below.

The composite atomic charges on ozone show only a rather benign and uninteresting total charge distribution, with slight negative charge (-0.0998) on each terminal oxygen and compensating positive charge at the center.

─────────── *I/O-3.9* ───────────

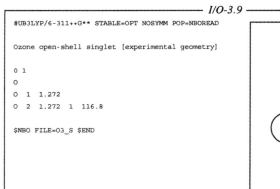

```
#UB3LYP/6-311++G** STABLE=OPT NOSYMM POP=NBOREAD

Ozone open-shell singlet [experimental geometry]

0 1
O
O  1  1.272
O  2  1.272  1  116.8

$NBO FILE=O3_S $END
```

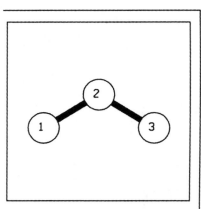

─────────── *I/O-3.10* ───────────

Summary of Natural Population Analysis:

Atom No	Natural Charge	Natural Population				Natural Spin Density
		Core	Valence	Rydberg	Total	
O 1	-0.09978	1.99990	6.07659	0.02328	8.09978	-0.52465
O 2	0.19954	1.99986	5.75371	0.04690	7.80046	0.00000
O 3	-0.09976	1.99990	6.07658	0.02328	8.09976	0.52465
* Total *	0.00000	5.99966	17.90688	0.09346	24.00000	0.00000

However, the final "natural spin density" column reveals the striking *spin polarization* in ozone, with ca. 0.5e excess β spin on O(1) and compensating excess α-spin on O(3), corresponding to significant "singlet diradical" character.

Details of the α-spin NAO populations (shown for NMB orbitals only) and atomic spin charges reveal other features of the surprising spin *a*symmetry in this species. As expected from the spin density values in I/O-3.10, the natural charges in I/O-3.11 differ by about 0.5e on the two ends [$+0.2124$ on O(1), -0.3122 on O(3)], corresponding to strong *left–right* spin polarization. (Of course, the β-spin NAOs show the "mirror image" of this asymmetric distribution, leading to the overall symmetric pattern in I/O-3.11.) In addition, one can see that the most highly occupied $2p^\uparrow$ spin-orbitals are $2p_x^\uparrow$ on O(1), but $2p_{y,z}^\uparrow$ on O(3), corresponding to additional *angular* spin polarization. These spin NAO descriptors well illustrate the concept of

```
──────────────────────────── I/O-3.11 ────────────
**********************************************************
*******          Alpha spin orbitals        *******
**********************************************************


NATURAL POPULATIONS:  Natural atomic orbital occupancies

  NAO Atom No lang   Type(AO)    Occupancy      Energy
  ----------------------------------------------------
    1   O  1   s    Cor( 1s)     0.99995     -19.10549
    2   O  1   s    Val( 2s)     0.92158      -0.98000
    6   O  1   px   Val( 2p)     0.96665      -0.36919
   10   O  1   py   Val( 2p)     0.38640      -0.28581
   14   O  1   pz   Val( 2p)     0.50131      -0.29405

   23   O  2   s    Cor( 1s)     0.99993     -19.27186
   24   O  2   s    Val( 2s)     0.83178      -1.02652
   28   O  2   px   Val( 2p)     0.70607      -0.45815
   32   O  2   py   Val( 2p)     0.72024      -0.46860
   36   O  2   pz   Val( 2p)     0.61876      -0.44368

   45   O  3   s    Cor( 1s)     0.99995     -19.11888
   46   O  3   s    Val( 2s)     0.92649      -1.02638
   50   O  3   px   Val( 2p)     0.61494      -0.32649
   54   O  3   py   Val( 2p)     0.88843      -0.42001
   58   O  3   pz   Val( 2p)     0.87079      -0.37275


Summary of Natural Population Analysis:

                                 Natural Population
                  Natural    -------------------------------------
  Atom No         Charge       Core    Valence   Rydberg   Total
  ----------------------------------------------------------------
    O  1          0.21244     0.99995  2.77594   0.01167  3.78756
    O  2          0.09977     0.99993  2.87685   0.02345  3.90023
    O  3         -0.31221     0.99995  3.30065   0.01161  4.31221
  ================================================================
  * Total *       0.00000     2.99983  8.95344   0.04673  12.00000
```

different spatial distributions for different spins (Sidebar 2.1), the signature feature of partial diradical character. Although quite rare in ground-state molecular singlet species, such partial diradical character is doubtless associated with the unusual reactivity and photochemistry of ozone.

Having detected and partially characterized the underlying spin polarization in ozone, we may well ask, "What causes that?" This is the type of question that NBO analysis is designed to answer, and we shall therefore return to this challenging example in the later chapters of this book.

SIDEBAR 3.2 *SOME ASPECTS OF RHF VERSUS UHF DESCRIPTION OF SINGLET SPECIES*

Most molecules of singlet spin symmetry are best described by a leading configuration of "restricted" Hartree–Fock (RHF) form (i.e., doubly occupied spatial orbitals) in near-equilibrium geometry. However, at a critical distance along any bond-breaking coordinate, the RHF-like portion of the singlet potential energy surface typically becomes unstable with respect to a lower-energy "unrestricted" Hartree–Fock (UHF) configuration of open-shell DODS form (different orbitals for different spins; Sidebar 2.1). (In the Gaussian program system, the "STABLE = OPT" keyword initiates a check for RHF instability and search for the lower-energy UHF solution, if available; this allows the RHF versus UHF character of a singlet species to be determined unambiguously.) Such open-shell "diradical" character is expected quite generally along any homolytic dissociation pathway, but the ozone molecule is exceptional in exhibiting UHF-type splitting even in its equilibrium geometry.

By definition, a spin "singlet" is an eigenfunction of the total squared spin angular momentum operator S_{op}^2 with eigenvalue $S(S + 1) = 0$ and spin multiplicity $2S + 1 = 1$. Although RHF-type wavefunctions automatically have this spin symmetry, UHF-type wavefunctions do not, and are therefore referred to as "broken symmetry" solutions. However, the broken-symmetry UHF configuration can always be corrected by adding its spin-flipped counterpart to obtain a double-configuration description. Nevertheless, the conceptual simplicity of the single-configuration UHF description and its effectiveness in describing the physical spin-polarization effect often make this the cost-effective choice, particularly when $\langle S_{op}^2 \rangle_{UHF} \cong 0$.

In this book, we focus primarily on how to obtain the NAO/NBO/NRT descriptors of a chosen wavefunction, rather than on how a wavefunction is chosen. The NAO/NBO/NRT descriptors of UHF-type description (as used throughout this book for open-shell systems) can be compared with the corresponding descriptors of more accurate wavefunctions for insights into the chemically significant differences, if any, that justify a more complex theoretical level.

PROBLEMS AND EXERCISES

3.1. Find the NAO electronic configuration for the O atom in each of the following open- and closed-shell ground-state species. For comparison purposes, treat each species as open shell (e.g., by using UB3LYP/6-311++G** method and NOSYMM, GUESS = MIX, STABLE = OPT keywords) to insure lowest-energy solution.

(a) An isolated atom (triplet)

(b) Dioxygen molecule, O_2 (triplet)

(c) Water molecule, H_2O (singlet)

(d) Superoxide anion, O_2^- (doublet)

(e) Ozone molecule, O_3 (open-shell singlet)

(f) Carbon monoxide, CO (singlet)

(g) Nitrosonium cation, NO^+ (singlet)

3.2. Compare the radial profiles of the effective oxygen 2s orbital (PNAO) for each species in Problem 3.1. For amusement, compare also with the form of the "2s basis AO" in your chosen basis set. (Try *any* of the s-type basis AOs if none are identified as the "2s" AO.)

3.3. The spin-orbital energies (ε_{2s}, ε_{2p}) of the oxygen 2s, $2p_x$, $2p_y$, $2p_z$ NAOs vary widely in the species of Problem 3.1, reflecting differences in occupancy, overall charge, and electron–electron repulsion in each configurational environment. Make an overall plot of NAO energy (vertical) versus occupancy (horizontal) for all the unique oxygen ε_{2s}, ε_{2p} spin NAOs of Problem 3.1, using different symbols to distinguish ε_{2s} versus ε_{2p} as well as values for ions versus neutrals.

(a) Unlike the simple hydrogenic case, the orbital energies ε_{2s}, ε_{2p} of many-electron atoms are expected to *differ* due to the effects of electron–electron repulsion, with s-orbitals lying *below* p-orbitals (due to their superior "penetration" to the nucleus despite the "screening" effect of other electrons). Can you see evidence (however faintly) for this tendency in your plotted values? Discuss briefly. Estimate the "typical" difference $\varepsilon_{2p} - \varepsilon_{2s}$ for highly occupied oxygen 2s, 2p spin-orbitals of neutral species.

(b) Orbital energies are also expected to vary with overall charge, becoming destabilized in anionic and stabilized in cationic environments. Can you see evidence (however faintly) for this tendency in your plotted values? Estimate the effect $\Delta\varepsilon_{2s}$, $\Delta\varepsilon_{2p}$ of changing species charge by ± 1 (in spin-orbitals of comparable occupancy).

(c) In self-consistent-field theories such as DFT, the "orbital energy" of an orbital depends on its occupancy, because an "occupying" electron reduces the effective nuclear screening for that orbital, which can only be screened by the other $N-1$ electrons, whereas an unoccupied orbital is screened by all N electrons. Can you see evidence (however faintly) for such occupancy dependence in your plotted values? From a best straight-line fit to your plotted values for neutral ε_{2p} versus occupancy, estimate the energetic shift ($\Delta\varepsilon_{2p}$) if spin-orbital occupancy is reduced by 0.5 electrons (all else being as equal as possible).

(d) Still other weaker dependencies can be seen in the scatter of ε_{2s}, ε_{2p} values with change in configurational environment, particularly in open-shell systems. Can you see evidence of the effect on an occupied α-ε_{2p} if the corresponding β-ε_{2p} is occupied or unoccupied? Comment on other general dependencies that may be present in this (limited) data set. (If desired, include other species in your data set to strengthen the generality of your conclusions.)

Chapter 4

Hybrids and Bonds in Molecules

Age-old questions concerning the nature of the "bonds" between atoms in molecules culminated in the remarkable Lewis structure model of G. N. Lewis (1916). The notion that such bonds were formed from directed hybrids was subsequently developed by Linus Pauling (1932), shortly after the discovery of quantum mechanics. Although many theoretical advances have ensued, it is fair to say that the underlying concepts of valence-shell hybridization, shared-electron pair bonds, and Lewis structural dot diagrams continue to dominate chemical thinking and pedagogy to this day.

Although localized Lewis structural hybrid and bonding concepts carry strong quantum mechanical overtones, these concepts achieved current textbook formulations long before accurate quantum mechanical wavefunctions were available to test their rather speculative underpinnings. The fact that these concepts still underlie modern chemical pedagogy testifies to the remarkable prescience of the theoretical pioneers who first achieved these initial formulations in the computational "dark ages." Nevertheless, we may expect that modern quantum mechanical wavefunctions should allow us the refine and extend these powerful concepts as originally envisioned. Even if textbooks may lag in this respect, a modern chemistry student is often fortunate to have web-accessible tools that now allow direct exploration of the more accurate and quantitative forms of hybrids, bonds, and Lewis structures in the best available modern wavefunctions. NBO analysis is currently the most general and widely used tool for "translating" modern quantum mechanical calculations into the qualitative language of localized bonding concepts.

In this chapter, we illustrate how to obtain the optimal "natural Lewis structure" (NLS) formulation of the wavefunction in terms of optimal NBOs for shared pairs (bonds) and lone pairs of the conventional Lewis structural dot diagram. We also describe how to assess the *accuracy* of the NLS representation, comparing it with alternative Lewis structural formulations (alternative "resonance structures") that might be suggested. In Sections 4.1–4.2, we first consider the relatively simple closed-shell molecules such as HF, CH_3OH, or H_2NCHO that conform to the octet rule. The

Discovering Chemistry With Natural Bond Orbitals, First Edition. Frank Weinhold and Clark R. Landis.
© 2012 John Wiley & Sons, Inc. Published 2012 by John Wiley & Sons, Inc.

more difficult cases presented by three-center bonding (e.g., B_2H_6), open-shell species (e.g., O_3), and Lewis-like bonding in transition metals (e.g., WH_6) will be described in Sections 4.4–4.6. In each case, the residual error of the localized NLS formulation naturally leads to consideration of the "resonance" (delocalization) corrections to the simple Lewis-like picture, to be discussed in Chapter 5.

4.1   BONDS AND LONE PAIRS IN MOLECULES

4.1.1   Hydrogen Fluoride

As a simple diatomic example, let us first consider the hydrogen fluoride molecule of Section 3.2. Following the NPA and natural electron configuration summaries (I/O-3.2), the NBO search summary appears as shown below (I/O-4.1).

As shown in the output, this particular NBO search terminated successfully after only a single "cycle," which satisfied the default search criteria. The search yielded a Lewis structure with one core (CR), one bond (BD), and three lone pair (LP) "Lewis-type" (L) NBOs, which described about 99.95% of the total electron density (i.e., 9.995 of the 10 electrons). These five L-type NBOs easily satisfied the default threshold (1.90e) for "pair" occupancy [the "0" under "Low occ (L)"] and the remaining 17 "non-Lewis" (NL) NBOs were all well below the 0.1e occupancy threshold ["High occ (NL)"] to be considered a satisfactory Lewis structure. [The "Dev" entry refers to "deviations" from the initial guess that steers multiple cycles of the search algorithm (if required), beyond the scope of this book; consult the NBO

I/O-4.1

```
NATURAL BOND ORBITAL ANALYSIS:

                    Occupancies      Lewis Structure   Low   High
             Occ.   ------------------  ----------------  occ   occ
   Cycle   Thresh.  Lewis    Non-Lewis  CR  BD  3C  LP   (L)   (NL)   Dev
===============================================================================
   1(1)     1.90    9.99526   0.00474    1   1   0   3    0     0    0.00
-------------------------------------------------------------------------------


Structure accepted: No low occupancy Lewis orbitals

-------------------------------------------------------------
   Core                  1.99995 ( 99.997% of    2)
   Valence Lewis         7.99531 ( 99.941% of    8)
==================    ==============================
   Total Lewis          9.99526 ( 99.953% of   10)
-------------------------------------------------------------
   Valence non-Lewis    0.00000 (  0.000% of   10)
   Rydberg non-Lewis    0.00474 (  0.047% of   10)
==================    ==============================
   Total non-Lewis      0.00474 (  0.047% of   10)
-------------------------------------------------------------
```

Bibliography website link (www.chem.wisc.edu/~nbo5) for further details.] The final lines quantify the overall accuracy of this NLS description (>99.95%), including the NMB (valence) versus Rydberg-type contributions to the NL remnant, confirming the high accuracy of the expected freshman-level dot diagram

$$H\!:\!\ddot{\underset{..}{F}}\!:$$

for this simple diatomic species.

Following the details of NBO composition, which are described in Section 4.2, the occupancies and energies of NBOs appear in the NBO summary table, as shown in I/O-4.2.

```
─────────────────────────── I/O-4.2 ───────────────────────────

  NATURAL BOND ORBITALS (Summary):


                                                   Principal Delocalizations
              NBO              Occupancy   Energy   (geminal,vicinal,remote)
  ==============================================================================
  Molecular unit  1   (HF)
    1. BD ( 1) F  1- H  2       2.00000   -0.88148
    2. CR ( 1) F  1             1.99995  -24.42481   26(v)
    3. LP ( 1) F  1             1.99938   -1.14280   25(v)
    4. LP ( 2) F  1             1.99796   -0.42162   24(v)
    5. LP ( 3) F  1             1.99796   -0.42162   23(v)
    6. RY*( 1) F  1             0.00000    1.59026
    7. RY*( 2) F  1             0.00000    1.16611
    8. RY*( 3) F  1             0.00000   65.88014
    9. RY*( 4) F  1             0.00000    1.26887
   10. RY*( 5) F  1             0.00000    1.86558
   11. RY*( 6) F  1             0.00000    4.54526
   12. RY*( 7) F  1             0.00000    1.26887
   13. RY*( 8) F  1             0.00000    1.86558
   14. RY*( 9) F  1             0.00000    4.54526
   15. RY*(10) F  1             0.00000    0.98223
   16. RY*(11) F  1             0.00000    2.76707
   17. RY*(12) F  1             0.00000    3.55886
   18. RY*(13) F  1             0.00000    3.78278
   19. RY*(14) F  1             0.00000    3.91547
   20. RY*(15) F  1             0.00000    3.91547
   21. RY*(16) F  1             0.00000    3.78278
   22. RY*(17) F  1             0.00000    4.29799
   23. RY*( 1) H  2             0.00204    1.44062
   24. RY*( 2) H  2             0.00204    1.44062
   25. RY*( 3) H  2             0.00063    1.75565
   26. RY*( 4) H  2             0.00004    1.20313
   27. RY*( 5) H  2             0.00000    0.89478
   28. RY*( 6) H  2             0.00000    2.04388
   29. BD*( 1) F  1- H  2       0.00000    0.40052
        ------------------------------
               Total Lewis     9.99526  ( 99.9526%)
           Valence non-Lewis   0.00000  (  0.0000%)
           Rydberg non-Lewis   0.00474  (  0.0474%)
        ------------------------------
               Total unit  1  10.00000  (100.0000%)
               Charge unit 1   0.00000
```

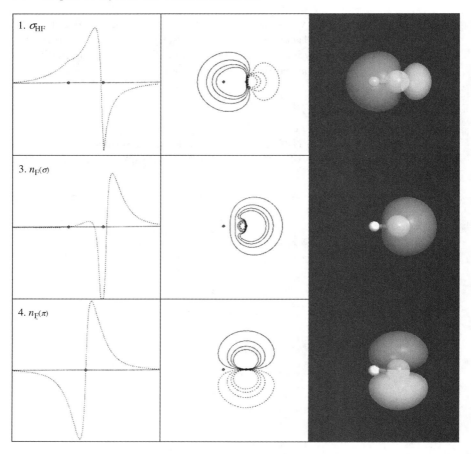

Figure 4.1   Distinct valence (P)NBOs of HF of bonding (BD:σ_{HF}) and nonbonding type (LP: on-axis $n_F^{(\sigma)}$ and off-axis $n_F^{(\pi)}$), shown in profile, contour, and surface plots. The profile of the p_y-type LP (NBO 4) is along a vertical line through the F nucleus, perpendicular to the equivalent p_x-type LP (NBO 5, not shown) that points out of the page. (See the color version of this figure in Color Plates section.)

As shown in the output, the five leading L-type NBOs 1–5, of near-double occupancy, are followed by the 24 remaining NL-type (starred) NBOs of Rydberg (RY*) or valence antibond (BD*) type, all of negligible occupancy (and safely ignorable for all practical purposes). The core (CR) NBO 2 is essentially identical to the F 1s NAO exhibited previously (Fig. 3.1). Profile, contour, and surface plots for the remaining valence NBOs of BD, LP type are shown in Fig. 4.1.

Perhaps the first surprise for some students is that the three fluorine lone pairs do not have the symmetric "tripod-like" shapes that are sometimes depicted in textbook cartoons of these orbitals. As shown clearly in Fig. 4.1 (and for reasons to be discussed in Sidebar 4.2), the on-axis "σ-type" $n_F^{(\sigma)}$ lone pair (NBO 3) is quite *distinct* from the two off-axis "π-type" $n_F^{(\pi x)}$, $n_F^{(\pi y)}$ lone pairs (NBOs 4, 5), as confirmed by their inequivalent occupancies and energies in I/O-4.2. On the scale of accuracy we are

――――――――――― *I/O-4.3* ―――――――――――

```
NATURAL BOND ORBITALS (Summary):


                                              Principal Delocalizations
            NBO              Occupancy   Energy  (geminal,vicinal,remote)
===============================================================================
Molecular unit  1  (CH4O)
   1. BD ( 1) H  1- C   2     1.99265   -0.52346  72(v),34(v)
   2. BD ( 1) C  2- O   3     1.99850   -0.81318
   3. BD ( 1) C  2- H   5     1.99703   -0.52365  33(v),69(g)
   4. BD ( 1) C  2- H   6     1.99703   -0.52365  33(v),69(g)
   5. BD ( 1) O  3- H   4     1.99033   -0.72065  16(v),68(v)
   6. CR ( 1) C  2           1.99930  -10.09312  69(g)
   7. CR ( 1) O  3           1.99980  -18.92886  16(v),53(v)
   8. LP ( 1) O  3           1.98417   -0.60841  16(v),68(v),51(v),18(v)
   9. LP ( 2) O  3           1.96534   -0.30134  70(v),71(v),50(v),17(v)
   :
  68. BD*( 1) H  1- C   2     0.01263    0.38816
  69. BD*( 1) C  2- O   3     0.00121    0.27623
  70. BD*( 1) C  2- H   5     0.01832    0.38295
  71. BD*( 1) C  2- H   6     0.01832    0.38295
  72. BD*( 1) O  3- H   4     0.00579    0.43161
             ------------------------------
                 Total Lewis    17.92417  ( 99.5787%)
           Valence non-Lewis     0.05628  (  0.3126%)
           Rydberg non-Lewis     0.01955  (  0.1086%)
             ------------------------------
                 Total unit  1  18.00000  (100.0000%)
                 Charge unit 1   0.00000
```

discussing, the tripod-like depiction is not even remotely "equivalent" to the optimal NBO depiction in Fig. 4.1. Some significant chemical consequences that follow from this distinction will be discussed in Chapter 9.

4.1.2 Methanol

A more representative polyatomic example is given by CH_3OH (cf. I/O-3.3). In this case, the default NBO search again required only one cycle and returned an optimal Lewis structure of very high accuracy (99.56% of the total electron density), containing two CR-type, five BD-type, and two LP-type NBOs. The NBO summary for occupancies and energies of these nine L-type orbitals and the final five (valence BD*) NL-type orbitals (i.e., neglecting the 58 RY*-type orbitals in this basis set) are shown in I/O-4.3.

(The "Principal Delocalizations" of this output section will be discussed in Chapter 5.)

The highly occupied L-type NBOs correspond to the expected Lewis structure dot diagram, which may be represented schematically as

$$
\begin{array}{c}
\text{H} \\
| \\
\text{H}-\text{C}-\ddot{\text{O}}-\text{H} \\
| \\
\text{H}
\end{array}
$$

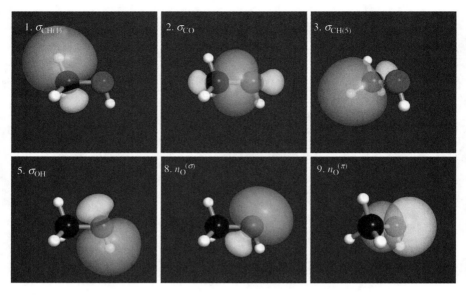

Figure 4.2   Lewis-type valence NBOs of CH_3OH (cf. I/O-4.3). (See the color version of this figure in Color Plates section.)

Once again, the two oxygen lone pairs, NBOs 8 and 9, are found to be *in*equivalent (not even faintly resembling "rabbit ears"), with the σ-type (in-plane) $n_O^{(\sigma)}$ (NBO 8) having significantly higher occupancy and lower energy than the π-type (out-of-plane) $n_O^{(\pi)}$ (NBO 9). Figure 4.2 displays surface plots of the L-type (P)NBOs of methanol, showing their general similarity to analogous NBOs of Fig. 4.1.

Although not visually apparent in Fig. 4.2, the three σ_{CH} bonds (NBOs 1, 3, 4) are also slightly inequivalent. As noted in Section 3.2, the proton charges of the out-of-plane C(2)−H(5) and C(2)−H(6) bonds differ slightly from the in-plane C(2)−H(1) bond (although the idealized methyl group was constrained to have perfect three-fold geometrical symmetry), and this inequivalence is also reflected in the slight differences between out-of-plane NBOs 3, 4 and in-plane NBO 1 shown in I/O-4.4. These subtle differences, as well as other aspects of the remaining small errors ($\sim$0.04%) of the NLS description (all to be discussed in Chapter 5), should not detract from admiration of the freshman-level Lewis structure concepts that so successfully account for the major features of the electronic distribution in this and many other polyatomic molecules.

4.1.3   Formamide

As a more challenging polyatomic species, we now consider the formamide molecule (H_2NCHO) with the Gaussian input geometry and atom numbering shown:

—————————— *I/O-4.4* ——————————

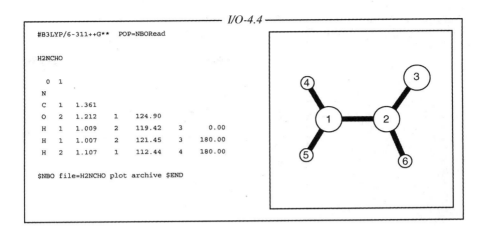

```
#B3LYP/6-311++G**   POP=NBORead

H2NCHO

  0  1
N
C   1   1.361
O   2   1.212    1    124.90
H   1   1.009    2    119.42    3      0.00
H   1   1.007    2    121.45    3    180.00
H   2   1.107    1    112.44    4    180.00

$NBO file=H2NCHO plot archive $END
```

The NBO search report (I/O-4.5) contains details of the evident difficulties in finding the "best" Lewis structure formula for this species. In this case, fully 19 cycles were required for the NBO search. In the search algorithm, the occupancy threshold ("Occ. Thresh.") is successively reduced from 1.90, 1.80, . . . , 1.50 while searching for the Lewis structure of lowest non-Lewis occupancy (highest possible Lewis occupancy). Finally, at cycle 10 (with threshold 1.80), a structure with six bonds and three lone pairs was found to have relatively low NL occupancy of 0.45516. Successive reductions of the threshold continued to return this structure as the best available, until it was finally accepted (cycle 19) as the final NLS. However, this NLS is identified as a "strongly delocalized structure" because of the one low-occupancy L-type NBO and one high-occupancy NL-type NBO found in the structure. Although the chosen NLS is indeed the "best possible" among all searched Lewis structures (i.e., all possible ways of drawing the bonds), one sees evidence in the search report for *alternative* Lewis structures of relatively low NL occupancy (0.73148 in cycle 11, 0.81531 in cycle 1, . . .). Such alternative Lewis structures indicate significant "resonance" in the formamide molecule, as will be discussed and quantified in Chapter 5.

The NLS metrics quoted at the end of I/O-4.5 document the reduced accuracy of the localized Lewis structure description in this case (98.10%, reduced from the >99.9% "typical" for CH_3OH and other common organic species). The residual 1.9% "delocalization error" (corresponding to $\sim$0.455e that could not be assigned to L-type NBOs of the best possible NLS) is seen to be primarily associated with *valence*-NL orbitals of BD* (valence antibond) type, whereas the corresponding contributions from Rydberg-NL (RY* type) orbitals are an order of magnitude smaller. Thus, the NBO search report points to significant chemical delocalization effects in formamide (and other amides) that underlie many of the interesting properties of proteins.

I/O-4.5

```
NATURAL BOND ORBITAL ANALYSIS:
```

		Occupancies		Lewis Structure				Low occ	High occ	
Cycle	Occ. Thresh.	Lewis	Non-Lewis	CR	BD	3C	LP	(L)	(NL)	Dev
1(1)	1.90	23.18469	0.81531	3	7	0	2	3	2	0.22
2(2)	1.90	22.90837	1.09163	3	7	0	2	3	2	0.18
3(3)	1.90	22.25366	1.74634	3	8	0	1	2	2	1.22
4(4)	1.90	22.90837	1.09163	3	7	0	2	3	2	0.18
5(5)	1.90	22.25366	1.74634	3	8	0	1	2	2	1.22
6(6)	1.90	22.90837	1.09163	3	7	0	2	3	2	0.18
7(7)	1.90	22.25366	1.74634	3	8	0	1	2	2	1.22
8(8)	1.90	22.90837	1.09163	3	7	0	2	3	2	0.18
9(9)	1.90	22.25366	1.74634	3	8	0	1	2	2	1.22
10(1)	1.80	23.54484	0.45516	3	6	0	3	1	1	0.22
11(2)	1.80	23.26852	0.73148	3	6	0	3	1	1	0.73
12(3)	1.80	23.54484	0.45516	3	6	0	3	1	1	0.22
13(1)	1.70	23.54484	0.45516	3	6	0	3	0	1	0.22
14(2)	1.70	23.54484	0.45516	3	6	0	3	0	1	0.22
15(1)	1.60	23.54484	0.45516	3	6	0	3	0	1	0.22
16(2)	1.60	23.54484	0.45516	3	6	0	3	0	1	0.22
17(1)	1.50	23.54484	0.45516	3	6	0	3	0	1	0.22
18(2)	1.50	23.54484	0.45516	3	6	0	3	0	1	0.22
19(1)	1.80	23.54484	0.45516	3	6	0	3	1	1	0.22

```
Strongly delocalized structure accepted
```

Core	5.99886 (99.981% of	6)
Valence Lewis	17.54597 (97.478% of	18)
Total Lewis	23.54484 (98.103% of	24)
Valence non-Lewis	0.40587 (1.691% of	24)
Rydberg non-Lewis	0.04929 (0.205% of	24)
Total non-Lewis	0.45516 (1.897% of	24)

The final summary of NBOs is shown in I/O-4.6, including the important BD*-type NL orbitals that are indicated to play the leading role in delocalization effects, but excluding the many remaining RY*-type orbitals. The L-type NBOs correspond to the Lewis structure diagram

which is indeed the best possible ("highest resonance weighting") for this molecule.

```
──────── I/O-4.6 ────────
NATURAL BOND ORBITALS (Summary):


                                         Principal Delocalizations
              NBO          Occupancy   Energy   (geminal,vicinal,remote)
===========================================================================
Molecular unit  1  (CH3NO)
   1. BD ( 1) N  1- C  2    1.99673   -0.86154   86(g),47(v)
   2. BD ( 1) N  1- H  4    1.99130   -0.69346   87(v),30(v),34(v),86(v)
                                                 72(v)
   3. BD ( 1) N  1- H  5    1.99200   -0.70092   86(v),33(v),65(v)
   4. BD ( 1) C  2- O  3    1.99816   -0.37904   85(g),13(v)
   5. BD ( 2) C  2- O  3    1.99684   -1.06911   84(v),82(g),30(g)
   6. BD ( 1) C  2- H  6    1.98379   -0.55427   83(v),47(v)
   7. CR ( 1) N  1          1.99949  -14.19130   31(v),30(v),33(v),34(v)
   8. CR ( 1) C  2          1.99961  -10.16605   50(v)
   9. CR ( 1) O  3          1.99976  -18.84744   30(v),31(v),82(v)
  10. LP ( 1) N  1          1.75091   -0.28548   85(v),32(v),71(v),66(v)
  11. LP ( 1) O  3          1.98437   -0.69759   30(v),45(v),82(v),31(v)
                                                 87(v),50(g),53(g)
  12. LP ( 2) O  3          1.85188   -0.25901   82(v),87(v),31(v),34(v)
   :
  82. BD*( 1) N  1- C  2    0.06363    0.44649
  83. BD*( 1) N  1- H  4    0.01244    0.39841
  84. BD*( 1) N  1- H  5    0.00842    0.39808
  85. BD*( 1) C  2- O  3    0.24274    0.00237   48(g),39(g),32(g)
  86. BD*( 2) C  2- O  3    0.00660    0.61241
  87. BD*( 1) C  2- H  6    0.07204    0.34075

         -----------------------------
            Total Lewis    23.54484   ( 98.1035%)
         Valence non-Lewis  0.40587   (  1.6911%)
         Rydberg non-Lewis  0.04929   (  0.2054%)
         -----------------------------
            Total unit  1  24.00000  (100.0000%)
           Charge unit  1   0.00000
```

Figure 4.3 depicts the leading L-type NBOs of this structure (omitting the hydride bonds, which play a secondary role in amide chemistry).

Several points are worthy of special attention in the NBO diagrams of Fig. 4.3:

(1) The carbonyl oxygen lone pairs, NBOs 11 and 12, are again seen to be of distinctly *in*equivalent form. The on-axis $n_O^{(\sigma)}$, NBO 11, is relatively inert, usually only weakly involved in carbonyl intra- and intermolecular interactions. In contrast, the off-axis $n_O^{(\pi)}$, NBO 12 (in-plane "p-π-type," labeled "p$_y$" in Zimmerman's terminology), is the primary "active" site of coordinative H-bonding (Chapter 9) and photochemical $n \rightarrow \pi^*$ excitation (Chapter 11). Thus, a "rabbit ears" depiction of carbonyl lone pairs is seriously erroneous and misleading with respect to important chemical properties of amide groups.

(2) The carbonyl double-bond NBOs are also seen to have distinctly *in*equivalent σ- (σ_{CO}, NBO 5) and π-bond (π_{CO}, NBO 4) forms, rather than the

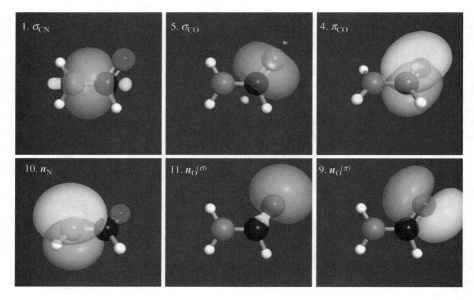

Figure 4.3   Leading Lewis-type valence NBOs of formamide (cf. I/O-4.6). (See the color version of this figure in Color Plates section.)

symmetrically equivalent "banana bond" forms that were sometimes advocated by Pauling.

(3) The nitrogen lone pair (n_N, NBO 10) is seen to be of highly unusual pure p form, consistent with the highly unusual *planar* structure of the amine group of amides, as discussed in Section 4.2. Together with the two p orbitals (p_C, p_O) of the carbonyl π_{CO} bond, the amine p_N orbital belongs to an *allylic*-like p_N-p_C-p_O arrangement of NAOs with strong p-π overlap, suggestive of the strong possibilities for π-type resonance as further explored in Chapter 5.

4.2 ATOMIC HYBRIDS AND BONDING GEOMETRY

Given the Lewis structural bonding patterns found in Section 4.1 for typical molecules (HF, CH_3OH, H_2NCHO), we now wish to investigate details of the bonding hybrids and their relationship to molecular geometry. The quantitative NBO hybridizations and directionalities are found to be in excellent agreement with the qualitative concepts of Pauling and other pioneer theorists, but with interesting subtleties that allow their original insights to be refined and extended. Modern wavefunctions testify eloquently to the aptness and accuracy of (most of) the simple hybrid and bonding concepts you learned in freshman chemistry.

It should be pointed out that in searching for the best possible hybrids and bonding pattern, the NBO program makes *no* use of molecular geometry information.

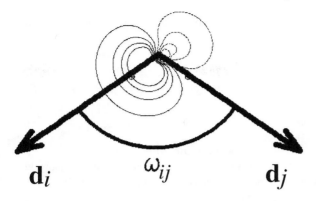

Figure 4.4 Geometry of directional vectors $\mathbf{d}_i$, $\mathbf{d}_j$ for two directional hybrids (with hybrid h_i shown for reference).

The extracted *Natural Hybrid Orbitals* (NHOs) are therefore not simply "encoded" forms of the molecular shape, as envisioned in "valence shell electron pair repulsions" (VSEPR)-type caricatures of hybridization theory. Instead, the NHOs represent optimal fits to the ESS-provided electronic occupancies (first-order density matrix elements; cf. *V&B*, p. 21ff) in terms of known angular properties of basis AOs. Thus, the NHOs *predict* preferred directional characteristics of bonding from angular patterns of electronic occupancy, and the deviations (if any) between NHO directions and the actual directions of bonded nuclei give important clues to bond "strain" or "bending" that are important descriptors of molecular stability and function.

The original concept of main-group "valence hybrids" refers simply to quantum mechanical mixing ("superposition") of the four atomic valence orbitals (s, p_x, p_y, p_z) to form four directed hybrid orbitals (h_1, h_2, h_3, h_4) that are variationally superior for chemical bonding; mathematically,

$$h_i = a_{i0}\, s + a_{ix}\, p_x + a_{iy}\, p_y + a_{iz}\, p_z \qquad (4.1)$$

Because the s orbital is isotropic, the direction of h_i is determined solely by its p-orbital mixing. Just as p_x, p_y, p_z point in the respective directions ($\mathbf{x}$, $\mathbf{y}$, $\mathbf{z}$) of unit vectors along the Cartesian axes, so does each new h_i point in a unique direction given by a unit vector $\mathbf{d}_i$ (see Fig. 4.4). If we define a "hybridization parameter" λ_i for each h_i as the ratio of squared p-type to s-type contributions:

$$\lambda_i \equiv (a_{ix}^2 + a_{iy}^2 + a_{iz}^2)/a_{i0}^2 \qquad (4.2)$$

and introduce modified (normalized) coefficients d_{ix}, d_{iy}, d_{iz} as

$$d_{ix} = a_{ix}/\lambda_i, \quad d_{iy} = a_{iy}/\lambda_i, \quad d_{iz} = a_{iz}/\lambda_i \qquad (4.3)$$

we can write the directional unit vector $\mathbf{d}_i$ for hybrid h_i as

$$\mathbf{d}_i = d_{ix}\mathbf{x} + d_{iy}\mathbf{y} + d_{iz}\mathbf{z} \qquad (4.4)$$

The normalized "p_i" orbital pointed in the $\mathbf{d}_i$ direction is given by the corresponding linear combination:

$$p_i = d_{ix}\, p_x + d_{iy}\, p_y + d_{iz}\, p_z \tag{4.5}$$

With these definitions, Equation (4.1) can finally be rewritten as (see *V&B*, p. 107ff)

$$h_i = (1 + \lambda_i)^{-1/2}[s + \lambda_i\, p_i] \tag{4.6}$$

which identifies h_i as an "$sp^{\lambda i}$ hybrid" oriented in direction $\mathbf{d}_i$.

The angle ω_{ij} between directed hybrids h_i, h_j is given by the usual dot product formula between their respective directional vectors (cf. Fig. 4.4), namely,

$$\mathbf{d}_i \cdot \mathbf{d}_j = \cos \omega_{ij} \tag{4.7}$$

However, general conservation principles of quantum mechanical wave mixing dictate that the final hybrids $\{h_i\}$ (like the atomic orbitals from which they originate) must be orthonormal:

$$\int h_i{}^*h_j\, d\tau = \delta_{ij} \tag{4.8}$$

Substitution of (4.6) into (4.8) for $sp^{\lambda i}$ hybrid h_i and $sp^{\lambda j}$ hybrid h_j leads to the important *Coulson directionality theorem* (see *V&B*, pp. 107–109)

$$\cos \omega_{ij} = \mathbf{d}_i \cdot \mathbf{d}_j = -(\lambda_i \lambda_j)^{-1/2} \tag{4.9}$$

which dictates the intrinsic angle ω_{ij} between hybrids h_i, h_j in terms of their respective hybridization parameters λ_i, λ_j. Equation (4.9) is the most important equation relating atomic sp^{λ} hybrids to molecular bonding geometry. (The analogous hybrid angles for sd^{μ}-type transition metal bonding are discussed in Section 4.6.)

Of course, the hybridization parameter λ_i (4.2) is merely a compact way of expressing the ratio of %-p character to %-s character in the hybrid, namely,

$$\lambda_i = \%\text{-p}/\%\text{-s} \tag{4.10}$$

which could vary anywhere between 0 (pure s) and ∞ (pure p). For example, the "standard" sp^2 and sp^3 hybrids have 66.7% and 75% p-character, respectively, but an $sp^{2.5}$ hybrid of 71.4% p-character or an $sp^{5.7}$ hybrid of 85% p-character are also possible. [If you were told that only sp^1, sp^2, and sp^3 hybrids are imaginable, you were misled.]

Alternatively, we can express the %-s, %-p character of the hybrid as

$$\%\text{-s} = 100*1/(1 + \lambda_i) \tag{4.11}$$

$$\%\text{-p} = 100*[\lambda_i/(1 + \lambda_i)] \tag{4.12}$$

The allowed values of the λ_i's are only constrained by the requirement that the total s-character and p-character from the four hybrids must sum properly to the total number of s orbitals (1) and p orbitals (3) available for their construction, namely, the "sum rules:"

$$\sum_{1 \leq i \leq 4} 1/(1 + \lambda_i) = 1 \text{ (s-orbital sum rule)} \tag{4.13}$$

$$\sum_{1 \leq i \leq 4} \lambda_i/(1 + \lambda_i) = 3 \text{ (p-orbital sum rule)} \tag{4.14}$$

which constrain the four hybrids to mutually consistent directions in three-dimensional space.

In actuality, the NHOs $\{h_i^{(A)}\}$ are obtained as linear combinations of *all* available NAOs $\{\theta_i^{(A)}\}$ on the atom:

$$h_i^{(A)} = \sum_j a_{ij} \, \theta_j^{(A)} \tag{4.15}$$

including (in principle) contributions from higher $d, f, \ldots$ orbitals. However, if we divide (4.15) into contributions from the NMB (valence shell $\theta_v^{(A)}$ NAOs) and NRB (Rydberg $\theta_r^{(A)}$ NAOs),

$$h_i^{(A)} = \sum_v a_{iv}\theta_v^{(A)} + \sum_r a_{ir} \, \theta_r^{(A)} \tag{4.16}$$

the NMB contributions [corresponding to the simple starting point (4.1)] are found to be overwhelmingly dominant. Hence, the elementary hybridization Equations (4.1)–(4.14) are found to provide *excellent* approximations to the quantitative NHOs found from the best available modern wavefunctions, and the student of chemistry should gain thorough familiarity with their usage. For further background on general hybridization theory, see *V&B*, Section 3.2.3.

Each of the NBOs $\{\Omega_i\}$ is expressed as a linear combination of constituent NHOs $\{h_j\}$, which in turn are composed of NAOs $\{\theta_k\}$. For example, a two-center σ_{AB} bond NBO between atoms A and B can be written as

$$\sigma_{AB} = c_A h_A + c_B h_B \tag{4.17}$$

where h_A, h_B [cf. (4.6)] are the respective hybrids on the atoms, and c_A, c_B are the *natural polarization coefficients* whose squares give the percentage contributions of h_A, h_B to the NBO. The polarity of each σ_{AB} bond can be quantified succinctly in terms of the *natural ionicity* parameter i_{AB}, defined as

$$i_{AB} \equiv (c_A^2 - c_B^2)/(c_A^2 + c_B^2) \tag{4.18}$$

The natural ionicity i_{AB} is zero for a pure covalent bond ($c_A = c_B$) but can achieve any value between -1 ($c_A = 0$; pure ionic hybrid on B) and $+1$ ($c_B = 0$; pure ionic hybrid on A), ranging smoothly between ionic and covalent limits. Do not even *think* about characterizing "ionic" and "covalent" as two distinct "types" of bonding; they are merely opposite limits of a *continuum* of ionicity values ($0 \leq |i_{AB}| \leq 1$) that exhibit *no* abrupt discontinuities or change of "type" as bond polarity shifts between the two ionic extremes.

With this background, we now proceed to examine details of the NBO output for the specific examples (HF, CH_3OH, H_2NCHO) chosen in Section 4.1.

4.2.1   Hydrogen Fluoride

The main NBO output for the hydrogen fluoride molecule is shown in I/O-4.7 which immediately follows the NBO search report I/O-4.1.

```
───────────────────────── I/O-4.7 ─────────────
   (Occupancy)   Bond orbital/ Coefficients/ Hybrids
-------------------------------------------------------------------

  1. (2.00000) BD ( 1) F  1- H  2
            ( 77.68%)   0.8814* F  1 s( 20.20%)p 3.94( 79.68%)d 0.01(  0.12%)
                                      0.0000 -0.4486  0.0287  0.0000  0.0000
                                      0.0000  0.0000  0.0000  0.0000  0.0000
                                      0.0000  0.0000  0.0000  0.8922 -0.0272
                                     -0.0019  0.0000  0.0000  0.0000  0.0000
                                      0.0000 -0.0341
            ( 22.32%)   0.4725* H  2 s( 99.88%)p 0.00(  0.12%)
                                     -0.9994  0.0000  0.0000  0.0000  0.0000
                                      0.0000 -0.0344
  2. (1.99995) CR ( 1) F  1          s(100.00%)
                                      1.0000  0.0000  0.0000  0.0000  0.0000
                                      0.0000  0.0000  0.0000  0.0000  0.0000
                                      0.0000  0.0000  0.0000  0.0000  0.0000
                                      0.0000  0.0000  0.0000  0.0000  0.0000
                                      0.0000  0.0000
  3. (1.99938) LP ( 1) F  1          s( 79.87%)p 0.25( 20.13%)d 0.00(  0.00%)
                                      0.0000  0.8936  0.0112  0.0000  0.0000
                                      0.0000  0.0000  0.0000  0.4485 -0.0073
                                      0.0045  0.0000  0.0000  0.0000  0.0000
                                      0.0000 -0.0039
  4. (1.99796) LP ( 2) F  1          s(  0.00%)p 1.00( 99.98%)d 0.00(  0.02%)
                                      0.0000  0.0000  0.0000  0.0000  0.0000
                                      0.0000  0.0000  0.0000  0.0000  0.9998
                                      0.0111 -0.0003  0.0000  0.0000  0.0000
                                      0.0000  0.0000  0.0000  0.0000 -0.0155
                                      0.0000  0.0000
  5. (1.99796) LP ( 3) F  1          s(  0.00%)p 1.00( 99.98%)d 0.00(  0.02%)
                                      0.0000  0.0000  0.0000  0.0000  0.0000
                                      0.9998  0.0111 -0.0003  0.0000  0.0000
                                      0.0000  0.0000  0.0000  0.0000  0.0000
                                      0.0000  0.0000  0.0000 -0.0155  0.0000
                                      0.0000  0.0000
  6. (0.00000) RY*( 1) F  1          s( 99.91%)p 0.00(  0.09%)d 0.00(  0.00%)
  :
 28. (0.00000) RY*( 6) H  2          s( 36.57%)p 1.73( 63.43%)
 29. (0.00000) BD*( 1) F  1- H  2
            ( 22.32%)   0.4725* F  1 s( 20.20%)p 3.94( 79.68%)d 0.01(  0.12%)
            ( 77.68%)  -0.8814* H  2 s( 99.88%)p 0.00(  0.12%)
```

The HF molecule output is similar to atomic NBO output seen previously (I/O-2.3), but with additional detail for the composition of each NBO from its constituent NHOs. As usual, we omit many of the uninteresting RY*-type NBOs ("leftovers" of the 6-311++G** basis) that make no significant contribution to molecular description.

As shown in I/O-4.7, NBO 1 is a two-center bond "BD (1)" (the first and only such bond in this species) between atoms F, H (in chosen input numbering) that is expressed as [cf. (4.17)]

$$\sigma_{FH} = c_F h_F + c_H h_H \qquad (4.19)$$

or more explicitly, as shown by entries for each atom,

$$\sigma_{FH} = 0.8814(sp^{3.94})_F + 0.4725(s)_H \qquad (4.20)$$

The squared polarization coefficients (given as parenthesized percentages before each hybrid listing) indicate that the σ_{HF} bond is rather strongly polarized toward F (77.68% on F, 22.32% on H), with corresponding ionicity parameter [cf. (4.18)]

$$i_{FH} = +0.5536 \qquad (4.21)$$

but still quite far from the "complete ionic" limit.

The displayed form of the "sp$^\lambda$ hybrid" shown in I/O-4.7 may initially seem somewhat confusing. The h_F hybrid for NBO 1 is more completely described as an "s^{1}p$^\lambda$d$^\mu$ hybrid" [the "1" on s is always understood as the "unit" against which p-character (λ) and d-character (μ) are measured], namely,

$$h_F = s^1 \, p^{3.94} \, d^{0.01} \qquad (4.22)$$

which corresponds to the parenthesized percentages given in the output (the more reliable way to "read" hybrid composition; see Sidebar 4.1)

$$h_F: \ 20.20\%\text{-s}, \ 79.68\%\text{-p}, \ 0.12\%\text{-d character} \qquad (4.23)$$

However, as shown in Equation (4.20), we shall generally neglect the weak contributions of d-type orbitals (which are confusingly called "polarization" orbitals by computational specialists) and concentrate on the dominant valence sp$^\lambda$ character of each bonding hybrid.

The 22 numbers printed below each hybrid are the NAO coefficients that form the hybrid (keyed to the 22 fluorine NAOs tabulated in I/O-3.1). For example, h_F is approximately described as

$$h_F \cong -0.45(2s)_F + 0.89(2p_z)_F \qquad (4.24)$$

As shown, for example, in I/O-4.8, each BD-type NBO includes a specification of NHO compositions in terms of a conventional sp^λ (exponential-type) label (which can be tricky to decipher) as well as the explicit percentages of s, p, d, . . . character (which can be read unambiguously). When in doubt, the latter should be trusted.

If the percentage s-character (%-s) is nonzero, no problems arise in converting %-p, %-s values to sp^λ form with Equations (4.10)–(4.12); effectively, the %-s becomes the "unit" for formal "s^1p^λ" labeling of relative %-p/%-s values. However, when %-s vanishes, it is necessary to choose a new "unit" to specify the angular ratios (e.g., in a hybrid label of "pd^μ" type, with p-character as the "unit"). In this case, the first nonvanishing angular component is given a "1.00" exponent, and remaining angular exponents are calculated by analogs of Equation (4.10) (e.g., with $\mu = \%\text{-d}/\%\text{-p}$, etc.).

An example of such relabeling is shown in I/O-4.8 for NBO 9, the $n_O^{(\pi)}$ lone pair, which is essentially a pure p-orbital as shown in the NHO specification:

```
9. (1.96534) LP ( 2) O 3 s( 0.00%)p 1.00( 99.95%)d 0.00( 0.05%)
```

The fact that this is "pure p" [with $\lambda = \infty$; cf. (4.10)] rather than "$sp^{1.00}$" (with $\lambda = 1.00$, as superficial reading of the label might suggest), is signaled by the "0.00%" value for %-s, which shifts the labeling scheme. For example, a pd^2 hybrid would be specified by

```
s(0.00%)p 1.00(33.33%)d 2.00(66.67%)
```

and a pure d hybrid would appear as

```
s(0.00%)p 0.00(0.00%)d 1.00(100.00%)
```

With a little care, the NHO output can be converted to an accurate exponent-type label even if the %-s, %-p, . . . values are ignored, but it's always wise to check how you have "read" the hybrid exponent(s) for consistency with the parenthesized percentage values.

and h_H is essentially the pure $(1s)_H$ NAO. Figure 4.5 depicts profile, contour, and surface plots of the overlapping (P)NHOs (upper panels) compared with the final σ_{HF} (P)NBO (lower panels), illustrating the chemical magic of the quantum mechanical wave-mixing phenomenon.

The remaining occupied NBOs in I/O-4.7 correspond to the fluorine core (NBO 2) and valence lone pairs (NBOs 3–5). As discussed in Sidebar 4.2, the on-axis $n_F^{(\sigma)}$ is basically an s-rich $sp^{0.25}$ hybrid, "opposite" to the p-rich $sp^{3.94}$ hybrid that was used for bonding in (4.20), and therefore quite distinct in composition, shape, and energetics from the two off-axis p-π lone pairs (NBOs 4, 5), which are basically of pure p_x, p_y form.

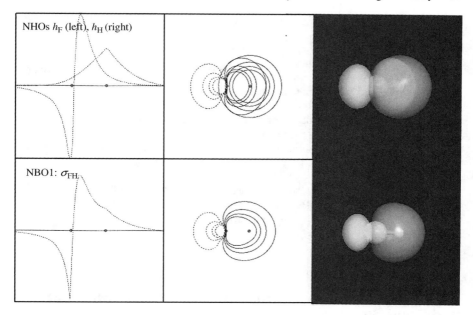

Figure 4.5 σ_{FH} bond of hydrogen fluoride, shown as overlapping NHOs (upper) or as final NBO (lower); (cf. I/O-4.7). (See the color version of this figure in Color Plates section.)

SIDEBAR 4.2   *"TRIPODS," "RABBIT EARS," AND OTHER ORBITAL ABSURDITIES*

Why the difference in F lone pairs in Fig. 4.1 (or O lone pairs in Figs. 4.2 and 4.3)? The answer is related to the basic quantum mechanical reason for hybridization itself (cf. *V&B*, pp. 52ff, 105ff).

An isolated atom has no reason to hybridize, because s and p orbitals differ in energy and symmetry, with no physical interaction to break symmetry nor variational incentive to "reward" s–p mixing. However, in the presence of a potential bonding partner along, say, the **z** direction, the spherical symmetry is broken and s–p_z interactions become possible. Because the perturbation occurs only along the **z** direction, the p_x, p_y orbitals remain unaffected, but s and p_z orbitals can mix to form two new "hybrid" orbitals h_1, h_2:

$$h_1 = c_s\, s + c_z\, p_z \qquad (4.25)$$

$$h_2 = c_z\, s - c_s\, p_z \qquad (4.26)$$

As shown, the mixing coefficients c_s, c_z of the in-phase hybrid (4.25) must be switched with a sign change in the out-of-phase hybrid (4.26) to maintain mutual orthogonality. [For example, if h_1 is primarily of p-character ($c_z > c_s$), then h_2 is primarily of s-character and points in the opposite direction.] In HF (Fig. 4.1), the on-axis lone pair $n_F^{(\sigma)}$ is the s-rich hybrid h_2, whereas the two off-axis $n_F^{(\pi)}$ lone pairs are essentially the atomic p_x, p_y orbitals that were left uninvolved in bond formation. Thus, in linear bonding the off-axis (atom-like) lone pairs must always remain distinct from the on-axis (hybrid) lone pair that shares the brunt of chemical bonding.

Similarly, in planar bonding (e.g., H_2O or related alcohols and ethers), hybridization can only involve the two p orbitals (e.g., p_x, p_y) that lie in the plane of bonding. Hybridization of oxygen s, p_x, p_y orbitals therefore results in three orthonormal hybrids in the bonding **x–y** plane, one of which becomes the in-plane $n_O^{(\sigma)}$ (as seen, e.g., in NBO 8 of Fig. 4.2 for CH_3OH), while the remaining unused p_z orbital becomes the out-of-plane $n_O^{(\pi)}$ (e.g., NBO 9 in Fig. 4.2). Analogous considerations apply to the inequivalent O lone pairs of carbonyl groups in aldehydes, amides, or ketones (cf. Fig. 4.3). Thus, the "rabbit ears" depiction of water lone pairs, no matter how impressively rendered in your textbook, makes *no* physical or chemical sense and should be eradicated from the thinking of all serious students of bonding theory.

[Rabbit-ears depictions are sometimes argued to be "mathematically equivalent" to the $n_O^{(\sigma)}$, $n_O^{(\pi)}$ forms, but such (Hückel-type) arguments cannot be justified at any level of theory that is relevant to the contemporary scale of chemical accuracy.]

4.2.2   Methanol and Formamide: Hybrid Directionality and Bond Bending

As a more representative three-dimensional molecular geometry, let us return to the CH_3OH molecule. Output details of the valence hybrids are shown for *L*-type NBOs in I/O-4.8 (with NAO coefficient tables omitted).

Recall (Section 3.2) that the methanol geometry was created with idealized tetrahedral bond angles, suggesting idealized sp^3 hybrids as in methane. Nevertheless,

```
─────────────────────────── I/O-4.8 ───────────────────────────
    (Occupancy)   Bond orbital/ Coefficients/ Hybrids
-----------------------------------------------------------------------
  1. (1.99265) BD ( 1) H  1- C  2              .
           ( 40.81%)   0.6388* H  1 s( 99.98%)p 0.00(  0.02%)
           ( 59.19%)   0.7694* C  2 s( 26.57%)p 2.76( 73.29%)d 0.01(  0.14%)
  2. (1.99850) BD ( 1) C  2- O  3
           ( 33.34%)   0.5774* C  2 s( 22.83%)p 3.37( 76.91%)d 0.01(  0.25%)
           ( 66.66%)   0.8165* O  3 s( 29.25%)p 2.42( 70.67%)d 0.00(  0.08%)
  3. (1.99703) BD ( 1) C  2- H  5
           ( 58.20%)   0.7629* C  2 s( 25.52%)p 2.91( 74.33%)d 0.01(  0.15%)
           ( 41.80%)   0.6465* H  5 s( 99.98%)p 0.00(  0.02%)
  4. (1.99703) BD ( 1) C  2- H  6
           ( 58.20%)   0.7629* C  2 s( 25.52%)p 2.91( 74.33%)d 0.01(  0.15%)
           ( 41.80%)   0.6465* H  6 s( 99.98%)p 0.00(  0.02%)
  5. (1.99033) BD ( 1) O  3- H  4
           ( 72.92%)   0.8539* O  3 s( 20.98%)p 3.76( 78.89%)d 0.01(  0.14%)
           ( 27.08%)   0.5204* H  4 s( 99.89%)p 0.00(  0.11%)
   .
   .
  8. (1.98417) LP ( 1) O  3              s( 49.84%)p 1.01( 50.12%)d 0.00(  0.03%)
  9. (1.96534) LP ( 2) O  3              s(  0.00%)p 1.00( 99.95%)d 0.00(  0.05%)
```

the C hybrid to O (sp$^{3.37}$, NBO 2) is seen to be slightly *richer* in p-character (76.91%) than those to the three H atoms, which also differ slightly from one another (sp$^{2.76}$ hybrid of 73.29%-p character in NBO 1, but sp$^{2.91}$ hybrids of 74.33%-p character in NBOs 3, 4). The four λ_i values for the carbon NHOs

$$\text{C hybrids: } \lambda_{H1} = 2.76, \ \lambda_{H5} = \lambda_{H6} = 2.91, \ \lambda_O = 3.37 \qquad (4.27)$$

are found to satisfy the sum rules (4.13 and 4.14) to high accuracy, and their average is very close to the "expected" $\lambda = 3.00$ (75%-p) for idealized tetrahedral sp^3 bonding. The same is found to be true for the four oxygen NHOs:

$$\text{O hybrids: } \lambda_C = 2.42, \ \lambda_H = 3.76, \ \lambda_{n(\sigma)} = 1.01, \ \lambda_{n(\pi)} = \infty \text{ (NBO 9)} \qquad (4.28)$$

(cf. Sidebar 4.1 for discussion of the "$\lambda = \infty$" for $n_O^{(\pi)}$ hybrid in NBO 9). Although the NHOs show "average" resemblance to idealized sp^3 hybridization, the deviations from ideality reflect subtle electronic influences that may be expected to *break* the idealized tetrahedral symmetry under complete geometry optimization.

Let us examine some structural consequences of these hybridizations in greater detail. From Equation (4.9), we can see that the hybrid angle ω_{13} for in-plane H$_1$−C−O carbon NHOs is given by

$$\omega_{13} = \cos^{-1}[-1/(2.76*3.37)^{1/2}] = 109.14° \qquad (4.29)$$

Similarly, ω_{56} for out-of-plane H$_5$−C−H$_6$ NHOs is

$$\omega_{56} = \cos^{-1}[-1/(2.91*2.91)^{1/2}] = 110.10° \qquad (4.30)$$

ω_{15} for H$_1$−C−H$_{5/6}$ NHOs is

$$\omega_{15} = \cos^{-1}[-1/(2.76*2.91)^{1/2}] = 110.66° \qquad (4.31)$$

and ω_{35} for O−C−H$_{5/6}$ NHOs is

$$\omega_{35} = \cos^{-1}[-1/(3.37*2.91)^{1/2}] = 108.62° \qquad (4.32)$$

The carbon NHOs are seen to be slightly "misaligned" with respect to the actual 109.47° angles between nuclei.

Some details of the "bond-bending" or "strain" in NHO (mis)alignments with nuclei are summarized in the "NHO directionality" output shown in I/O-4.9. For cases in which the NHO direction differs from the line of nuclear centers by 1° or more (and other threshold criteria noted at the top of the output are satisfied), the table gives the polar (theta) and azimuthal (phi) angles that specify the "line of centers" direction in spherical polar coordinates for the coordinate system chosen by the host ESS, together with the corresponding hybrid direction and the angular deviation (Dev) at each center. For CH$_3$OH, the maximum deviations are seen to be rather small ($\sim$2°), suggesting (as expected) that bond strain is minimal in this acyclic species.

─── *I/O-4.9* ───

```
NHO DIRECTIONALITY AND BOND BENDING (deviations from line of nuclear centers)


        [Thresholds for printing:  angular deviation  >  1.0 degree]
                                   p- or d-character  > 25.0%
                                   orbital occupancy  >  0.10e


                    Line of Centers      Hybrid 1            Hybrid 2
                    ---------------   -------------------  -------------------

            NBO       Theta  Phi     Theta  Phi   Dev      Theta  Phi   Dev
        ========================================================================
    1. BD ( 1) H 1- C 2    90.0  340.5     --     --    --      90.0  162.2  1.6
    2. BD ( 1) C 2- O 3    90.0  270.0    90.0  272.1  2.1      90.0   88.1  1.9
    3. BD ( 1) C 2- H 5    35.3   35.3    35.0   33.5  1.1       --     --    --
    4. BD ( 1) C 2- H 6   144.7   35.3   145.0   33.5  1.1       --     --    --
    8. LP ( 1) O 3          --     --     90.0    0.0   --       --     --    --
    9. LP ( 2) O 3          --     --      5.7    0.0   --       --     --    --
```

─── *I/O-4.10* ───

```
    (Occupancy)   Bond orbital/ Coefficients/ Hybrids
----------------------------------------------------------------------------

  1. (1.99673) BD ( 1) N  1- C  2
              ( 62.10%)   0.7881* N  1 s( 39.03%)p 1.56( 60.87%)d 0.00(  0.10%)
              ( 37.90%)   0.6156* C  2 s( 33.41%)p 1.99( 66.47%)d 0.00(  0.12%)
  2. (1.99130) BD ( 1) N  1- H  4
              ( 70.18%)   0.8377* N  1 s( 30.40%)p 2.29( 69.50%)d 0.00(  0.10%)
              ( 29.82%)   0.5461* H  4 s( 99.96%)p 0.00(  0.04%)
  3. (1.99200) BD ( 1) N  1- H  5
              ( 69.73%)   0.8350* N  1 s( 30.49%)p 2.28( 69.41%)d 0.00(  0.10%)
              ( 30.27%)   0.5502* H  5 s( 99.96%)p 0.00(  0.04%)
  4. (1.99816) BD ( 1) C  2- O  3
              ( 29.82%)   0.5461* C  2 s(  0.00%)p 1.00( 99.59%)d 0.00(  0.41%)
              ( 70.18%)   0.8377* O  3 s(  0.00%)p 1.00( 99.88%)d 0.00(  0.12%)
  5. (1.99684) BD ( 2) C  2- O  3
              ( 35.71%)   0.5975* C  2 s( 35.23%)p 1.83( 64.62%)d 0.00(  0.15%)
              ( 64.29%)   0.8018* O  3 s( 39.75%)p 1.51( 60.04%)d 0.01(  0.21%)
  6. (1.98379) BD ( 1) C  2- H  6
              ( 56.83%)   0.7538* C  2 s( 32.10%)p 2.11( 67.77%)d 0.00(  0.13%)
              ( 43.17%)   0.6570* H  6 s( 99.96%)p 0.00(  0.04%)
   :
 10. (1.75091) LP ( 1) N  1           s(  0.00%)p 1.00( 99.99%)d 0.00(  0.01%)
 11. (1.98437) LP ( 1) O  3           s( 60.38%)p 0.66( 39.59%)d 0.00(  0.03%)
 12. (1.85188) LP ( 2) O  3           s(  0.01%)p 1.00( 99.91%)d 0.00(  0.09%)
   :

                    Line of Centers      Hybrid 1            Hybrid 2
                    ---------------   -------------------  -------------------

            NBO       Theta  Phi     Theta  Phi   Dev      Theta  Phi   Dev
        ========================================================================
    1. BD ( 1) N 1- C 2    90.0   46.4     --     --    --      90.0  229.6  3.1
    2. BD ( 1) N 1- H 4    90.0  287.0    90.0  285.6  1.4       --     --    --
    4. BD ( 1) C 2- O 3    90.0  351.3     5.7  349.0 84.3       5.7  171.6 84.3
    5. BD ( 2) C 2- O 3    90.0  351.3    90.0  350.0  1.3       --     --    --
    6. BD ( 1) C 2- H 6    90.0  114.0    90.0  112.2  1.7       --     --    --
   10. LP ( 1) N 1          --     --      5.7   63.3   --       --     --    --
   11. LP ( 1) O 3          --     --     90.0  171.1   --       --     --    --
   12. LP ( 2) O 3          --     --     90.0   80.9   --       --     --    --
   85. BD*( 1) C 2- O 3    90.0  351.3     5.7  349.0 84.3       5.7  171.6 84.3
```

For comparison, we show some details of the formamide NBOs and the corresponding NHO directionality table in I/O-4.10. The apparent large "Dev" of 84.3° for NBO 4 is merely a consequence of its π-bond nature (cf. orbital plots in Fig. 4.3), in which the constituent p-π-type NHOs point $\sim$90° from the line of nuclear centers. The remaining deviations of σ-type hybrids are rather small, suggesting that bond strain is again minimal in the formamide species. However, much stronger examples of bond bending can be found in cyclopropane and similar "strained" cyclic species (cf. *V&B*, p. 146ff for further discussion).

4.3 BOND POLARITY, ELECTRONEGATIVITY, AND BENT'S RULE

One can see from Equation (4.10) that hybrid p-character (and associated λ hybridization parameter) strongly affects hybrid direction and molecular shape. But what affects hybrid p-character? The answer to this question gives one of the deepest insights into molecular shape, and is expressed in simple and intuitive terms by Bent's rule (cf. *V&B*, p. 138ff), the deeper principle that underlies success of the valence shell electron pair repulsions (VSEPR) model (Sidebar 4.3).

SIDEBAR 4.3 | *"FAT" LONE PAIRS, "SKINNY" POLAR BONDS, AND OTHER VSEPR MISCONCEPTIONS*

Although the strange "steric demand" concepts of VSEPR models, skillfully deployed, can "give the right answer" in a surprising number of cases, the same steric-style VSEPR reasoning proves unreliable in what might be considered reasonable extensions to more complex molecules. (Quantitation of steric size is discussed in Chapter 6.)

For example, if one considers ethane-like molecules in which one of the hydride bonds on each end is replaced by a "fat" lone pair (e.g., in hydrazine, H_2N-NH_2) versus a "skinny" polar bond (e.g., in difluoroethane, H_2FC-CH_2F), one might reasonably expect that the "fat" lone pairs would "repel" as far as possible, to the *anti* conformation, whereas the "skinny" polar bonds would prefer being adjacent to one another in a tilted *syn* conformation, namely,

anti   *syn*

However, neither expectation is correct; both species adopt twisted *gauche* conformers by significant energetic margins. If instead one puts a lone pair at one end and polar bond at the other, as in $H_2\ddot{N}-CH_2F$, the preferred *anti* conformer separates the lone pair and polar bond as *far* as possible, contrary to what seems "reasonable" from VSEPR assumptions.

Another simple example is presented by the methanol (CH_3OH) molecule, as previously considered in Section 3.2. Instead of the idealized tetrahedral methyl geometry imposed in I/O-3.3, VSEPR-style logic might lead one to expect that the methyl group should tilt *away from* the "fat rabbit ears" and into the "skinnier" polar OH bond if allowed to optimize its geometry. In fact, the *opposite* occurs, with the methyl group seemingly "tilting into the lone pairs" by ca. 2–3°. [Similar VSEPR-defiant methyl tilting effects are seen in methylamine (H_3C-NH_2) and related species.]

For those who continue chemistry studies beyond freshman level, it is not surprising that VSEPR-type concepts play no significant role in more advanced quantum mechanical theories of organic or inorganic molecular structure. Judicious replacement of VSEPR concepts with equivalent Bent's rehybridization concepts (e.g., replacing "fat" by "more s-like," and "skinny" by "more p-like") could significantly improve the accuracy of current freshman-level pedagogy.

Bent's rule describes how a central main-group atom A allocates the percentage s/p-character of its bonding hybrids toward bonding partners X, Y of *un*equal electronegativity. (It would be quantum mechanically unreasonable to expect A to use equivalent hybrids to form *in*equivalent A—X and A—Y bonds.) Specifically, Bent's rule for main-group atoms can be stated as follows:

A central atom tends to direct hybrids of higher p-character (higher λ) toward more electronegative substituents (4.33)

or equivalently,

Atomic s-character (lower λ) tends to accumulate in hybrids directed toward the least electronegative substituents (4.34)

[The corresponding generalization of Bent's rule for transition metals is described in *V&B*, p. 421ff.]

The quantum mechanical rationalization for Bent's rule can be readily understood by beginning chemistry students, based on the realization that valence s orbitals always lie *beneath* valence p orbitals in energy ($\varepsilon_s < \varepsilon_p$). The electrons that remain "close" to A (i.e., in bonds strongly polarized toward A, with ionicity $i_{AX} > 0$) will therefore demand s-rich hybrids, to keep their energy as low as possible, while those that are "far" from A (i.e., in bonds strongly polarized toward X, with ionicity $i_{AX} < 0$) can be allocated the remaining p-rich hybrids, preserving lowest overall energy. In particular, *lone pair* hybrids ("bonds to atoms of *zero* electronegativity") should acquire highest s-character, whereas *vacant* hybrids ("bonds to atoms of infinite electronegativity") correspondingly acquire highest p-character according to Bent's rule.

Despite its simple and intuitive character, Bent's rule is surprisingly successful in anticipating the subtle variations of s/p-character found in quantitative NHOs. For example, one can see in I/O-4.8 that the methanol C hybrid to O has higher p-character (76.9%) than those to H's (73.3–74.3%), consistent with the higher electronegativity of O versus H. Similarly, one can see that the O hybrid to H has higher p-character (78.9%) than the corresponding C hybrids to H (consistent with the greater electronegativity difference between central atom and ligand in O−H versus C−H bonding). Alternatively, if we compare the three in-plane oxygen hybrids, we see that the hybrids to H ($sp^{3.76}$, 78.9%-p) or C ($sp^{2.42}$, 70.7%-p) are far richer in p-character than the oxygen $n_O^{(\sigma)}$ lone pair ($sp^{1.01}$, 50.1%-p), as Bent's rule anticipates.

When combined with the Coulson directionality theorem (4.9), Bent's rule shows the far-reaching connection between molecular shape and atom electronegativity differences. For example, if the idealized tetrahedral sp^3 hybrids of methane (CH_4) are perturbed by replacing an H by F to make fluoromethane (CH_3F), one anticipates from (4.33) and (4.34) that the electronegative F substituent will draw a carbon hybrid of *higher* p-character (actually, $sp^{3.84}$) than those to H ($sp^{2.75}$), which results, according to (4.9), in slightly reduced F−C−H angles (to 108.6°) and expanded H−C−H angles (110.3°), as though the polar C−F bond "occupies smaller angular volume." In the qualitative form stated above, Bent's rule sometimes leads to ambiguities or conflicts with other hybridization constraints (see Sidebar 4.4), and its quantitative formulation involves considerable complexity (see *V&B*, p. 139ff). Nevertheless, judicious application of Bent's rehybridization concept allows one to draw useful predictive inferences concerning many subtle variations of molecular shape, based on known electronegativity differences.

SIDEBAR 4.4   *BORDERLINE VIOLATIONS OF BENT'S RULE*

Bent's rule is usually well satisfied by the quantitative NHO hybridizations. However, exceptional cases of intrinsic ambiguity or conflict with other hybridization constraints sometimes lead to apparent violations of Bent's rule that are worthy of special note.

Ambiguous cases often arise in ionic species, where there is intrinsic uncertainty concerning the charge state of each atom. For example, the σ_{CO} bond of neutral carbon monoxide (CO) reflects the greater p-character of the C hybrid ($s^{2.59}$) than the O hybrid ($sp^{1.19}$), as suggested by Bent's rule. However, the corresponding radical cation (CO^+) shows a *reversal* of hybrid p-character ($sp^{0.38}$ on C versus $sp^{2.93}$ on O) that seemingly contradicts the "known" higher electronegativity of O compared to C. However, it is ambiguous in the latter case whether one should compare electronegativities of neutral C versus neutral O, or of cationic C^+ versus neutral O (or some alternative charge partitioning, for example, $C^{+0.75}$ versus $O^{+0.25}$). Natural population analysis indicates that C^+ versus O is indeed the relevant electronegativity difference for application of Bent's rule, and cationic

C^+ is indeed more electronegative than neutral O. Similar ambiguities arise in open-shell neutral species (where atoms may have different effective spin-charges in the two spin sets), ylidic species (where "formal charges" alter the usual neutral atom properties), and other "unusual" atomic charge states.

A more surprising exception occurs in ammonia (NH_3), whose geometry is superficially in good accord with Bent's rule. The optimized geometry exhibits slightly reduced H$-$N$-$H angles ($107.9°$) and correspondingly increased formal lp$-$N$-$H angles, seemingly consistent with greater s-character in the n_N lone pair and greater p-character in the σ_{NH} hybrids, as anticipated by Bent's rule. However, the actual NBOs of NH_3 show an unexpected result,

$$NH_3: \sigma_{NH} = 0.82(sp^{2.82})_N + 0.57(s)_H$$
$$NH_3: n_N = (sp^{3.65})_N$$

with *higher* p-character in the lone pair (78.4%, $\lambda_n = 3.65$) than in the N$-$H bond hybrid (73.7%, $\lambda_H = 2.82$) and evidence of unusual bond strain. Ammonia is apparently an unusual outlier compared to related species such as NF_3

$$NF_3: \sigma_{NF} = 0.60(sp^{6.78})_N + 0.80(s)_H$$
$$NF_3: n_N = (sp^{0.59})_N$$

or PH_3

$$PH_3: \sigma_{PH} = 0.70(sp^{5.85})_P + 0.71(s)_H$$
$$PH_3: n_P = (sp^{0.77})_P$$

that exhibit the common Bent-compliant hybridization pattern of p-rich polar bonds and s-rich lone pairs. The anomalous "anti-Bent" behavior of NH_3 is associated with unusually large bond-bending deviations ($\sim 4°$) of *opposite* sense to those in PH_3 (see *V&B*, p. 147ff), and with unusually low barrier to umbrella inversion and relatively unpuckered C_{3v} equilibrium geometry, all exceptional compared to other Group-15 hydrides.

Why does Bent's rule seem to fail in this case? One can see that a near-planar C_{3v} geometry brings Bent's rule into direct conflict with the planar bonding symmetry restriction (Sidebar 4.2), which requires the lone pair to be of *pure* p character (rather than "s-rich," as Bent's rule suggests). Apparently, the NH bonds of NH_3 (strongly polarized toward the central N atom) allow the planar-symmetry limit to dominate (by a narrow margin; see below), leading to weak pyramidalization, low-inversion barrier, p-rich lone pair, and H nuclei *lagging* progressively behind the NHOs as pyramidalization proceeds. In contrast, the PH bonds of PH_3 allow Bent's rule to dominate, leading to pronounced nonplanarity, high inversion barrier, s-rich lone pair, and H nuclei *leading* the NHOs as pyramidalization proceeds.

Even NH_3 is only marginally anti-Bent. If one examines how the NHO hybridizations vary with pyramidalization angle or R_{NH} (Fig. 4.6), one can see that only a miniscule change in equilibrium geometry ($\sim 2°$ pyramidalization increase or ~ 0.05 Å NH bond length increase) would restore NH_3 to "normal" Bent-compliant hybridization (and remove much of its exceptional bond strain). Thus, ammonia appears to be a borderline "exception that proves the rule."

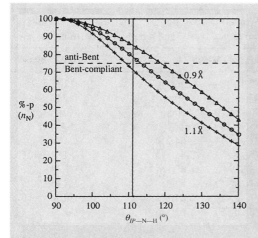

Figure 4.6 Percentage p-character of nitrogen lone pair n_N in ammonia as a function of pyramidalization angle θ_{lp-N-H}, shown for R_{NH} at equilibrium (1.015 Å; circles), slightly elongated (1.1 Å; plusses), and slightly contracted (0.9 Å; triangles). The vertical line marks the equilibrium pyramidalization angle (111.0°), and the horizontal dashed line (ideal sp^3 hybrid, 75%-p) marks the boundary between planar-limit ("anti-Bent") and Bent's rule hybridization. Note that only a slight increase in pyramidalization angle (by ~2°) or R_{NH} (by ~0.05 Å) would restore Bent-compliant hybrids.

Despite occasional exceptions such as NH_3, Bent's rule generally provides more fundamental and accurate rationalizations for substituent-induced rehybridization and geometry changes than does the VSEPR model.

Of course, the attentive student will recognize that these and many similar structural inferences are usually given freshman-level rationalizations in terms of the VSEPR model (Sidebar 4.3). The VSEPR model gives "the right answer for the wrong reason" in selected main-group examples, which are generally understood more satisfactorily in terms of Bent's rehybridization concepts. However, VSEPR-type concepts fail *spectacularly* for many transition metal species (cf. *V&B*, p. 389ff), whereas Bent's rehybridization rule, suitably generalized to sd^μ-type bonding (*V&B*, p. 421ff), continues to account successfully for molecular shape changes. Moreover, when extended to other main-group applications (beyond a narrow domain of fortuitous agreement with rehybridization concepts), the VSEPR-type "steric demand" concepts fail conspicuously (Sidebar 4.3). [A beginning chemistry student who was indoctrinated with VSEPR-style rationalizations, but never introduced to Bent's rule, may therefore wish to consider a request for tuition refund!]

The most direct measure of bond polarity and atom electronegativity difference is provided by the natural ionicity parameter of Equation (4.18). As examples, Table 4.1 displays the σ_{CX}-type NBOs and i_{CX} parameters for carbon bonds to X = H, N, O in H_2NCHO, CH_3OH, and simple hydrocarbons. In the C–X competition for bonding electrons, positive $i_{CX} > 0$ signals that C is more successful ("more electronegative") than X, whereas $i_{CX} < 0$ ($c_X^2 > c_A^2$) signals that X gained the greater share of the electron pair. The H_2NCHO i_{CX} entries of Table 4.1 show that carbon is more electronegative than hydrogen ($i_{CH} = 0.14$), but distinctly less electronegative than nitrogen ($i_{CN} = -0.24$) or oxygen [$i_{CO} = -0.29(\sigma), -0.40(\pi)$].

However, we should recognize that the actual i_{AB} competition is between the *groups* at either end of the A-B bond. The "effective electronegativity" of A is altered

Table 4.1 C–X bonding NBOs and associated natural ionicity values (i_{CX}) for various bonding partners (X = H, N, O) in formamide, methanol, and simple hydrocarbons.

Molecule	X	C–X bonding NBO	i_{CX}
H_2NCHO	H	$\sigma_{CH} = 0.75(sp^{2.11})_C + 0.66(s)_H$	+0.137
	N	$\sigma_{CN} = 0.62(sp^{1.99})_C + 0.79(sp^{1.56})_N$	−0.242
	O	$\sigma_{CO} = 0.60(sp^{1.83})_C + 0.80(sp^{1.51})_O$	−0.286 (σ)
	O	$\pi_{CO} = 0.55(p)_C + 0.84(p)_O$	−0.404 (π)
CH_3OH	H	$\sigma_{CH} = 0.77(sp^{2.76})_C + 0.64(s)_H$	+0.184 (H_1)
	H	$\sigma_{CH} = 0.76(sp^{2.91})_C + 0.65(s)_H$	+0.164 (H_5,H_6)
	O	$\sigma_{CO} = 0.58(sp^{3.37})_C + 0.82(sp^{2.42})_O$	−0.334
$H_3C{-}CH_3$	H	$\sigma_{CH} = 0.77(sp^{3.25})_C + 0.63(s)_H$	+0.194
$H_2C{=}CH_2$	H	$\sigma_{CH} = 0.77(sp^{2.37})_C + 0.64(s)_H$	+0.186
HCCH	H	$\sigma_{CH} = 0.78(sp^{1.09})_C + 0.62(s)_H$	+0.224

by other substituent groups as well as by resonance delocalization effects. For example, the C–O ionicity in methanol ($i_{CO} = -0.33$) differs somewhat from that in formamide ($i_{CO} = -0.29$), because the groups at the carbon end (H_3C- versus H_2NC-) and oxygen end ($-OH$ versus $-O$) differ in the two cases. One important aspect of this difference is the substituent-induced rehybridization at each center (in accordance with Bent's rule), with greater s-character tending to confer greater effective electronegativity. This is illustrated in Table 4.1 by the noticeably higher i_{CH} for acetylene (0.22) compared to other CH bonds in the table (0.14–0.19). The additional influence of hyperconjugative polarization shifts (Chapter 5) is evident in differences between in-plane versus out-of-plane CH bonds of methanol or the slightly irregular sp^λ-dependence in the hydrocarbons.

The intuitive connection between bond ionicity and atomic electronegativity differences suggests a definite relationship between these quantities (cf. V&B, p. 131ff). The relationship can be made explicit by defining a "natural" scale of atomic electronegativity (Ξ_A) by the following equation:

$$i_{AH} = 1 - \exp[-0.45(\Xi_A - \Xi_H)] \tag{4.35}$$

or equivalently,

$$\Xi_A = \Xi_H - \ln[(1 - i_{AH})/0.45] \tag{4.36}$$

The H-atom electronegativity Ξ_H is fixed to match the assigned value on the Pauling or Allred–Rochow electronegativity scales ($\Xi_H = 2.10$), and the exponential scale factor "0.45" in (4.35) is chosen to express the resulting Ξ-scale values as nearly as possible in "Pauling units." The required i_{AH} values for each atom A are calculated from the simplest possible Lewis-like AH_n species in equilibrium ground-state geometry.

As defined in this manner, the natural electronegativity scale agrees closely with empirical Pauling, Mulliken, or Allred–Rochow scales (as closely as any of these scales agree with one another). Such "natural" scale is based on a more methodical and firmly grounded theoretical procedure than the empirical scales, and can be systematically improved as more accurate theoretical methods are developed, particularly for heavier elements of the periodic table.

One can see from Table 4.1 that the σ_{CO} and π_{CO} NBOs of the carbonyl double bond have *distinct* ionicity values (-0.29 versus -0.40). Such differences suggest that a π-bonding atom may be described as possessing both "σ-electronegativity" ($\Xi_A^{(\sigma)}$) and "pi-electronegativity" ($\Xi_A^{(\pi)}$) values that are generally *un*equal. The natural π-electronegativity scale can be defined by an equation analogous to (4.35):

$$i_{AC}^{(\pi)} = 1 - \exp[-0.45(\Xi_A^{(\pi)} - \Xi_C^{(\pi)})] \tag{4.37}$$

or equivalently,

$$\Xi_A^{(\pi)} \equiv \Xi_C^{(\pi)} - \ln[(1 - i_{AC}^{(\pi)})/0.45] \tag{4.38}$$

where the "reference" C atom is assigned (rather arbitrarily) to equal σ- and π-electronegativity values:

$$\Xi_C^{(\pi)} = \Xi_C^{(\sigma)} = 2.60 \tag{4.39}$$

and the $i_{AC}^{(\pi)}$ values are taken from prototype $H_nA=CH_2$ species of simplest double-bonded form.

Table 4.2 presents a comprehensive list of natural σ-electronegativities for elements 1–120, and Table 4.3 presents corresponding π-electronegativities for some Group 14–16 elements. [Note from these tables that oxygen has slightly lower π-electronegativity ($\Xi_O^{(\pi)} = 3.43$) than σ-electronegativity ($\Xi_O^{(\sigma)} = 3.48$), so the formamide σ_{CO} versus π_{CO} ionicity differences shown in Table 4.1 suggest interesting conjugative effects of the amide environment (Chapter 5), rather than simple inductive effects of neighboring electronegativity differences.] The natural electronegativity values of Tables 4.2 and 4.3, combined with Equations (4.35) and (4.37), allow one to make reasonably intelligent guesses of bond polarities for many species, particularly when hyperconjugative and conjugative perturbations are minimal.

Table 4.2 Natural σ-electronegativities $\Xi_A^{(\sigma)}$ of elements 1-120, from higher-level B3LYP/6-311++G** (or relativistic LAC3p++) theory for normal-valent hydrides (unparenthesized) or lower-level estimates based on monohydride bond polarities, NAO energies, or other descriptors (parenthesized). (Tabulated values for f-Group elements correspond to a specific electron configuration and may vary widely for other low-lying configurations employed in bonding.)

Z	Atom	$\Xi_A^{(\sigma)}$	Z	Atom	$\Xi_A^{(\sigma)}$	Z	Atom	$\Xi_A^{(\sigma)}$	Z	Atom	$\Xi_A^{(\sigma)}$
1	H	[2.10]	31	Ga	1.39	61	Pm	(0.96)	91	Pa	(1.06)
2	He	(4.04)	32	Ge	1.74	62	Te	(0.97)	92	U	(0.99)
3	Li	0.79	33	As	1.93	63	Eu	(0.80)	93	Np	(1.15)
4	Be	1.02	34	Se	2.21	64	Gd	(0.96)	94	Pu	(1.07)
5	B	1.86	35	Br	2.47	65	Tb	(0.96)	95	Am	(0.90)

(continued)

Table 4.2 (*Continued*)

Z	Atom	$\Xi_A^{(\sigma)}$	Z	Atom	$\Xi_A^{(\sigma)}$	Z	Atom	$\Xi_A^{(\sigma)}$	Z	Atom	$\Xi_A^{(\sigma)}$
6	C	2.60	36	Kr	(2.73)	66	Dy	(0.97)	96	Cm	(1.04)
7	N	3.07	37	Rb	0.83	67	Ho	(0.93)	97	Bk	(1.04)
8	O	3.48	38	Sr	0.83	68	Er	(0.81)	98	Cf	(1.04)
9	F	3.89	39	Y	1.09	69	Tm	(0.79)	99	Es	(0.98)
10	Ne	(4.44)	40	Zr	1.43	70	Yb	(0.82)	100	Fm	(0.85)
11	Na	0.88	41	Nb	1.67	71	Lu	(1.01)	101	Md	(1.10)
12	Mg	1.04	42	Mo	2.16	72	Hf	1.34	102	No	(0.96)
13	Al	1.35	43	Tc	2.25	73	Ta	1.54	103	Lr	(1.00)
14	Si	1.78	44	Ru	2.31	74	W	1.94	104	Rf	(1.25)
15	P	2.06	45	Rh	2.23	75	Re	2.20	105	Db	(1.42)
16	S	2.42	46	Pd	2.04	76	Os	2.17	106	Sg	(1.72)
17	Cl	2.76	47	Ag	1.48	77	Ir	2.22	107	Bh	(2.15)
18	Ar	(3.12)	48	Cd	(1.18)	78	Pt	2.30	108	Hs	(2.10)
19	K	0.82	49	In	1.32	79	Au	2.01	109	Mt	(2.21)
20	Ca	0.87	50	Sc	1.58	80	Hg	(1.51)	110	Ds	(2.40)
21	Sc	1.16	51	Sb	1.72	81	Tl	1.43	111	Rg	(2.26)
22	Ti	1.55	52	Te	1.95	82	Pb	1.64	112	Uub	(1.84)
23	V	1.79	53	I	2.19	83	Bi	1.70	113	Uut	(1.48)
24	Cr	(2.10)	54	Xe	(2.40)	84	Po	(1.92)	114	Uuq	(1.60)
25	Mn	(2.03)	55	Cs	0.81	85	At	(2.16)	115	Uup	(1.68)
26	Fe	2.03	56	Ba	0.78	86	Rn	(2.28)	116	Uuh	(1.88)
27	Co	1.96	57	La	(0.88)	87	Fr	(0.74)	117	Uus	(2.12)
28	Ni	1.87	58	Ce	(0.82)	88	Ra	(0.81)	118	Uuo	(2.20)
29	Cu	1.47	59	Pr	(0.80)	89	Ac	(0.79)	119	?	(0.67)
30	Zn	(1.17)	60	Nd	(0.93)	90	Th	(0.95)	120	?	(0.79)

Table 4.3 Natural π-electronegativities $\Xi_A^{(\pi)}$ for selected group 14–16 elements (B3LYP/6-311++G** level).

Z	Atom	$\Xi_A^{(\pi)}$	Z	Atom	$\Xi_A^{(\pi)}$	Z	Atom	$\Xi_A^{(\pi)}$
6	C	[2.60]	14	Si	2.11	32	Ge	2.12
7	N	2.85	15	P	2.44	33	As	2.43
8	O	3.43	16	S	2.86	34	Se	2.79

4.4 HYPOVALENT THREE-CENTER BONDS

To this point, we have considered only default features of NBO analysis, those performed on every input species without keywords or other user intervention. In this section, we encounter the first of many keyword options that allow further exploration of specialized molecular species or properties.

In this section, we wish to explore the bonding challenges presented by *hypovalent* ("electron deficient") species such as diborane (B_2H_6) and other boron hydrides. A general feature of these molecules is their apparent lack of sufficient valence electrons to form the necessary electron-pair bonds between atoms. (For example, the 12 valence electrons of B_2H_6 are inadequate to form the 7 bonds that are presumably needed to connect 8 atoms.)

The resolution of this bonding paradox was achieved with the "three-center bond" (τ-bond) concept, symbolically represented by a "Y-bond" connector between τ_{ABC}-bonded atoms A, B, C, namely,

$$\overset{\textstyle B}{\underset{\textstyle A \quad\; C}{\diagup\;\diagdown}}$$

Such novel 3c/2e (three-center/two-electron) connectors allow two of the H's to form τ_{BHB} "bridges" between boron atoms, while the remaining four H atoms are linked by ordinary 2c/2e σ_{BH} bonds as depicted below:

$$\text{(see structure)} \qquad (D_{2h} \text{ symmetry})$$

Introduction of the 3c/2e τ-bond as a novel structural element of Lewis structure diagrams allowed Lipscomb and others to successfully rationalize the bonding and geometry of many hypovalent species.

Even if a novice NBO user were unaware of the conceptual problems presented by hypovalency, there would be ample numerical evidence that something is drastically wrong with a conventional (1c,2c) Lewis structure description of diborane or other electron-deficient species. Suppose a default NBO analysis is attempted for the diborane input geometry file and atom numbering shown in I/O-4.11 below.

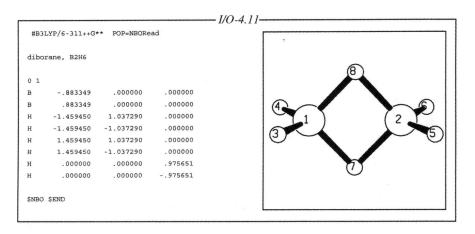

```
———————————————— I/O-4.11 ————————————————

   #B3LYP/6-311++G**   POP=NBORead


   diborane, B2H6

   0 1
   B      -.883349     .000000     .000000
   B       .883349     .000000     .000000
   H     -1.459450    1.037290     .000000
   H     -1.459450   -1.037290     .000000
   H      1.459450    1.037290     .000000
   H      1.459450   -1.037290     .000000
   H       .000000     .000000     .975651
   H       .000000     .000000    -.975651


   $NBO $END
```

─────*I/O-4.12*─────

```
--------------------------------------------------------
  Core                  3.99646 ( 99.911% of    4)
  Valence Lewis         9.74570 ( 81.214% of   12)
  =================     ===========================
  Total Lewis          13.74215 ( 85.888% of   16)
--------------------------------------------------------
  Valence non-Lewis     2.24756 ( 14.047% of   16)
  Rydberg non-Lewis     0.01029 (  0.064% of   16)
  =================     ===========================
  Total non-Lewis       2.25785 ( 14.112% of   16)
--------------------------------------------------------
```

In contrast to the usual >99% success of the optimal NLS, the NBO search for this job (I/O-4.12) reports that only 85.9% of the electron density could be accommodated by the "best" structure (corresponding to about 2.25 "missing" electrons)—a dismal result.

However, by merely inserting the "3CBOND" keyword in the $NBO keylist to request the three-center bond search, as shown below:

```
$NBO 3CBOND $END
```

one obtains a greatly improved NLS description that now contains two three-center (3C) bonds as well as four ordinary two-center (BD) bonds (and the usual CR pairs), as summarized in I/O-4.13.

The new 3c-extended NLS now accounts for >99.6% of the electron density, fully comparable to the ordinary molecules considered previously. Note that adding

─────*I/O-4.13*─────

```
Natural  Bond  Orbital   Analysis:

                     Occupancies        Lewis structure    Low   High
             Occ.   ------------------  ----------------    occ   occ
  Cycle     thresh.  Lewis  Non-Lewis   CR  BD  3C  LP      (L)   (NL)   Dev
  ===================================================================
   1(1)      1.90   15.93641  0.06359    2   4   2   0       0     0    0.15
    :
--------------------------------------------------------
  Core                  3.99645 ( 99.911% of    4)
  Valence Lewis        11.93995 ( 99.500% of   12)
  =================     ===========================
  Total Lewis          15.93641 ( 99.603% of   16)
--------------------------------------------------------
  Valence non-Lewis     0.05323 (  0.333% of   16)
  Rydberg non-Lewis     0.01037 (  0.065% of   16)
  =================     ===========================
  Total non-Lewis       0.06359 (  0.397% of   16)
--------------------------------------------------------
```

the 3CBOND keyword would have no effect on the NLS for previous molecules, except to lengthen the overhead of computer time for the NBO search. For hypovalent species, however, the 3CBOND search leads to *qualitative* (not just incremental) NLS improvement, showing that three-center τ-bonds deserve to be recognized along with one-center lone pairs and two-center σ, π bonds as rightful members of the small arsenal of localized bonding motifs that electron pairs employ to build molecules. In cases of doubt, the student explorer should always launch the 3CBOND search as a possible means for repairing major defects of a standard 1c/2c NLS description.

A general three-center τ_{ABC} NBO is built from three contributing hybrids h_A, h_B, h_C

$$\tau_{ABC} = c_A h_A + c_B h_B + c_C h_C \tag{4.40}$$

Each Lewis-type τ_{ABC} must therefore be complemented by two remaining three-center *anti*bond NBOs (labeled 3C* in NBO output) to conserve basis completeness and orthonormality. In many cases, these valence antibond NBOs (as optimally chosen by the NBO program) correspond to two-center "π-type" $\tau_{ABC}^{(\pi)*}$ and three-center "Δ-type" $\tau_{ABC}^{(\Delta)*}$ linear combinations (cf. *V&B*, p. 306ff),

$$\tau_{ABC}^{(\pi)*} = \mathscr{N}_\pi [c_C h_A - c_A h_C] \tag{4.41}$$

$$\tau_{ABC}^{(\Delta)*} = \mathscr{N}_\Delta [c_B c_A h_A - (c_A^2 + c_B^2) h_B + c_B c_C h_C] \tag{4.42}$$

($\mathscr{N}_\pi$, $\mathscr{N}_\Delta$ = normalization constants). A portion of the NBO listing for the three-center $B_1-H_8-B_2$ NBOs of B_2H_6 is displayed in I/O-4.14, showing the Lewis-type τ_{BHB} bond (NBO 2) and non-Lewis-type $\tau_{BHB}^{(\Delta)*}$ (NBO 25) and $\tau_{BHB}^{(\pi)*}$ (NBO 26) antibonds for this case.

As shown in I/O-4.14, the three orthonormal $B_1-H_8-B_2$ NBOs may be expressed approximately as

$$\tau_{B1H8B2} = 0.53(sp^{4.47})_{B1} + 0.67(s)_{H8} + 0.53(sp^{4.47})_{B2} \tag{4.43}$$

$$\tau_{B1H8B2}^{(\pi)*} = 0.71(sp^{4.47})_{B1} - 0.71(sp^{4.47})_{B2} \tag{4.44}$$

$$\tau_{B1H8B2}^{(\Delta)*} = 0.47(sp^{4.47})_{B1} - 0.75(s)_{H8} - 0.47(sp^{4.47})_{B2} \tag{4.45}$$

Figure 4.7 shows the Lewis-type NBO (4.43) in contour diagrams of overlapping NHOs (left) and final NBO (center), or as a surface plot (right). Figure 4.8 similarly shows the non-Lewis-type NBOs (4.44), (4.45) in contour and surface plots.

```
————————————I/O-4.14————————————

   (Occupancy)   Bond orbital/ Coefficients/ Hybrids
------------------------------------------------------------------------
  2. (1.98537) 3C ( 1) B  1- B  2- H  8
          ( 27.85%)   0.5277* B  1 s( 18.30%)p 4.47( 81.70%)
                             0.0004  0.4276  0.0128  0.5631  0.0011
                             0.0000  0.0000 -0.7066 -0.0262
          ( 27.85%)   0.5277* B  2 s( 18.30%)p 4.47( 81.70%)
                             0.0004  0.4276  0.0128 -0.5631 -0.0011
                             0.0000  0.0000 -0.7066 -0.0262
          ( 44.30%)   0.6656* H  8 s(100.00%)
                             1.0000  0.0067
 25. (0.01535) 3C*( 1) B  1- B  2- H  8
          ( 22.15%)   0.4707* B  1 s( 18.30%)p 4.47( 81.70%)
                             0.0004  0.4276  0.0128  0.5631  0.0011
                             0.0000  0.0000 -0.7066 -0.0262
          ( 22.15%)  -0.4707* B  2 s( 18.30%)p 4.47( 81.70%)
                            -0.0004 -0.4276 -0.0128  0.5631  0.0011
                             0.0000  0.0000  0.7066  0.0262
          ( 55.70%)  -0.7463* H  8 s(100.00%)
                             1.0000  0.0067
 26. (0.00043) 3C*( 1) B  1- B  2- H  8
          ( 50.00%)   0.7071* B  1 s( 18.30%)p 4.47( 81.70%)
                            -0.0004 -0.4276 -0.0128 -0.5631 -0.0011
                             0.0000  0.0000  0.7066  0.0262
          ( 50.00%)  -0.7071* B  2 s( 18.30%)p 4.47( 81.70%)
                            -0.0004 -0.4276 -0.0128  0.5631  0.0011
                             0.0000  0.0000  0.7066  0.0262
          (  0.00%)   0.0000* H  8 s(  0.00%)
                             0.0000  0.0000
```

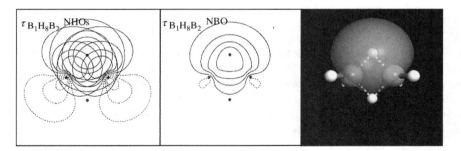

Figure 4.7 Lewis-type three-center τ_{BHB} bond of B_2H_6, showing contour plots for overlapping NHOs (left) and final NBO (center), and corresponding surface plot (right). (See the color version of this figure in Color Plates section.)

4.5 OPEN-SHELL LEWIS STRUCTURES AND SPIN HYBRIDS

Open-shell NBO hybridization and bonding patterns present some of the starkest conflicts with freshman textbook concepts. Indeed, elementary textbooks often give no hint of the open-shell (partial diradical) character of open-shell singlet systems

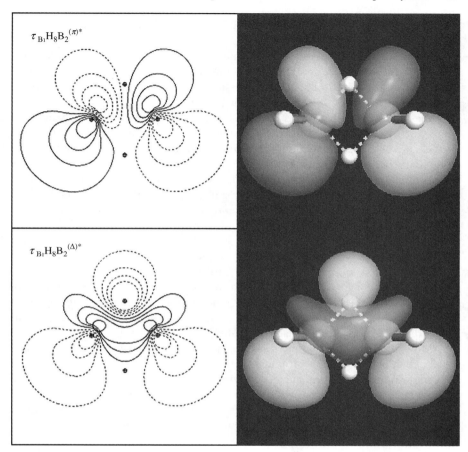

Figure 4.8   Non-Lewis-type three-center antibonds $\tau_{BHB}^{(\pi)*}$ (upper) and $\tau_{BHB}^{(\Delta)*}$ (lower) of B_2H_6, shown in contour and surface plots. (See the color version of this figure in Color Plates section.)

such as ozone (O_3) (Section 3.3.2), nor its underlying role in the unusual structural, reactive, and photochemical properties of these species. While simple radical cation and anion species (e.g., HF^+, Section 3.3.1) may conform tolerably to the freshman-level picture of "double occupancy" for all but one orbital of a parent "perfect paired" species, this description is often deeply misleading, and should only be considered one possible limit of the more general "different orbitals for different spins" picture that is needed to accurately describe open-shell species. More complex open-shell species such as molecular O_2 or O_3 can actually be described reasonably well by a localized Lewis-like structural representation (see below), but *only* if one adopts the generalized concept of "different Lewis structures for different spins" (DLDS). In effect, we need to envision *different* hybridization and bonding patterns ("spin Lewis structures") for α and β electrons as a result of the differing Coulomb and exchange forces in the two spin sets, thereby generalizing our structure–function intuitions to the DLDS spin hybrids that such spin Lewis structures suggest.

What is meant by a spin Lewis structure? The notion of "Lewis structure" inherently refers to *localized* (1c,2c) assignments of electrons to nonbonding (1c) or bonding (2c) spin-orbitals. For open-shell species, these (1c,2c) spin-orbital patterns will generally differ in the two spin sets. For such cases, a "lone pair" becomes a "lone particle" (1c spin-NBO) and an "electron-pair bond" becomes a "one-electron bond" (2c spin-NBO) of definite spin. While the α NBOs and β NBOs may be closely matching in certain NBO regions (corresponding to partial compliance with the elementary "double occupancy" concept), in general the α and β electrons are free to adopt *distinct* (1c,2c) Lewis-like bonding patterns throughout the molecule, with a resulting spin-density distribution (difference of α and β spin density) that extends over multiple "unpaired" electrons and (1c,2c) regions of the molecule.

Open-shell Lewis-like bonding patterns can be depicted by simple modification of the usual Lewis-type bonding diagrams for closed-shell species. For this purpose, one might choose to replace the (1c,2c) "strokes" of the conventional closed-shell diagram by explicit up arrow ($\uparrow$) or down arrow ($\downarrow$) symbols in separate diagrams for each spin. But alternatively, and much more simply, one can merely agree to understand the strokes of each spin Lewis diagram as representing *single* electrons rather than the "pairs" of a usual closed-shell Lewis diagram.

Molecular oxygen (O_2) provides a simple illustration of this altered stroke-type depiction for DLDS structures. As can be inferred from general NBO *Aufbau* principles for homonuclear diatomic molecules (*V&B*, p. 157ff), O_2 is a ground-state triplet species, exhibiting paramagnetic *attraction* to an external magnetic field. The optimal Lewis-like bonding patterns for α spin (single-bonded) and β spin (triple-bonded) structures are depicted below in ordinary stroke-type notation

$$
\begin{array}{cc}
|\overline{O}-\overline{O}| & |O\equiv O| \\
\alpha & \beta
\end{array}
\qquad (4.46)
$$

where each stroke denotes a *one*-electron 1c (nonbonded) or 2c (bonded) Lewis structural feature. Though unconventional, such NBO Lewis-like description is of high overall accuracy, accounting for 99.87% (α) and 99.97% (β) of total electron density in the two spin manifolds. [In contrast, the best possible "maximum spin-paired" description (obtained by including the MSPNBO keyword in the $NBO keylist) accounts for only 87.47% of total electron density, corresponding to *hundred-fold*-larger errors than those of the elegantly compact DLDS description (4.46).] Table 4.4 summarizes some details of the composition and occupancy of the optimal open-shell NBOs for O_2 in each spin set.

Given the two spin-Lewis structures in (4.46), one can envision the composite "spin hybrid" as having average bond order of *two*, correctly indicative of bond

Table 4.4 Optimal spin-NBOs for spin-Lewis structures of molecular O_2 [Text $(4.46\alpha,\beta)$], showing occupancy (and parenthesized degeneracy) for each distinct 1c (n_O) or 2c (σ_{OO}, π_{OO}) feature of the open-shell Lewis structure.

α Spin		β Spin	
Occ.	α-NBO	Occ.	β-NBO
1.0000(1)	$\sigma_{OO} = 0.71(sp^{4.10})_1 + 0.71(sp^{4.10})_2$	1.0000(1)	$\sigma_{OO} = 0.71(sp^{3.53})_1 + 0.71(sp^{3.53})_2$
0.9994(2)	$n_O^{(\sigma)} = (sp^{0.22})_{1,2}$	1.0000(2)	$\pi_{OO}^{(x,y)} = 0.71(p_{x,y})_1 + 0.71(p_{x,y})_2$
0.9975(4)	$n_O^{(x,y)} = (p_{x,y})_{1,2}$	0.9990(2)	$n_O^{(\sigma)} = (sp^{0.26})_{1,2}$

length and strength intermediate between those of standard single or triple bonds. Superficially, such spin averaging resembles the averaging of resonance structures in closed-shell species. However, the differences between spin hybrids and resonance hybrids are more significant than their similarities. Unlike ordinary resonance structures, the individual spin-Lewis structures in $(4.46\alpha,\beta)$ "live" in different spin spaces, and their wave mechanical mixing is spin *forbidden* in nonrelativistic theory. Thus, the O_2 spin hybrid carries none of the connotations of "mixing," "delocalization," or "stabilization" that are commonly associated with resonance-type phenomena. Instead, the spin-Lewis structures $(4.46\alpha,\beta)$ merely represent a more specific and accurate formulation of the "single Lewis structure" concept for open-shell species, taking account of the *differing* Lewis-like (1c,2c) patterns that are generally needed for the two spin sets.

A more complex and interesting illustration of DLDS behavior is provided by the ozone molecule (Section 3.3.2), an open-shell *singlet* species whose optimal Lewis-like bonding patterns are depicted in $(4.47\alpha,\beta)$:

$$|\overline{O}{=}\overline{O}{-}\underline{\overline{O}}| \qquad |\underline{\overline{O}}{-}\overline{O}{=}\overline{O}|$$
$$\alpha \qquad\qquad \beta \qquad\qquad\qquad (4.47)$$

As shown in (4.47), the spin-Lewis structures of ozone resemble allylic-like single- and double-bond structures. Each O—O linkage of the ozone spin hybrid is thereby associated with formal $1{-}1/2$ bond order that is correctly indicative of bonding character *intermediate* between ordinary single and double bonds. However, as in the case of O_2, such spin averaging carries no connotations of special allylic-like stability or "resonance mixing." The spin separation depicted in $(4.47\alpha,\beta)$ instead reflects the *in*stability of open-shell spin sets and their mutual tendency to spin *polarize*, thereby avoiding opposite-spin electrons, which fail to provide the spin-allowed Hamiltonian interactions to reward "mixing." Nevertheless, *same*-spin interactions of resonance-delocalization type may still contribute ayllic-like stabilization *within* each spin manifold, as discussed in Chapter 5.

```
─────────────────────────── I/O-4.15 ──────────
 Natural  Bond  Orbital  Analysis , alpha spin orbitals:
 -----------------------------------------------------------------------
 .
 .
 .
   (Occupancy)    Bond orbital/ Coefficients/ Hybrids
 -----------------------------------------------------------------------
   1. (0.99930) BD ( 1) O  1- O  2
               ( 31.22%)  0.5587* O  1 s( 0.00%)p 1.00( 99.78%)d 0.00( 0.22%)
               ( 68.78%)  0.8294* O  2 s( 0.00%)p 1.00( 99.88%)d 0.00( 0.12%)
   2. (0.99807) BD ( 2) O  1- O  2
               ( 41.78%)  0.6463* O  1 s( 13.85%)p 6.20( 85.86%)d 0.02( 0.30%)
               ( 58.22%)  0.7630* O  2 s( 22.34%)p 3.47( 77.46%)d 0.01( 0.20%)
   3. (0.99810) BD ( 1) O  2- O  3
               ( 56.76%)  0.7534* O  2 s( 21.62%)p 3.62( 78.18%)d 0.01( 0.20%)
               ( 43.24%)  0.6576* O  3 s( 13.42%)p 6.43( 86.31%)d 0.02( 0.28%)
   7. (0.99896) LP ( 1) O  1          s( 81.90%)p 0.22( 18.10%)d 0.00( 0.00%)
   8. (0.96665) LP ( 2) O  1          s( 4.82%)p19.72( 95.13%)d 0.01( 0.05%)
   9. (0.99808) LP ( 1) O  2          s( 57.09%)p 0.75( 42.88%)d 0.00( 0.03%)
  10. (0.99918) LP ( 1) O  3          s( 81.93%)p 0.22( 18.06%)d 0.00( 0.02%)
  11. (0.97187) LP ( 2) O  3          s( 5.19%)p18.25( 94.76%)d 0.01( 0.05%)
  12. (0.88900) LP ( 3) O  3          s( 0.00%)p 1.00( 99.94%)d 0.00( 0.06%)
```

Abridged NBO output for the ozone α-spin structure (4.47α) (cf. I/O-3.9 for input and numbering) is shown in I/O-4.15. (The corresponding β-spin output differs only in interchange of atom labels 1, 3.) Each spin-Lewis structure describes only 98.5% of the associated electron spin density, indicating significant same-spin resonance-type NBO delocalizations (e.g., of $n_O^{\uparrow} \rightarrow \pi_{OO}^{*\uparrow}$ type; cf. Chapter 5) that significantly affect ozone structure and reactivity. Further aspects of the interplay between spin hybridization (involving Lewis-type NBOs) and resonance hybridization (involving non-Lewis NBOs) will be discussed in Chapter 5.

4.6   LEWIS-LIKE STRUCTURES IN TRANSITION METAL BONDING

The remarkable Lewis-like bonding of transition metals (see *V&B*, pp. 365–387) is based on the primacy of sd$^\mu$ hybridization and the associated 12-electron ("duodectet rule") modification of Lewis structure diagrams. The idealized bond angles ω_{ij} between sd$^{\mu i}$ hybrid h_i and sd$^{\mu j}$ hybrid h_j are found to satisfy an equation analogous to the Coulson orthogonality theorem (4.9) for the geometric mean hybridization parameter $\mu \equiv (\mu_i \mu_j)^{1/2}$, namely,

$$\cos \omega_{ij} = \pm[(\mu - 2)/3\,\mu]^{1/2} \tag{4.48}$$

(with $\cos \omega_{ij} = 0$ for $\mu \leq 2$). Note that the two allowed signs in (4.48) lead to distinct supplementary acute (+) and obtuse (−) hybrid angles for any chosen μ. Any student who has seriously contemplated the startling idealized geometries associated with

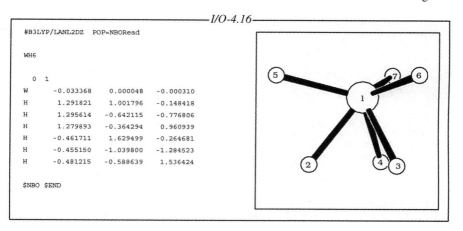

I/O-4.16

```
#B3LYP/LANL2DZ   POP=NBORead

WH6

  0  1
W       -0.033368     0.000048    -0.000310
H        1.291821     1.001796    -0.148418
H        1.295614    -0.642115    -0.776806
H        1.279893    -0.364294     0.960939
H       -0.461711     1.629499    -0.264681
H       -0.455150    -1.039800    -1.284523
H       -0.481215    -0.588639     1.536424

$NBO $END
```

sd^μ hybridization (see *V&B*, Figs. 4.2–4.7) will never again give serious credence to VSEPR-type structural concepts.

As a simple illustration of transition metal hybridization and Lewis-like structures, let us consider tungsten hexahydride (WH_6) as a prototype of idealized sd^5 hybridized bonding. The input file and chosen atomic numbering for the optimized (B3LYP/LANL2DZ-level) geometry of WH_6 are given in I/O-4.16.

Although the displayed WH_6 geometry (with "open" versus "closed" tripod-like features) looks weird from a VSEPR viewpoint, it is indeed the most stable equilibrium form, *far* lower in energy than any imagined octahedral or other VSEPPR-compliant alternative [none of which are stable with respect to deformation to isomers consistent with Equation (4.48)]. The optimized bond angles of the displayed WH_6 geometry are all within 3–4° of the idealized sd^5 angles (63.4°, 116.6°) given by Equation (4.48), showing clearly that the intrinsic directions of valence hybrids (not VSEPR-type "repulsions") are controlling the molecular shape.

NBO analysis of WH_6 proceeds routinely, leading to the expected six W-H bonds (and no lone pairs) consistent with the formal Lewis-like diagram

$$
\begin{matrix}
 & H & \\
H & | & H \\
 & W & \\
H & | & H \\
 & H &
\end{matrix}
\qquad (4.49)
$$

for duodectet-compliant bonding. With the six σ_{WH} NBOs displayed in abridged form in I/O-4.17, the Lewis-like structure (4.49) provides an excellent description of total electron density (99.55%, rivaling the Lewis structural accuracy for common main-group species).

```
─────────────────────────── I/O-4.17 ───────────────
NATURAL BOND ORBITAL ANALYSIS:
  :
Total Lewis              79.63829  ( 99.548% of  80)
  :
  (Occupancy)    Bond orbital/ Coefficients/ Hybrids
-----------------------------------------------------------------------
  1. (1.91904) BD ( 1)  W  1- H  2
               ( 50.11%)   0.7079* W  1 s( 13.56%)p 0.02(  0.22%)d 6.36( 86.22%)
               ( 49.89%)   0.7063* H  2 s(100.00%)
  2. (1.91925) BD ( 1)  W  1- H  3
               ( 50.11%)   0.7079* W  1 s( 13.57%)p 0.02(  0.22%)d 6.35( 86.21%)
               ( 49.89%)   0.7063* H  3 s(100.00%)
  3. (1.91928) BD ( 1)  W  1- H  4
               ( 50.11%)   0.7079* W  1 s( 13.57%)p 0.02(  0.22%)d 6.35( 86.21%)
               ( 49.89%)   0.7063* H  4 s(100.00%)
  4. (1.97608) BD ( 1)  W  1- H  5
               ( 43.50%)   0.6596* W  1 s( 19.79%)p 0.01(  0.29%)d 4.04( 79.93%)
               ( 56.50%)   0.7516* H  5 s(100.00%)
  5. (1.97601) BD ( 1)  W  1- H  6
               ( 43.51%)   0.6597* W  1 s( 19.79%)p 0.01(  0.29%)d 4.04( 79.93%)
               ( 56.49%)   0.7516* H  6 s(100.00%)
  6. (1.97601) BD ( 1)  W  1- H  7
               ( 43.51%)   0.6596* W  1 s( 19.79%)p 0.01(  0.29%)d 4.04( 79.92%)
               ( 56.49%)   0.7516* H  7 s(100.00%)
```

As shown in I/O-4.17, the three σ_{WH} bonds of the "open" tripod (NBOs 4–6) have slightly different hybridizations (sd$^{6.36}$, 86.2% d-character) than those (sd$^{4.04}$, 79.9% d-character) of the "closed" tripod (NBOs 1–3), namely,

$$\sigma_{WH}(\text{"closed"}) = 0.71(\text{sd}^{6.36})_W + 0.71(\text{s})_H \tag{4.50}$$

$$\sigma_{WH}(\text{"open"}) = 0.66(\text{sp}^{4.04})_W + 0.75(\text{s})_H \tag{4.51}$$

However, the average hybridization conforms closely to the expected sd^5 (83.3% d-character) of equivalent idealized hybrids, consistent with the overall close match to idealized 63.4°, 116.6° valence angles as noted above. Figure 4.9 displays contour and surface plots of NBO (4.51) to illustrate how such sd$^{\mu}$-based σ_{WH} NBOs differ from corresponding sp$^{\lambda}$-based main-group hydride bonds considered previously.

Despite differences of detail, the Lewis-like structures for transition metals present a highly satisfying analogy to main-group Lewis bonding. As shown in I/O-4.17, the Lewis-type NBOs of WH$_6$, like their main-group counterparts, exhibit slight deviations from exact double occupancy, reflecting the role of resonance-type departures from the idealized Lewis-structure picture that will be considered in the following chapter. Nevertheless, students of chemistry should rejoice that the elementary Lewis-type bonding and hybridization picture, as suitably generalized to transition metals, continues to exhibit remarkable accuracy and efficacy for describing chemical bonding phenomena across the periodic table. The Lewis picture

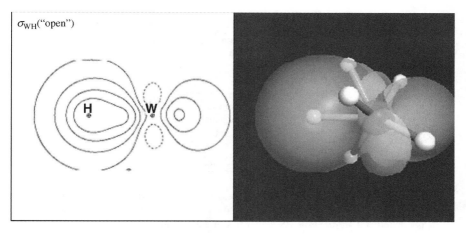

σ_{WH}("open")

Figure 4.9 "Open" σ_{WH} NBO (4.51) of WH_6, shown in contour and surface plots. (See the color version of this figure in Color Plates section.)

therefore serves as the natural starting point for deeper exploration of resonance-type corrections and the associated subtleties of chemical behavior.

PROBLEMS AND EXERCISES

4.1. For a system composed of one C atom, one O atom, and two H atoms, various (local) equilibrium isomeric species are possible on the lowest singlet potential energy surface. The Gaussian input deck shown below

```
#UB3LYP/6-311++G** NOSYMM GUESS=MIX POP=NBOREAD

isomeric local equilibrium COH2 species on the
singlet potential energy surface: Isomer-1

0 1
C
O   1   co
H   1   ch    2   och
H   1   chp   2   ochp   3   180.

co    1.2019
ch    1.1080
chp   1.1080
och   121.97
ochp  121.97

$NBO file=Isomer_1 $END
```

includes a z-matrix for a general planar COH_2 species (with atom numbering C_1, O_2, H_3, H_4), specified by five variables (co, ch, chp, och, ochp) as defined below:

Variable	Definition
co	Bond distance C_1-O_2
ch	Bond distance C_1-H_3
chp	Bond distance C_1-H_4
och	Bond angle $O_2-C_1-H_3$
ochp	Bond angle $O_2-C_1-H_4$

The table below gives numerical values of these geometrical variables for the lowest five (near-)equilibrium isomers on the ground-state singlet surface, together with the associated B3LYP/6-311++G** energies of each isomer:

Isomer	E (a.u.)	co (Å)	ch (Å)	chp (Å)	och (°)	ochp (°)
1	−114.541849	1.2019	1.1080	1.1080	121.97	121.97
2	−114.528568	1.1277	(5.0)	(5.0)	175.73	175.73
3	−114.458136	1.3113	1.1150	1.8689	102.24	29.33
4	−114.393611	1.1741	1.1251	(5.0)	124.52	113.34
5	−114.301544	(5.0)	5.6370	5.6370	7.79	7.79

(Note that the input deck contains pre-entered numerical values for isomer 1, but you can readily substitute values for any other desired isomer. Note also that the parenthesized "5.0" value is an arbitrarily chosen large separation between molecular units that may chemically react if brought into closer proximity.)

(a) Describe the chemical species present in each isomeric geometry 1–5 by giving the best-possible Lewis structure representation for each species (if necessary, using different Lewis diagrams for α and β spin) and a verbal description in reasonable chemical language.

(b) For each isomeric species 1–5, determine the maximum deviation from the s orbital sum rule (Eq. 4.13) and p orbital sum rule (Eq. 4.14). Similarly, determine the maximum occupancy of non-NMB atomic orbitals ("polarization orbitals") for any species to test the accuracy of the NMB approximation.

(c) Where three or more atoms are bonded together, the NHO hybrid directions may exhibit significant "bond bending." For any isomeric species where such bond angles are present, determine the angle between the corresponding NHOs from the Coulson directionality theorem, and identify the species and bond angle that appear to exhibit the greatest angular strain (noncylindrical NBO symmetry). Do you find evidence of greater occupancy of polarization orbitals at such strained nuclei?

(d) Among these isomeric species, determine the maximum deviation of bond ionicity from the value predicted by the natural electronegativity values of Table 4.2.

(e) Can you find evidence for any deviations from Bent's rule in the NHO hybridizations, polarizations, or bond angles for these species?

(f) If possible with your ESS, test the stability of the NBO descriptors in Problems 4.1.a–e with respect to changes in method (e.g., UHF, UMP2, UCCSD, alternative DFT functionals) or basis (e.g., 6-31G*, 6-31+G*, aug-cc-pVTZ) and summarize your conclusions.

4.2. Isomers 1 and 3 correspond to a formal "intramolecular hydride shift" reaction. Find the "linear synchronous transit" (LST) pathway between these two isomers by evaluating each variable v at the successive intermediate values

$$v_\lambda = \lambda v_3 + (1-\lambda)v_1$$

for $\lambda = 0, 0.1, 0.2, \ldots, 0.9, 1.0$ (so the species is at isomer 1 for $\lambda = 0$ and at isomer 3 for $\lambda = 1$). Find the approximate "transition state" $\lambda = \lambda_{TS}$ where the "best" Lewis structure description *switches* from one isomeric form to the other. How does this compare with the apparent high-energy point (if any) along the LST pathway?

(a) From your NBO output, find the NL-occupancy (rho* value) for the two best Lewis structures found for each λ. Plot your results in a graph of rho* versus λ to show how the relative rho* "errors" apparently *cross* at λ_{TS}. (This problem anticipates the *continuous* description of such cross-over transitions in terms of Natural Resonance Theory, Chapter 10.)

(b) Does the transferring H exhibit "hydridic" character near the transition state? Plot the natural charge on this atom for each point on the reaction pathway, and comment on the overall atomic charge pattern at λ_{TS}.

(c) Carry out analogs of Problems 4.1b–f to investigate the accuracy of NMB sum rules (Problem 4.1b), NHO hybridization angles (Problem 4.1c), bond ionicity estimates (Problem 4.1d), Bent's rule deviations (Problem 4.1e), or methodological stability (Problem 4.1f) along the $1 \rightarrow 3$ pathway.

(d) Can you see evidence of how the transferring H is altering the bond geometry, hybridization, and polarity in the manner suggested by Bent's rule? Plot the NHO angular deviations for the "moving" hydride bond at each end of the pathway and comment briefly on regularities you can recognize either at the $O_2-C_1-H_4$ or $C_1-O_2-H_4$ limit.

4.3. Repeat Problem 4.2 for the $3 \rightarrow 4$ bond-dissociation reaction to find the λ_{TS} at which the best NLS description switches from one structure to the other. Is there a corresponding energy-barrier that identifies a "transition state" along the reaction path in this case?

4.4. The LST pathway is only a crude approximation to the presumed "intrinsic reaction coordinate" (IRC, minimum-energy pathway) that crosses through the true transition state (TS) saddle point. If possible with your ESS, find the true TS and IRC for hydride transfer reaction $1 \rightarrow 3$ (Problem 4.2), and compare the optimal NLS description, charge distribution, NHO angle deviations, or other features of interest with the corresponding LST-TS features found previously.

4.5. The optimal NBO structure is usually in excellent agreement with textbook representations, but exceptions are still commonly found in the representation of second-row oxyanions (such as sulfates, phosphates, or perchlorates; see *V&B*, p. 302ff) and other cases of apparent "hypervalency" (such as phosphine oxide, H_3PO; see *V&B*, p. 179ff). How many such exceptions can you find in your freshman chemistry textbook? How many can you find in the latest issue of *Journal of the American Chemical Society*? (Section 5.5 describes how you can test which Lewis structural formulation is more accurate, and by how much.)

Chapter 5

Resonance Delocalization Corrections

What would the world be like if the Lewis structure picture were *exact*, and resonance effects were absent? Such a "world without resonance" is the essential defining characteristic of the *Natural Lewis Structure wavefunction* $\Psi^{(L)}$, a well-defined starting point for the systematic NBO-based exploration of chemical behavior.

The previous chapter has given considerable evidence for the accuracy of the $\Psi^{(L)}$-based picture in a variety of open- and closed-shell species, based on the high percentage of electron density that is accounted for in Lewis-type NBOs alone. The complete NBO basis set $\{\Omega_i\}$ naturally separates into Lewis and non-Lewis components,

$$\{\Omega_i\} = \{\Omega_i^{(L)}\} + \{\Omega_j^{(NL)}\} \tag{5.1}$$

and total electron density (ρ) can be similarly divided into Lewis (ρ_L) and non-Lewis (ρ_{NL}) contributions,

$$\rho = \rho_L + \rho_{NL} \tag{5.2}$$

In a similar vein, we can envision the total wavefunction Ψ to be composed of the dominant Lewis-type contribution $\Psi^{(L)}$ with secondary non-Lewis "correction" $\Psi^{(NL)}$:

$$\Psi = \Psi^{(L)} + \Psi^{(NL)} \tag{5.3}$$

The high %-ρ_L (or low %-ρ_{NL}) exhibited by numerous open- and closed-shell species gives strong (but indirect) evidence that the "resonance-free world" described by $\Psi^{(L)}$ must closely resemble the full solution Ψ of Schrödinger's equation, at least in some average or overall sense. Nevertheless, we expect that the "small correction" $\Psi^{(NL)}$ will play the *dominant* role in certain chemical phenomena of interest, such as aromaticity. In this chapter, we wish to characterize L-type versus NL-type contributions to chemical properties in more direct fashion, seeking to understand the subtle influences of resonance-type delocalization corrections to the localized $\Psi^{(L)}$-based picture. The NBO program includes a powerful array of perturbative and

Discovering Chemistry With Natural Bond Orbitals, First Edition. Frank Weinhold and Clark R. Landis.
© 2012 John Wiley & Sons, Inc. Published 2012 by John Wiley & Sons, Inc.

variational analysis tools for this purpose, including $DEL, $CHOOSE, and natural resonance theory (NRT) options that will be introduced in this chapter.

5.1 THE NATURAL LEWIS STRUCTURE PERTURBATIVE MODEL

Equation (5.3) suggests a general perturbation theoretic approach to analyzing the quantum mechanical Schrödinger equation:

$$H_{op}\Psi = E\Psi \qquad (5.4)$$

whose solution Ψ (with associated energy E) provides a complete description of the chemical system described by Hamiltonian operator H_{op}. The idealized Lewis model wavefunction $\Psi^{(L)}$ may be envisioned as satisfying a corresponding *model* Schrödinger equation (see Sidebar 5.1 for mathematical details):

$$H_{op}{}^{(L)}\Psi^{(L)} = E^{(L)}\Psi^{(L)} \qquad (5.5)$$

where $H_{op}{}^{(L)}$ is the Lewis-type Hamiltonian operator for an idealized "model chemistry" in which resonance-type ($\Psi^{(NL)}$) effects are *absent*. $E^{(L)}$ is the associated NLS energy eigenvalue, which can also be expressed as

$$E^{(L)} = \int \Psi^{(L)} {}^{*}H_{op}{}^{(L)}\Psi^{(L)}d\tau \qquad (5.6)$$

where $d\tau$ denotes integration over all space-spin coordinates of the N-electron wavefunction $\Psi^{(L)}$.

Given the model Lewis-type Schrödinger equation (5.5) as a starting point, we now introduce the difference operator $H_{op}{}^{(NL)}$ and energy $E^{(NL)}$ such that the system Hamiltonian H_{op} can be rewritten as

$$H_{op} = H_{op}{}^{(L)} + H_{op}^{(NL)} \qquad (5.7)$$

and the system energy E as

$$E = E^{(L)} + E^{(NL)} \qquad (5.8)$$

In this formulation, the model Schrödinger equation (5.5) describes the model chemistry of an idealized resonance-free world, whereas $E^{(NL)}$ describes the energetic corrections due to resonance delocalization (departures from the idealized chemistry of a single localized Lewis structure).

Equations (5.3), (5.7), and (5.8) form the starting point for a systematic "perturbation theory" analysis, whose deeper details need not concern us here (see *V&B*, p. 16ff). In this approach, the NLS model $H_{op}{}^{(L)}$ is regarded as the unperturbed Hamiltonian, with known eigenfunction $\Psi^{(L)}$ and energy eigenvalue $E^{(L)}$ that are assumed to be well understood. The resonance-type corrections to energy ($E^{(NL)}$), density (ρ_{NL}), or other properties can then be expressed (analyzed or evaluated) in orderly fashion from the known properties of the model Lewis system. The NBO program contains powerful algorithms (invoked by keywords to be described below), which perform the perturbative decompositions described by Equations (5.2)–(5.8), so it is only necessary to understand the general outlines of the NLS perturbative model in order to begin analyzing interesting chemical effects.

As indicated in the previous chapters, the unperturbed $\Psi^{(L)}$ corresponds to an idealized single-configuration picture (represented as a single-determinant SCF-type wavefunction; cf. Sidebar 5.1) in which each Lewis-type NBO has exact double occupancy (or single occupancy in open-shell case). In this single-determinant limit, the model N-electron Schrödinger equation (5.5) leads to a corresponding Lewis-type *one*-electron eigenvalue equation,

$$h_{\mathrm{op}}^{(0)}\,\Omega_i^{(L)} = \varepsilon_i^{(L)}\Omega_i^{(L)}, \ i = 1, 2, \dots, N \tag{5.9}$$

whose first N eigenfunctions $\{\Omega_i^{(L)}\}$ (counting spin NBOs separately) are the filled Lewis-type NBOs ($\sigma_{AB}, \sigma_{CD}, \dots$), with corresponding orbital energies $\varepsilon_i^{(L)}$. However, the eigenfunctions of $h_{\mathrm{op}}^{(0)}$ also include the remaining non-Lewis-type NBOs $\Omega_j^{(NL)}(\sigma_{AB}^{\ *}, \sigma_{CD}^{\ *}, \dots)$,

$$h_{\mathrm{op}}^{(0)}\Omega_j^{(NL)} = \varepsilon_j^{(NL)}\Omega_j^{(NL)}, j = N+1, \dots \tag{5.10}$$

that are formally vacant (unused in $\Psi^{(L)}$). We refer to the filled (Lewis-type) NBOs of (5.9) as "donor" orbitals and the vacant (non-Lewis-type) NBOs of (5.10) as "acceptor" orbitals.

In the resonance-free world of $h_{\mathrm{op}}^{(0)}$, the donor and acceptor NBOs have no interaction (due to their mutual orthogonality), i.e.,

$$\int \Omega_i^{(L)}*h_{\mathrm{op}}^{(0)}\Omega_j^{(NL)}d\tau = 0, \text{ for all } i,j \tag{5.11}$$

However, the corresponding *real*-world effective 1e-Hamiltonian operator F_{op} (i.e., of Fock, Kohn–Sham, or related type for other theory levels) has *non*-vanishing donor–acceptor interactions,

$$F_{ij} = \int \Omega_i^{(L)*}F_{\mathrm{op}}\Omega_j^{(NL)}d\tau \neq 0 \tag{5.12}$$

and hence will lead to real-world donor–acceptor *mixings* ("delocalizations") that bring in contributions from non-Lewis NBOs, or equivalently, from configurations ("resonance structures") other than $\Psi^{(L)}$. The leading perturbative corrections due to such resonance-type donor–acceptor mixings will be evaluated in the following section.

SIDEBAR 5.1 | *NATURAL LEWIS STRUCTURE WAVEFUNCTION AND HAMILTONIAN*

The mathematical keys to Lewis-based perturbative reformulation of the Schrödinger equation are (1) definition of the Lewis wavefunction $\Psi^{(L)}$ in terms of an associated variational functional; and (2) recasting of the variational definition into an equivalent eigenvalue equation (Euler equation of the variational functional). These steps will be sketched here rather schematically, using somewhat more advanced mathematical concepts than required elsewhere in this book. However, these concepts are all within the province of a proper mathematical introduction to quantum mechanics that an aspiring student of quantum chemistry is expected to master. For simplicity, we focus on the conventional closed-shell case based on localized (1c,2c) electron pairs, but the arguments are readily generalized to open-shell species, 3c bonds, and other extensions of Lewis structure concepts.

Starting from a given Lewis structural diagram, described by a localized configurational assignment

$$(\sigma_{AB})^2(\sigma_{CD})^2 \ldots \tag{5.13}$$

we can envision the associated variational trial function $\Psi^{(L)}$ in which each localized σ_{AB} is built from an arbitrary variational combination of orthonormal bonding hybrids h_A, h_B,

$$\sigma_{AB} = c_A h_A + c_B h_B \tag{5.14}$$

and each hybrid is an arbitrary variational combination of orthonormal atomic orbitals (NAOs) on atomic centers A, B. The variational *Ansatz* for (5.13) can then be expressed as

$$\Psi^{(L)} = A_{op}\{(\sigma_{AB})^2(\sigma_{CD})^2 \ldots\} = \det|(\sigma_{AB})^2(\sigma_{CD})^2 \ldots| \tag{5.15}$$

where A_{op} is the "antisymmetrizer operator" that guarantees compliance with the Pauli exclusion principle, leading to the "Slater determinant" det|. . .| at the right (expressed somewhat schematically, suppressing details of the singlet spin function associated with each localized pair function). $\Psi^{(L)}$ can therefore be more precisely characterized as the Slater determinant of doubly occupied NBOs of the given Lewis structural diagram, where all remaining details of the Lewis-type NBOs (5.14) are to be determined by the quantum mechanical variational principle for the system with known Hamiltonian operator H_{op}.

In terms of the formal "calculus of variations," the variational determination of $\Psi^{(L)}$ can be expressed as a *variational functional* satisfying the stationary condition:

$$\delta \int \Psi^{(L)*} H_{op} \Psi^{(L)} d\tau = 0 \tag{5.16}$$

which corresponds to the requirement that the variational integral $I = \int \Psi^{(L)*} H_{op} \Psi^{(L)} d\tau$ be *minimized* with respect to all possible variations of coefficients and hybrids in (5.14). [If $I = I(x)$ depended on only a single variable x, so that "δ" corresponds simply to "d/dx" variation, then (5.15) would be equivalent to the usual stationary condition of differential calculus, $dI/dx = 0$, to find where $I(x)$ is minimized.]

However, as pointed out by Löwdin (see *V&B*, p. 7ff and references therein), *any* variational procedure such as (5.16) can be formally recast as an eigenvalue equation ("Euler equation" of the variational functional)

$$H_{op}^{(L)} \Psi^{(L)} = E^{(L)} \Psi^{(L)} \tag{5.17}$$

where $H_{op}^{(L)}$ is a model Hamiltonian for the model Schrödinger equation (5.17) that emulates the full Schrödinger equation (5.4) in the variational subspace. In effect, solving a restrictive form (5.16) of the variational principle for the full system Hamiltonian H_{op} is equivalent to solving the *exact* Schrödinger equation (5.17) for a restrictive model Hamiltonian $H_{op}^{(L)}$.

Further details of constructing $H_{op}^{(L)}$ from the optimized $\Psi^{(L)}$ are beyond the scope of this book. However, the single-determinant form of $\Psi^{(L)}$, Equation (5.15), allows Equation (5.17) to be factored into simple *one*-electron (SCF-like) eigenvalue equations that make this construction straightforward for the NBO program.

5.2   SECOND-ORDER PERTURBATIVE ANALYSIS OF DONOR–ACCEPTOR INTERACTIONS

The unperturbed Lewis-type description (5.9 and 5.10) neglects the real-world interactions (5.12) between donor (Lewis-type) and acceptor (non-Lewis-type) NBOs of a parent Lewis structure $\Psi^{(L)}$. However, the typical high accuracy of this description (often $>99\%$-ρ_L) leads us to expect that low-order perturbative corrections may be adequate to capture the donor–acceptor (resonance) effects of greatest chemical interest. This is the reason experienced NBO users usually turn first to the "second-order perturbative analysis" section of NBO output.

The perturbation theory of NBO donor–acceptor interactions can be expressed quite simply in graphical or equation form for the leading (second-order) correction $\Delta E_{ij}^{(2)}$ for each $\Omega_i^{(L)}$-$\Omega_j^{(NL)}$ donor–acceptor pair. The schematic perturbation diagram for doubly occupied donor NBO $\Omega_i^{(L)}$ interacting with vacant acceptor NBO $\Omega_j^{(NL)}$ is depicted in Fig. 5.1. The unperturbed energy levels (outer) have respective NBO energies $\varepsilon_i^{(L)}$, $\varepsilon_j^{(NL)}$ on the vertical energy scale. However, in the presence of the perturbation, these levels mix (dashed lines) and split to become the final perturbed levels (center), with the lower level (marked ε_-) now *below* $\varepsilon_i^{(L)}$ while the higher level (marked ε_+) rises correspondingly above $\varepsilon_j^{(NL)}$. (Such a perturbative splitting pattern is dictated by the arcane wave-mixing rules of quantum superposition; see *J. Chem. Ed.* **76**, 1141, 1999.) Because only two electrons are involved in this interaction, they naturally occupy the lower ε_- level. By the magic of quantum mechanics, this results in overall energy *lowering* ("2e-stabilization") of the electron pair as a result of perturbative mixing with (and partial delocalization into) the higher-energy $\Omega_j^{(NL)}$ orbital. (Only in quantum mechanics can you lower the energy by partially mixing in a contribution from a *higher*-energy orbital!)

As shown in Fig. 5.1, the net perturbative energy lowering, $\Delta E_{ij}^{(2)}$, can be expressed by the following simple equation:

$$\Delta E_{ij}^{(2)} = -q_i |F_{ij}|^2 / (\varepsilon_j^{(NL)} - \varepsilon_i^{(L)}) \qquad (5.18)$$

where q_i is the occupancy of the donor orbital (~ 2), F_{ij} is given by (5.12) and $\varepsilon_i^{(L)}$, $\varepsilon_j^{(NL)}$ are the respective donor and acceptor orbital energies. For a given chemical species, the NBO program evaluates the second-order energies (5.18) for *all possible* donor–acceptor combinations, then prints the table of "E(2)" values for those deemed sufficiently large to be of chemical interest.

To capitalize on the chemical magic of donor–acceptor interactions, electrons must find acceptor orbitals. We know that atoms that lack empty valence orbitals are generally too "noble" for the rewards of electron pair sharing, but fortunately,

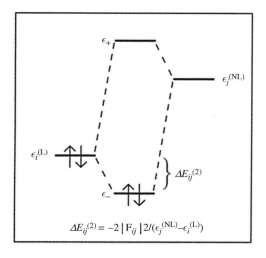

$$\Delta E_{ij}^{(2)} = -2\,|\,F_{ij}\,|\,2/(\epsilon_j^{(NL)} - \epsilon_i^{(L)})$$

Figure 5.1. 2e-stabilizing interaction between a filled donor orbital $\Omega_i^{(L)}$ and vacant acceptor orbital $\Omega_j^{(NL)}$, leading to energy lowering $\Delta E_{ij}^{(2)}$.

practically all molecules offer a variety of such valence acceptor orbitals in the form of their *valence antibonds*. Superficially, "antibonds" might seem to be antithetical to molecular stabilizations or intermolecular attractions, but nothing could be farther from the truth. (Antibond NBOs are sometimes also confused with "virtual" orbitals of SCF theory, but this too is superficial and erroneous.) Indeed, when atoms A, B unite to form a chemical bond through unfilled valence hybrids h_A, h_B, only the bonding (*in*-phase, Lewis-type) superposition

$$\sigma_{AB} = c_A h_A + c_B h_B \tag{5.19}$$

is filled to capacity to become a "donor," whereas the complementary "antibonding" (*out*-of-phase, non-Lewis) superposition

$$\sigma^*_{AB} = c_B h_A - c_A h_B \tag{5.20}$$

remains an available "acceptor," the unsaturated molecular vestige of parent atomic valence shell vacancies. Although extravalent (Rydberg-type RY*) orbitals also remain available as potential acceptors, the valence antibonds (BD* NBOs) commonly provide by far the most important source of acceptor orbitals [and stabilizing E(2) interactions] for molecular species.

Let us illustrate E(2) output for the formamide molecule, whose Lewis structure and donor NBOs were previously described in Section 4.1.3. The second-order perturbation theory analysis for H_2NCHO is shown in abridged form in I/O-5.1. In this case, 40 donor–acceptor E(2) values were found that exceed the 0.5 kcal/mol threshold, but only the 13 valence-shell entries (i.e., excluding CR, RY* NBOs) are included in the abridged listing. (One could reset the default E(2) threshold, e.g., to 10 kcal/mol by inserting the keyword "E2PERT = 10" in the $NBO keylist, thereby suppressing all but the largest E(2) entries.)

---------- *I/O-5.1* ----------

SECOND ORDER PERTURBATION THEORY ANALYSIS OF FOCK MATRIX IN NBO BASIS

Threshold for printing: 0.50 kcal/mol

Donor NBO (i)	Acceptor NBO (j)	E(2) kcal/mol	E(j)-E(i) a.u.	F(i,j) a.u.
within unit 1				
1. BD (1) N 1- C 2	86. BD*(2) C 2- O 3	0.81	1.47	0.031
2. BD (1) N 1- H 4	86. BD*(2) C 2- O 3	0.63	1.31	0.026
2. BD (1) N 1- H 4	87. BD*(1) C 2- H 6	2.12	1.03	0.042
3. BD (1) N 1- H 5	86. BD*(2) C 2- O 3	3.61	1.31	0.061
4. BD (1) C 2- O 3	85. BD*(1) C 2- O 3	1.09	0.38	0.019
5. BD (2) C 2- O 3	82. BD*(1) N 1- C 2	0.92	1.52	0.034
5. BD (2) C 2- O 3	84. BD*(1) N 1- H 5	1.05	1.47	0.035
6. BD (1) C 2- H 6	83. BD*(1) N 1- H 4	4.44	0.95	0.058
10. LP (1) N 1	85. BD*(1) C 2- O 3	59.61	0.29	0.117
11. LP (1) O 3	82. BD*(1) N 1- C 2	1.29	1.14	0.035
11. LP (1) O 3	87. BD*(1) C 2- H 6	1.01	1.04	0.029
12. LP (2) O 3	82. BD*(1) N 1- C 2	23.51	0.71	0.117
12. LP (2) O 3	87. BD*(1) C 2- H 6	22.08	0.60	0.105

As shown in the $|\Delta E_{ij}^{(2)}|$ entries (third column), a few delocalizations leap out for special attention. Most conspicuous is the 59.6 kcal/mol stabilization associated with $n_N \rightarrow \pi^*_{CO}$ delocalization. (NBOs 10 → 85), as well as the two delocalizations from the second ("p_y-type") oxygen lone pair $n_O^{(y)}$ into σ^*_{CN} (12 → 82; 23.5 kcal/mol) and σ^*_{CH} (10 → 87; 22.1 kcal/mol), which we can single out for special attention.

How do these large stabilization values originate? From Equation (5.18) one can see that $|\Delta E_{ij}^{(2)}|$ stabilizations are increased by (1) a small "energy gap" $\Delta\varepsilon_{ij} = \varepsilon_j^{(NL)} - \varepsilon_j^{(L)}$ in the denominator, and/or (2) a strong $|F_{ij}|$ interaction element in the numerator. Although textbooks commonly emphasize the energy gap factor, one can see from the numerical "$E(j)–E(i)$" values (column 4) that $\Delta\varepsilon_{ij}$ is commonly *large* and of *limited* variability with values ranging from 0.29 to 1.52 a.u. (ca. 180–950 kcal/mol) in I/O-5.1, accounting for only a small fraction of the actual ca. 90-fold range of E(2) values. In most cases, far more important is the strength of $|F_{ij}|$ interaction (column 5), whose *squared* $|F_{ij}|^2$ values exhibit ca. 40-fold variations in I/O-5.1. The largest delocalizations clearly benefit from both factors, but $|F_{ij}|$ is generally the more important in terms of "chemical interest."

What leads to a large $|F_{ij}|$ value? The simplest and most powerful way to think about $|F_{ij}|$ interactions is in terms of orbital "overlap," making implicit use of the *Mulliken approximation*, i.e.,

$$|F_{ij}| \propto S_{ij}^{(PNBO)} \tag{5.21}$$

where

$$S_{ij}^{(PNBO)} = \int P\Omega_i^{(L)*} P\Omega_j^{(NL)} d\tau \tag{5.22}$$

is the overlap integral of nonorthogonal PNBOs $^P\Omega_i^{(L)}$ and $^P\Omega_j^{(NL)}$. (More explicitly, F_{ij} and $S_{ij}^{(PNBO)}$ are generally of opposite sign, with proportionality constant of order unity.) Even if only crudely approximate, the Mulliken approximation (5.21) allows one to effectively *visualize* from PNBO overlap diagrams the orbital features that most strongly control $|F_{ij}|$ interaction strength, and thus to guide creative thinking about donor–acceptor stabilization. The powerful visual imagery of (P)NBO overlap diagrams, combined with the accuracy of the NBOs themselves, often allows rich chemical insights to be gained even from "eyeball accuracy" estimates of orbital overlap.

Let us first consider the dominant $n_N \to \pi^*_{CO}$ (NBO 10 $\to$ 85) delocalization, as pictured in Fig. 5.2. As shown in the upper panels, the out-of-plane n_N, π^*_{CO} NBOs are favorably aligned for strong π-type overlap, displayed in contour and surface plots in the lower panels.

The formal 2e-promotion in the π system can be represented by the configurational and Lewis structural changes shown in Fig. 5.3. As shown in the figure, $(n_C)^2 \to (\pi^*_{CO})^2$ NBO delocalization corresponds to formal "breaking" of the π_{CO} bond [because $(\pi_{CO})^2(\pi^*_{CO})^2$, with zero net bond order, is equivalent to

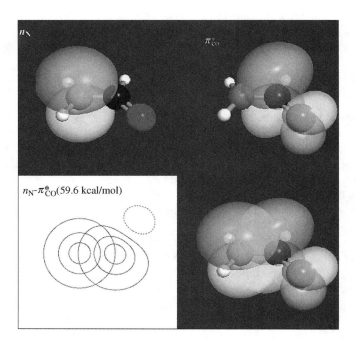

Figure 5.2. Formamide n_N and π^*_{CO} NBOs, shown individually (upper panels) and in interaction (lower panels) as contour and surface plots. (The contour plot is a top–down view of the π system, with chosen contour plane slicing through the π orbitals 1 Å above the molecular plane.) (See the color version of this figure in Color Plates section.)

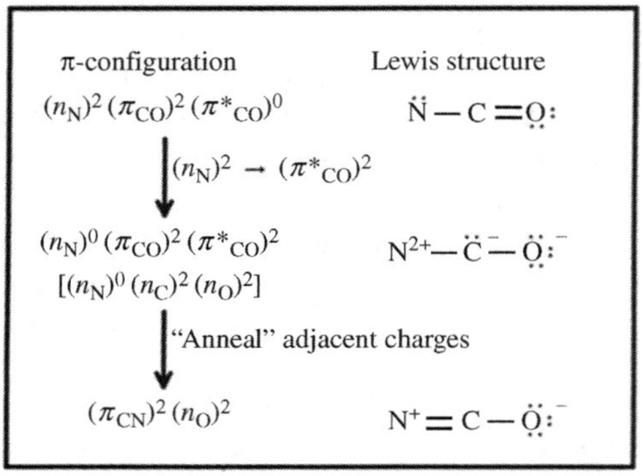

π-configuration	Lewis structure

$(n_N)^2 (\pi_{CO})^2 (\pi^*_{CO})^0$ $\ddot{N} - C = \underset{..}{O}:$

$\bigg\downarrow (n_N)^2 \rightarrow (\pi^*_{CO})^2$

$(n_N)^0 (\pi_{CO})^2 (\pi^*_{CO})^2$ $N^{2+} - \ddot{C} - \ddot{O}: ^-$
$[(n_N)^0 (n_C)^2 (n_O)^2]$

$\bigg\downarrow$ "Anneal" adjacent charges

$(\pi_{CN})^2 (n_O)^2$ $N^+ = C - \ddot{\underset{..}{O}}: ^-$

Figure 5.3. Configurational and Lewis (resonance) structure changes associated with formal $(n_N)^2 \rightarrow (\pi^*_{CO})^2$ NBO delocalization corrections in the π system of formamide, showing the formal equivalence to amide resonance shift.

nonbonding configuration $(n_C)^2 (n_O)^2]$ and "annealing" of the adjacent filled (C) and empty (N) p-orbitals to form a dative π_{CN} bond. The overall effect of the 2e-delocalization is therefore an admixture of the alternative Lewis (resonance) structure:

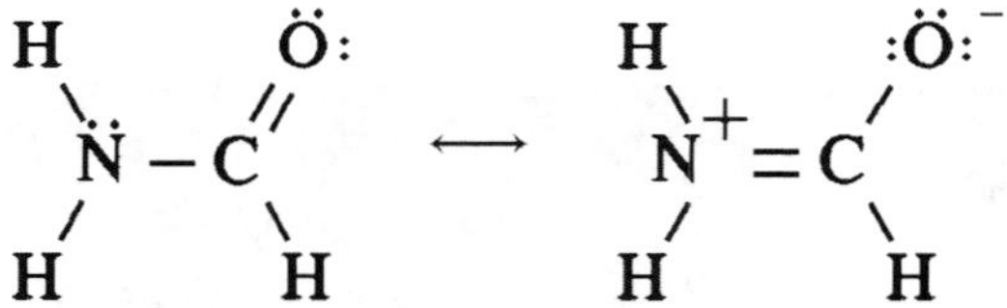

$$\begin{array}{ccc} \overset{H}{\underset{H}{\diagdown}} \ddot{N} - C \overset{\ddot{O}:}{\underset{H}{\diagup}} & \longleftrightarrow & \overset{H}{\underset{H}{\diagdown}} N^+ = C \overset{:\ddot{O}: ^-}{\underset{H}{\diagup}} \end{array} \tag{5.23}$$

which is indeed the expected strong "resonance delocalization" in this species. As the dipolar resonance mnemonic in (5.23) suggests, partial admixture of N=C−O resonance character leads to weakened (lengthened and red-shifted) CO bonding, strengthened CN bonding, and reduced N−C torsional flexibility (partial double-bond character), all well-known electronic signatures of amide groups. The allylic-type resonance in (5.23) identifies the strong $n_N \rightarrow \pi^*_{CO}$ delocalization as representative of typical "conjugative" interaction phenomena, with strong associated effects on geometry and reactivity.

Next most important in I/O-5.1 are the two strong delocalizations from $n_O^{(y)}$ (NBO 12) into vicinal antibonds σ^*_{CN} (NBO 82; 23.5 kcal/mol) and σ^*_{CH} (NBO 87; 22.1 kcal/mol), as pictured in Fig. 5.4. As shown in the contour and surface overlap diagrams, the in-plane p_y-type oxygen lone pair $n_O^{(y)}$ is perpendicular to the CO bond, well-positioned to interact strongly with the "backside" lobes of each antibond $(\sigma^*_{CN}, \sigma^*_{CH})$ at the neighboring vicinal positions. Because the σ_{CN}, σ_{CH} bonds have slightly different polarizations (due to the greater electronegativity difference in the former case), the σ^*_{CN} antibond has slightly greater amplitude at its backside carbon lobe, leading to slightly stronger $n_O^{(y)}$-σ^*_{CN} (versus $n_O^{(y)}$-σ^*_{CH}) interaction. As in Fig. 5.3, each of these NBO delocalizations can be equivalently expressed as an admixture of an alternative resonance structure, namely,

$$n_O{}^{(y)} \rightarrow \sigma^*_{CN} : \quad \begin{array}{c} H \\ \diagdown \\ \\ H \diagup \end{array} \ddot{N}{:}^- \quad C \overset{\ddot{O}^+}{\underset{H}{\diagup}} \tag{5.24}$$

$$n_O{}^{(y)} \rightarrow \sigma^*_{CH} : \quad \begin{array}{c} H \\ \diagdown \\ \\ H \diagup \end{array} \dot{N} - C \overset{\ddot{O}^+}{\underset{H{:}^-}{\diagup}} \tag{5.25}$$

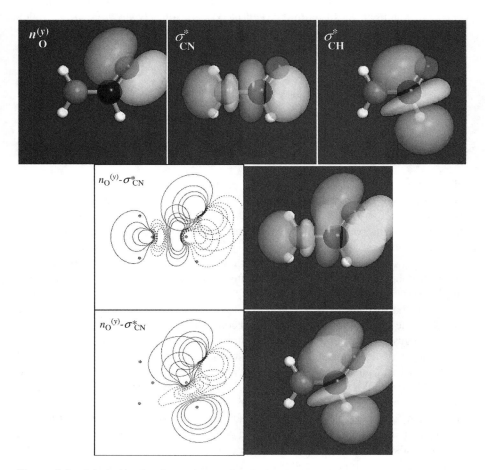

Figure 5.4. Principal in-plane lone-pair → antibond delocalizations of formamide, showing individual donor $\left(n_O{}^{(y)} \right)$ and acceptor $\left(\sigma^*_{CN}, \sigma^*_{CH} \right)$ NBOs (upper panels), and overlapping donor–acceptor pairs (lower panels) in contour and surface plots. (See the color version of this figure in Color Plates section.)

Although the relative weightings of these resonance structures (to be evaluated in Section 5.6) are expected to be somewhat weaker than the principal resonance delocalization of (5.23), they can nevertheless contribute to appreciable structural and reactive effects, such as weakening of CN (5.24) and CH (5.25) or strengthening of CO to partially compensate the bond-order shifts due to (5.23). Further aspects of net bond-order changes for multiple resonance contributions are discussed in Section 5.6.

Because the delocalizations depicted in Fig. 5.4 involve only the saturated σ skeleton, they are formally classified as "hyperconjugative" (rather than "conjugative") in character. Nevertheless, one can see that the $n_O^{(y)}$ donor of carbonyl compounds is a powerful hyperconjugator, and that each of its primary σ-delocalizations is only about a factor of 2–3 weaker than the famous amide n_N-π^*_{CO} π-delocalization. [For a recent authoritative review of hyperconjugative phenomena, see I. V. Alabugin, K. M. Gilmore, and P. W. Peterson, *Hyperconjugation*, Wiley Interdisciplinary Reviews: Computational Molecular Science **1**, 109–141, 2011.]

Still other hyperconjugative σ-delocalizations are seen in I/O-5.1 that appear to be of chemically significant strength, such as

$$\sigma_{CH} \rightarrow \sigma^*_{NH(a)}\,(\text{NBOs }6 \rightarrow 83, 4.4\,\text{kcal/mol}) \tag{5.26}$$

$$\sigma_{NH} \rightarrow \sigma^*_{CO(a)}\,(\text{NBOs }3 \rightarrow 86, 3.6\,\text{kcal/mol}) \tag{5.27}$$

$$\sigma_{NH} \rightarrow \sigma^*_{CH(a)}\,(\text{NBOs }2 \rightarrow 87, 4.4\,\text{kcal/mol}) \tag{5.28}$$

all involving hyperconjugatively coupled *vicinal* bond–antibond NBOs in *antiperiplanar* ("trans") orientation. It is apparent from these and other examples that the *anti* bond–antibond orientation typically leads to stronger hyperconjugation than the corresponding *syn* orientation. This difference is illustrated in Fig. 5.5 for CH−NH* hyperconjugations, comparing $\sigma_{CH} \rightarrow \sigma^*_{NH(a)}$ (left, 4.4 kcal/mol) versus $\sigma_{CH} \rightarrow \sigma^*_{NH(s)}$ (right, <0.05 kcal/mol) both in contour and surface overlap diagrams. A glance at Fig. 5.5 shows that the *anti*-arrangement (left) indeed offers more favorable in-phase overlap (blue with blue, yellow with yellow), whereas *syn* incurs unfavorable phase mismatches (blue with yellow) on one side or the other of the nodal plane bisecting the NH* antibond. The advantageous σ-delocalizations that occur in *anti*-arrangements (staggered conformers) compared to *syn* arrangements (eclipsed conformers) are the essential electronic origin of the famous *ethane rotation barrier* and related torsional phenomena that favor conformational staggering in single-bonded molecules (see *V&B*, p. 234ff). Of course, 2c-bond NBOs are generally *weakened* donors compared to 1c-lone pair NBOs [which have a ca. twofold advantage in the $|F_{ij}|^2$ interaction factors in (5.18)], so that lone-pair delocalizations such as shown in Fig. 5.4 are typically more "controlling" than bond delocalizations such as shown in Fig. 5.5.

It is also apparent that *geminal* delocalizations (i.e., of $\sigma_{AX} \rightarrow \sigma^*_{AY}$ form, involving X−A−Y bonding pattern) are generally much weaker than vicinal delocalizations (i.e., of $\sigma_{AX} \rightarrow \sigma^*_{BY}$ form, involving X−A−B−Y bonding pattern). Only one such above-threshold interaction [$\sigma_{CN} \rightarrow \sigma^*_{CO}$, NBOs 1 $\rightarrow$ 86 (0.81 kcal/mol)] appears in I/O-5.1 for formamide. The dominance of vicinal (*v*) over geminal (*g*) or

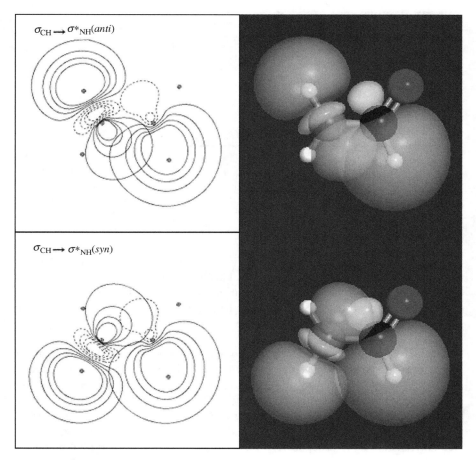

Figure 5.5.   Comparison contour and surface plots of vicinal $\sigma_{CH}-\sigma^*_{NH}$ interactions in *anti* (upper) versus *syn* (lower) orientations, showing the far more favorable NBO overlap in *anti*periplanar arrangement [consistent with the stronger hyperconjugative stabilization evaluated in the E(2) table]. (See the color version of this figure in Color Plates section.)

more remote-type (*r*) delocalizations is also evident in the NBO summary table, I/O-4.5, which identifies the *v/g/r* classification of non-Lewis acceptor NBOs for "principal delocalizations" of each Lewis-type NBO, ordered according to numerical entries of the E(2) table. Although geminal hyperconjugation is usually anticipated to be "negligible" in near-equilibrium geometry of acyclic hydrocarbons and other simple main-group compounds, these interactions exhibit complex dependence on angular and polarity variations that sometimes lead to surprising stabilizations in strained cyclic geometries (see *V&B*, p. 263ff). As shown in Fig. 5.6, both vicinal and geminal bond–antibond delocalizations have simple mappings onto corresponding "arrow-pushing" or resonance diagrams, analogous to those given previously in (5.24) and (5.25).

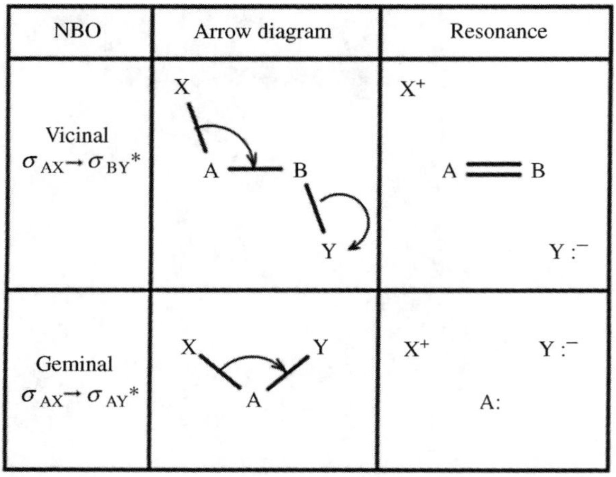

NBO	Arrow diagram	Resonance
Vicinal $\sigma_{AX} \to \sigma_{BY}^*$	X A —— B Y	X⁺ A ≡ B Y :⁻
Geminal $\sigma_{AX} \to \sigma_{AY}^*$	X Y A	X⁺ Y :⁻ A:

Figure 5.6. Generic "arrow pushing" diagram (left) and secondary resonance structure (right) for vicinal (upper) and geminal (lower) NBO donor–acceptor interactions.

[A final type of NBO donor–acceptor interaction shown in I/O-5.1 is the curious $\pi_{CO} \to \pi_{CO}^*$ delocalization (NBOs 4-85; 1.1 kcal/mol) of the carbonyl π-bond into its *own* antibond, a form of left–right "electron correlation" effect. Such correlation effects (generally absent in Hartree–Fock-level wavefunctions) are relatively weak compared to other entries of the E(2) table, and lie outside the scope of discussion in this book.]

The E(2) delocalization values are also reflected in occupancy shifts from donor to acceptor NBOs. As suggested in the perturbation diagram, Fig. 5.1, interaction of $\Omega_i^{(L)}$ and $\Omega_j^{(NL)}$ (with energy lowering $\Delta E_{ij}^{(2)}$) involves superposition mixing of donor and acceptor orbitals, with consequent partial delocalization of the electron pair from its parent Lewis-type $\Omega_i^{(L)}$ into the non-Lewis $\Omega_j^{(NL)}$ orbital. The initial unperturbed occupancy (2e) of donor $\Omega_i^{(L)}$ is therefore reduced by a small quantity of charge ($q_{i \to j}$) that is "transferred" to acceptor $\Omega_j^{(NL)}$. From general low-order perturbation theory formulas, one can estimate that the charge transfer $q_{i \to j}$ is approximately *proportional to* the associated stabilization energy $\Delta E_{ij}^{(2)}$

$$q_{i \to j} \propto |\Delta E_{ij}^{(2)}| \tag{5.29}$$

with a proportionality constant (essentially, the "$E(j)-E(i)$" energy difference in I/O-5.1) that is of order *unity* if all quantities are expressed in atomic units (see *V&B*, p. 58ff). Given the energy conversion factor 1 a.u. = 627.51 kcal/mol, one can see that even 0.01e delocalization (i.e., 1% of an electron) corresponds to about 0.01*627 ≅ 6 kcal/mol of stabilization, potentially significant on the usual scale of "chemical interest."

For example, from the NBO summary, I/O-4.5, one can see that π_{CO}^* (NBO 85) gained about 0.24e occupancy, close to the amount that was lost by n_N (NBO 10, occupancy 1.75) in the powerful $n_N \to \pi_{CO}^*$ interaction ($|\Delta E_{ij}^{(2)}| \cong 60$ kcal/mol). Similarly, the total charge transfer from donor $n_O^{(y)}$ (NBO 12, occupancy = 1.852) is approximately equal to the summed occupancy of its two principal acceptor NBOs (Fig. 5.4), σ_{CN}^* (occupancy 0.064) and σ_{CH}^* (occupancy 0.072), each with

$|\Delta E_{ij}^{(2)}| \cong 20 \, \text{kcal/mol}$, all roughly consistent with the crude proportionality (5.24). These relationships tell us that NPA population shifts as small as 0.001e may signal potential effects of chemical interest. (They also tell us that alternative population measures with uncertainties greater than ca. 0.001e are unlikely to provide reliable analysis of chemically significant effects.)

For electron correlation methods that lack an effective 1e-Hamiltonian operator to evaluate orbital energetics, the E(2) table is unavailable. However, in such cases the user can often "read" the important donor–acceptor interactions indirectly from a variety of alternative NBO descriptors, such as:

(i)   The occupancies of antibond NBOs (as above).

(ii) The associated overlap integrals [cf. (5.21)] or density matrix elements (Appendix C).

(iii) The delocalization "tails" of natural localized molecular orbitals (Section 5.4) and associated dipolar (Section 6.2), NMR (Chapter 7), and other properties.

(iv) NRT resonance weightings (Section 5.6), combined with the general mnemonic relationship (Fig. 5.6) between resonance structure and NBO donor–acceptor interactions.

The close connection between basic $|\Delta E_{ij}^{(2)}|$ stabilization energies and other wave-function properties insures that general patterns of the E(2) table will be reflected in many analysis details. The student should check NBO descriptors (i)–(iv) and associated experimental properties to verify overall consistency with the delocalization pattern displayed in the E(2) table.

Even without consulting the numerical entries of the E(2) table, an alert chemistry student will generally look first for antibonds in the *vicinal anti*-positions around each lone pair (or other strong donor NBO) as principal sites for resonance delocalizations, based on general considerations discussed above. In order to go beyond the elementary Lewis structure picture, the first step is to identify details (occupancy, shape, and location) of the important valence antibonds.

5.3 $DEL ENERGETIC ANALYSIS [INTEGRATED ESS/NBO ONLY]

For those fortunate to have a fully integrated (linked) ESS/NBO5 program (Section 1.1), the $DEL keylist and associated keyword options provide powerful "deletions" methods of energetic analysis, based on quasi-variational (rather than perturbation theoretic) assessment of donor–acceptor interactions and their structural conse-quences. In effect, the $DEL options allow one to *delete* single or multiple donor–acceptor interactions and recalculate the energy, geometry, and other molecular properties as though the world was created without such interactions. By comparing the $DEL properties with those of the full calculation, one identifies by difference the specific energetic and structural consequences of the deleted interaction(s). This approach often allows one to isolate the "smoking gun" that is most responsible for a

```
─────────────────────────── I/O-5.2 ───────────────────────────
#N b3lyp/6-311++g**  NoSymm POP=NBODel IOp(5/48=10000)


H2NCHO


  0  1
  N
  C   1   1.361
  O   2   1.212    1    124.90
  H   1   1.009    2    119.42    3      0.00
  H   1   1.007    2    121.45    3    180.00
  H   2   1.107    1    112.44    4    180.00


$NBO file=h2ncho $END
$DEL
  delete 2 elements
    12 82
    12 87
$END
```

particular structural or energetic feature of interest. It also allows one to take partial account of higher-order coupling effects—cooperative synergism versus anticooperative competition of multiple donor–acceptor interactions—that are beyond second-order perturbative description. Because such $DEL deletions require complete energy recalculation (involving all 1e, 2e integrals of the original wavefunction calculation), they involve intimate (linked) cooperation with the host ESS program and cannot be performed by a stand-alone GENNBO version of the NBO program.

The desired list of orbital or F_{ij} deletions is specified in a $DEL ... $END keylist, appended after the main $NBO ... $END keylist at the end of the input file, as illustrated for formamide (cf. Section 4.1.3) in I/O-5.2. For Gaussian input (as shown), the "POP = NBODEL" keyword must be included in the route card (line 1) in order to process the attached $DEL keylist requests, and the "NOSYMM" keyword is recommended to avoid errors when the chosen $DEL deletions break molecular symmetry. [The "IOp(5/48 = 10000)" entry (unnecessary in pre-G03 or current G09 versions), corrects for a DFT coding error in initial release of Gaussian 03.] In the input file shown, the $DEL keylist requests deletion of two F_{ij} "elements," namely, the interactions between NBOs 12 and 82 ($F_{12,82}$) and NBOs 12 and 87 ($F_{12,87}$), the hyperconjugative $n_O^{(y)} \rightarrow \sigma^*_{CN}$ and $n_O^{(y)} \rightarrow \sigma^*_{CH}$ interactions depicted in Fig. 5.4.

The menu of available $DEL selections is extensive, grouped into nine distinct deletion types (see *NBO Manual*, p. B-16ff for a comprehensive listing, and p. B-48ff for illustrations). The simplest and most general deletion type, as illustrated in I/O-5.2, is that for individual F_{ij} elements, using command syntax of the form

$$\text{DELETE } n \text{ ELEMENTS } i_1 j_1 \ldots i_n j_n \qquad (5.30)$$

where the n index pairs (i,j) follow the command on the same line or subsequent lines. (Deleting F_{ij} also implies deletion of F_{ji}, so each index pair can be specified in either

```
———————————————— I/O-5.3 ————————————————
$DEL
  delete 1 element 10 85
  delete 2 elements 12 82  12 87
  delete 3 orbitals 83 86 87
  nostar
$END
```

order.) A second $DEL command type deletes entire non-Lewis orbitals (tantamount to complete removal of the corresponding $\Omega_j^{(NL)}$ NBOs from the SCF variational basis set), and is of the form

$$\text{DELETE } n \text{ ORBITALS } j_1 j_2 \ldots j_n \tag{5.31}$$

Most dramatic is the NOSTAR deletion type, which deletes *all* non-Lewis ("starred") NBOs, hence reducing the SCF variational basis set to the Lewis-type $\{\Omega_i^{(L)}\}$ NBOs alone (i.e., that of the NLS wavefunction $\Psi^{(L)}$). Still other command types allow more complex deletions for selected blocks of F_{ij} matrix elements, selected bonding relationships (vicinal, geminal) or chemical groupings, and so forth. Note that it is generally permissible to include *multiple* deletion commands in the same $DEL keylist, each of which will be processed sequentially (as described below). An illustrative example is given in the $DEL keylist of I/O-5.3, which successively checks the effects of deleting the main conjugative interaction $F_{10,85}$ (Fig. 5.2), the primary $n_O^{(y)}$-type hyperconjugations $F_{12,82}$ and $F_{12,87}$ (Fig. 5.4), and the secondary hyperconjugating antibonds 83, 86, 87 (5.26–5.28), as well as determining the basic NOSTAR energy $E^{(L)}$ (5.6) that underlies the NLS perturbative model. By creative use of these command types, the student explorer can usually "zero in" on the one or few delocalization interactions most responsible for a given structural or energetic anomaly of interest.

Let us first illustrate $DEL output for a job that contains the single deletion command:

$$\text{DELETE 1 ELEMENT 10 85} \tag{5.32}$$

to delete the primary $n_N \to \pi^*_{CO}$ (NBO 10 → 85) delocalization of amide resonance. This leads first to the output shown in I/O-5.4, which echoes the deletion task and prints the NBO occupancies for the new deletion density.

As seen in the output, the only significant effect of this deletion is to back-transfer about 0.23e from the acceptor π^*_{CO} antibond (NBO 85) to the donor n_N orbital, "undoing" the principal effect of $F_{10,85}$ interaction. The modified $DEL density is then employed for one-cycle energy evaluation (quasi-variational expectation value) with the original Fock operator to give the modified E($DEL) value, as shown in I/O-5.5.

─────────────────────── *I/O-5.4* ───────────────────────

```
Deletion of the following NBO Fock matrix elements:
   10, 85;


Orbital occupancies:


     Orbital                No deletions  This deletion  Change
----------------------------------------------------------------------
 1. BD ( 1) N 1- C 2          1.99673       1.99673     0.00000
 2. BD ( 1) N 1- H 4          1.99130       1.99130     0.00000
 3. BD ( 1) N 1- H 5          1.99200       1.99200     0.00000
 4. BD ( 1) C 2- O 3          1.99816       1.98933    -0.00883
 5. BD ( 2) C 2- O 3          1.99684       1.99684     0.00000
 6. BD ( 1) C 2- H 6          1.98379       1.98379     0.00000
 7. CR ( 1) N 1              1.99949       1.99949     0.00000
 8. CR ( 1) C 2              1.99961       1.99961     0.00000
 9. CR ( 1) O 3              1.99976       1.99976     0.00000
10. LP ( 1) N 1              1.75091       1.98942     0.23851
11. LP ( 1) O 3              1.98437       1.98437     0.00000
12. LP ( 2) O 3              1.85188       1.85188     0.00000
  :
82. BD*( 1) N 1- C 2          0.06363       0.06363     0.00000
83. BD*( 1) N 1- H 4          0.01244       0.01244     0.00000
84. BD*( 1) N 1- H 5          0.00842       0.00842     0.00000
85. BD*( 1) C 2- O 3          0.24274       0.00912    -0.23362
86. BD*( 2) C 2- O 3          0.00660       0.00660     0.00000
87. BD*( 1) C 2- H 6          0.07204       0.07204     0.00000
```

─────────────────────── *I/O-5.5* ───────────────────────

```
NEXT STEP:  Evaluate the energy of the new density matrix
            that has been constructed from the deleted NBO
            Fock matrix by doing one SCF cycle.


-----------------------------------------------------------------------
Requested convergence on RMS density matrix=1.00D-04 within  1 cycles.
Requested convergence on MAX density matrix=1.00D-02.
Requested convergence on          energy=1.00D-02.
No special actions if energy rises.
Keep R1 integrals in memory in canonical form, NReq=     9267406.
>>>>>>>>>> Convergence criterion not met.
SCF Done:  E(RB+HF-LYP) =  -169.856275512    A.U. after   2 cycles
           Convg  =    0.6106D-02           -V/T =  2.0006
           S**2  =    0.0000


-----------------------------------------------------------------------
 Energy of deletion :     -169.856275512
   Total SCF energy :     -169.954780172
                      --------------------
      Energy change :      0.098505 a.u.,        61.813 kcal/mol
-----------------------------------------------------------------------
```

Because the single-pass evaluation method interrupts the usual SCF iterative sequence (which would simply restore the density to its original self-consistent form), a Gaussian warning message is issued ("convergence criterion not met"; ignore it) and the quasi-variational "energy of deletion" is given along with the original "total SCF energy." As seen at the bottom of I/O-5.5, the recalculated $DEL energy value ($-169.856276$ a.u.) is *higher* than the former SCF energy, because E($DEL) was variationally raised by *loss* of the stabilizing $n_N \rightarrow \pi^*_{CO}$ interaction. In this case, the $DEL estimate of $n_N - \pi^*_{CO}$ stabilization, $\Delta E(\$DEL) = 61.8$ kcal/mol, agrees reasonably with the perturbative E(2) estimate, $|\Delta E_{ij}^{(2)}| = 59.6$ kcal/mol (I/O-5.1).

Table 5.1 compares the variational $\Delta E(\$DEL)$ estimates with corresponding perturbative $\Delta E_{ij}^{(2)}$ estimates for all the donor–acceptor stabilizations included in I/O-5.1, and Fig. 5.7 displays the excellent correlation between these two estimates over the full range of conjugative and hyperconjugative interactions. The two types of estimates are seen to be mutually consistent in all qualitative respects, but differences of the order of 15–20% (and sometimes larger) are commonly found for individual entries. Such differences are intrinsic to approximations made in either method, and may be taken as representative uncertainties to be assigned to either estimate.

A deeper level of uncertainties may arise from DFT evaluations of $\Delta E(\$DEL)$ (see *NBO 5.0 Manual*, p. B-20), because the $DEL densities appear "unusual" compared with those used to guide semi-empirical DFT construction. Table 5.1 includes comparison B3LYP versus HF values (all at the same geometry and basis level) for both $\Delta E_{ij}^{(2)}$ and $\Delta E(\$DEL)$ estimates in formamide, showing that DFT and *ab initio* HF values agree sensibly (within the expected differences of correlated versus uncorrelated description) in this case. Such DFT versus HF "reality checks" can provide useful warnings of DFT artifacts in $\Delta E(\$DEL)$ evaluations.

Table 5.1   Comparison of perturbative $[\Delta E_{ij}^{(2)}]$ versus variational deletion $[\Delta E(\$DEL)]$ estimates of donor–acceptor stabilization (kcal/mol) for leading donor (*i*) and acceptor (*j*) NBOs of formamide (cf. I/O-5.1).

NBOs		DFT(B3LYP)			HF		
i	*j*	$\Delta E_{ij}^{(2)}$	$\Delta E(\$DEL)$	% difference	$\Delta E_{ij}^{(2)}$	$\Delta E(\$DEL)$	% difference
10	85	59.61	61.81	$+3.7$	85.80	51.08	-40.5
12	82	23.51	24.07	$+2.4$	30.67	23.02	-24.9
12	87	22.08	23.71	$+7.4$	29.74	22.09	-25.7
6	83	4.44	5.15	$+16.0$	5.38	4.97	-7.6
3	86	3.61	3.98	$+10.2$	3.96	3.72	-6.1
2	87	2.12	2.29	$+8.0$	2.58	2.32	-10.1
11	82	1.29	1.48	$+14.7$	1.40	1.25	-10.7
4	85	1.09	1.88	$+72.5$	0.66	0.53	-19.7
5	84	1.05	1.26	$+20.0$	1.21	1.22	$+0.8$
11	87	1.01	1.21	$+20.0$	1.15	1.03	-10.4
5	82	0.92	1.02	$+10.9$	1.01	0.86	-14.9
1	86	0.81	0.90	$+11.1$	0.91	0.78	-14.3
2	86	0.63	0.65	$+3.2$	0.75	0.67	-10.7

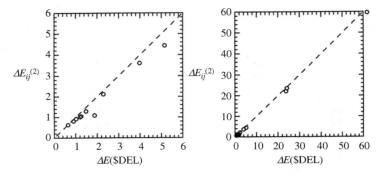

Figure 5.7.   Correlation of perturbative $\left(\Delta E_{ij}^{(2)}\right)$ versus $DEL-variational ($\Delta E(\$DEL)$) estimates (kcal/mol) of donor–acceptor stabilization, for weaker hyperconjugative interactions (left) and full range of conjugative and hyperconjugative interactions (right) in formamide (cf. Table 5.1).

Instead of a single deletion, let us now consider NOSTAR deletion of *all* interactions with non-Lewis NBOs. In this case, we obtain the occupancy and energy changes shown in I/O-5.6, in which all Lewis-type NBOs are restored to exact double occupancy and all non-Lewis NBOs are completely vacant, corresponding to the idealized NLS limit $\Psi^{(L)}$:

```
───────────────── I/O-5.6 ─────────────────
Orbital occupancies:


     Orbital                No deletions  This deletion   Change
 ----------------------------------------------------------------------
  1. BD ( 1) N 1- C 2          1.99673      2.00000      0.00327
  2. BD ( 1) N 1- H 4          1.99130      2.00000      0.00870
  3. BD ( 1) N 1- H 5          1.99200      2.00000      0.00800
  4. BD ( 1) C 2- O 3          1.99816      2.00000      0.00184
  5. BD ( 2) C 2- O 3          1.99684      2.00000      0.00316
  6. BD ( 1) C 2- H 6          1.98379      2.00000      0.01621
  7. CR ( 1) N 1               1.99949      2.00000      0.00051
  8. CR ( 1) C 2               1.99961      2.00000      0.00039
  9. CR ( 1) O 3               1.99976      2.00000      0.00024
 10. LP ( 1) N 1               1.75091      2.00000      0.24909
 11. LP ( 1) O 3               1.98437      2.00000      0.01563
 12. LP ( 2) O 3               1.85188      2.00000      0.14812
   :
 82. BD*( 1) N 1- C 2          0.06363      0.00000     -0.06363
 83. BD*( 1) N 1- H 4          0.01244      0.00000     -0.01244
 84. BD*( 1) N 1- H 5          0.00842      0.00000     -0.00842
 85. BD*( 1) C 2- O 3          0.24274      0.00000     -0.24274
 86. BD*( 2) C 2- O 3          0.00660      0.00000     -0.00660
 87. BD*( 1) C 2- H 6          0.07204      0.00000     -0.07204
   :
 -----------------------------------------------------------------------

   Energy of deletion :    -169.674185334
      Total SCF energy :   -169.954780172
                          --------------------
         Energy change :     0.280595 a.u.,         176.076 kcal/mol
 -----------------------------------------------------------------------
```

From the calculated energy [$E(L)$] of this NLS wavefunction and the difference [$E(NL)$] from total SCF energy, we therefore obtain [cf. Eq. (5.8)]

$$E(L) = -169.674185 \ a.u., E(NL) = -0.280595 \ \text{a.u.} \qquad (5.33)$$

Although the non-Lewis energy contribution is quite appreciable in absolute terms [$E(NL) = 176.08 \ \text{kcal/mol}$], it is a relatively small correction (0.16%) in percentage terms, consistent with the high accuracy of $\Psi^{(L)}$ as a perturbative starting point.

The estimate of $E(NL)$ in Equation (5.33) is still somewhat misleading, because the $DEL evaluation was carried out at the final equilibrium geometry (which includes the strong effects of resonance delocalization) rather than the preferred geometry of the idealized "resonance-free" NLS species itself. To assess the NLS energy $E_{opt}(L)$ in relaxed geometry, and determine specific geometrical effects of resonance delocalization, we can *reoptimize* the NLS geometry as described in Sidebar 5.2. The NOSTAR reoptimization is found to lead to considerable lowering of $E(L)$ [and corresponding reduction of $E(NL)$]

$$E_{opt}^{(L)} = -169.690666 \ \text{a.u.}, E_{opt}^{(NL)} = -0.167825 \ \text{a.u.} \qquad (5.34)$$

as well as pronounced geometry changes; including pyramidalization and reorientation of the NH_2 group, increased R_{CN} and decreased R_{CO} toward "standard" C−N and C=O bond lengths. These geometry changes accompanying *loss* of $E(NL)$ stabilization are consistent with the view that the planarity and other extraordinary structural features of amide groups are directly attributable to powerful resonance-type $n_N - \pi^*_{CO}$ interactions.

SIDEBAR 5.2   *$DEL-OPTIMIZATIONS WITH INTEGRATED GAUSSIAN/NBO: FORMAMIDE NLS*

Users with integrated Gaussian/NBO programs have the opportunity to combine the powerful Gaussian OPT (optimization) keyword options with $DEL options to find the geometric consequences of specific NBO deletions. Gaussian $DEL optimizations are rather restrictive and time consuming because they require numerical (rather than analytic) gradients, mandating use of numerical eigenvalue-following (EF) search algorithms based on z-matrix (rather than redundant internal) coordinates. Despite the technical difficulties and limitations, $DEL optimization techniques can provide a goldmine of information concerning geometrical effects of specific donor–acceptor interactions, as illustrated in the *NBO website* tutorial (www.chem.wisc.edu/~nbo5/tut_del.htm) on $DEL optimizations. Here, we use these techniques to determine the optimal NLS structure (NOSTAR deletion) of formamide, as though the resonance-type $E(NL)$ stabilizations did not exist in nature.

A sample Gaussian input file to perform the formamide NOSTAR optimization is shown below:

```
#B3LYP/6-311++G** POP=NBODEL NOSYMM OPT

$DEL(NOSTAR) optimization of formamide

0 1
  C    1    cn
  O    2    co      1    nco
  H    1    nh      2    cnh      3    ocnh
  H    1    nhp     2    cnhp     3    ocnhp
  H    2    ch      1    nch      4    hnch

     cn         1.26
     co         1.11
     nco        134.
     nh         1.09
     cnh        129.
     ocnh        10.
     nhp        1.01
     cnhp       121.
     ocnhp      158.
     ch         1.10
     nch        112.
     hnch       160.

$NBO PRINT=0 $END
$DEL
   NOSTAR
$END
```

Note that the POP = NBODEL keyword must be included (along with OPT) on the Gaussian route card. Gaussian optimization requires the $DEL keylist to have the form shown, with NOSTAR (or other deletion commands) separated from delimiter $DEL and $END lines. The NOSYMM keyword should generally be included on the Gaussian route card (in case low-symmetry deletions are desired), and the PRINT = 0 keyword should generally be included in the $NBO keylist (to minimize unwanted NBO output from intermediate optimization steps). Note also the (required) use of symbolic names for all variables to be optimized (up to 50 in number), with input geometry in traditional z-matrix format.

The fully optimized NLS geometry of "resonance-free" formamide resulting from this job is shown below:

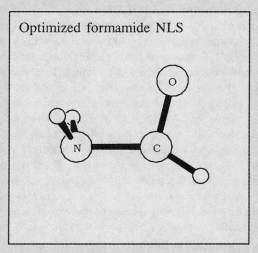

Optimized formamide NLS

As shown in the diagram, loss of E(NL) resonance stabilization dramatically alters the NLS geometry of the amide group. The NH_2 group is now pyramidalized in the manner common to amine groups in other molecules. The NLS bond lengths R_{CN} (1.57 Å) and R_{CO} (1.35 Å) are both elongated compared to their values in the physical molecule (1.36 and 1.21 Å, respectively), as expected when vicinal hyperconjugative interactions are absent. The NLS difference $R_{CN} - R_{CO}$ (0.22 Å) is also similar to that expected for idealized single versus double bonds, whereas the corresponding difference in the physical molecule (0.15 Å) is diminished by amide resonance. Despite the relatively small percentage contribution of E(NL) to total energy [text Eq. (5.33)], its deletion evidently results in *qualitative* changes in formamide structural and vibrational properties, as Pauling's classical resonance concepts suggest.

5.4 DELOCALIZATION TAILS OF NATURAL LOCALIZED MOLECULAR ORBITALS

As depicted in Fig. 5.1, the interaction of Lewis $\left(\Omega_i^{(L)}\right)$ and non-Lewis $\left(\Omega_j^{(NL)}\right)$ NBOs results in perturbative "mixing" of the parent orbitals to form new in-phase $\left(^{sl}\Omega_i^{(L)}\right)$ and out-of-phase $\left(^{sl}\Omega_j^{(NL)}\right)$ superpositions of *semi*-localized form, namely,

$$^{sl}\Omega_i^{(L)} = (1-t_{ij}^2)^2\Omega_i^{(L)} + t_{ij}\Omega_j^{(NL)} \tag{5.35}$$

$$^{sl}\Omega_j^{(NL)} = (1-t_{ij}^2)^2\Omega_j^{(NL)} - t_{ij}\Omega_i^{(L)} \tag{5.36}$$

where t_{ij} is a weak mixing coefficient ($t_{ij} \ll 1$). As shown in Equation (5.35), the in-phase (occupied, lower-energy) $^{sl}\Omega_i^{(L)}$ orbital is predominantly the parent

Lewis-type NBO $\Omega_i^{(L)}$ with a weak non-Lewis "tail" from $\Omega_j^{(NL)}$, whereas the out-of-phase (vacant, higher-energy) $^{sl}\Omega_j^{(NL)}$ (Eq. 5.36) is predominantly the non-Lewis $\Omega_j^{(NL)}$ NBO with weak Lewis-type tail from $\Omega_i^{(L)}$.

In the simple perturbative model of Fig. 5.1, the final perturbed "doubly occupied orbital" of the system is the lower-energy superposition $^{sl}\Omega_i^{(L)}$ (labeled "ε_-") rather than the unperturbed $\Omega_i^{(L)}$. Formally, the exact double occupancy identifies $^{sl}\Omega_i^{(L)}$ as a (semi-localized) molecular orbital, called a "natural localized molecular orbital" (NLMO). Although NLMOs look nothing like the "canonical" molecular orbitals (CMOs) you were probably shown in elementary textbooks, they are actually *unitarily equivalent* to textbook CMOs. This is a fancy way of saying that NLMOs and CMOs are equally valid single-configuration descriptors of the molecular wavefunction. Compared to CMOs, however, the NLMOs are generally far more recognizable and transferable from molecule to molecule, thus offering considerable pedagogical advantages as conceptual building blocks of molecular electronic structure. (Further aspects of the relationship between NLMO-based versus CMO-based descriptions of molecular systems are discussed in Chapter 11.)

More generally, each occupied NLMO may be expressed as a parent L-type $\Omega_i^{(L)}$ with weak NL-type delocalization tails (governed by coefficients t_{ij}) from each $\Omega_j^{(NL)}$:

$$^{sl}\Omega_i^{(L)} = t_{ii}\Omega_i^{(L)} + \sum_j t_{ij}\Omega_j^{(NL)}, i = 1, 2, \ldots, N/2 \qquad (5.37)$$

The residual "virtual" NLMOs are correspondingly written as

$$^{sl}\Omega_j^{(NL)} = t_{jj}\Omega_j^{(NL)} + \sum_i t_{ji}\Omega_i^{(L)}, j = N/2 + 1, \ldots \qquad (5.38)$$

The tails of the $^{sl}\Omega_i^{(L)}$'s represent the intrinsic contribution (nonvanishing occupancy) of each NL-type $\Omega_j^{(NL)}$ in "delocalizing" the parent L-type NBO in the molecular environment. The tails of the $^{sl}\Omega_j^{(NL)}$'s represent "unused" portions of $\Omega_i^{(L)}$ (the slight differences from full double occupancy), corresponding to remaining vacancies that accompany $\Omega_i^{(L)}$-$\Omega_j^{(NL)}$ charge delocalization in the final molecule. Like the NBOs from which they are formed, as well as the CMOs to which they are unitarily equivalent, the NLMOs form a complete orthonormal set.

At a single-configuration Hartree–Fock or DFT level the Lewis-based $^{sl}\Omega_i^{(L)}$'s (5.37) are exactly doubly occupied, and the virtual $^{sl}\Omega_j^{(NL)}$'s (5.38) make *no* contribution to the energy, wavefunction, or other physical properties of the system. From this viewpoint the weak non-Lewis tails of the $^{sl}\Omega_i^{(L)}$'s are the *only* true delocalization effects of physical significance, and all remaining "delocalization" of CMOs is mere window dressing, tending to erode the simplicity, familiarity, and transferability of NLMOs. [At correlated multiconfigurational levels, the NL-type NLMOs (5.38) gain slight occupancy and the L-type NLMOs (5.37) have slightly less than full double occupancy, but the $^{sl}\Omega_i^{(L)}$'s still remain far the most important orbitals from a pedagogical viewpoint.] In this section, we therefore wish to explore the compositions and properties of occupied NLMOs $\{^{sl}\Omega_i^{(L)}\}$, focusing particularly on the delocalization tails that distinguish these NLMOs from their parent NBOs.

To obtain printed details of NLMOs, one merely inserts the "NLMO" keyword into the $NBO keylist, namely:

$$\text{\$NBO NLMO \$END} \tag{5.39}$$

which leads to printout as shown (in abridged form) for formamide in I/O-5.7:

```
───────────────────────── I/O-5.7 ─────────────────────────
NATURAL LOCALIZED MOLECULAR ORBITAL (NLMO) ANALYSIS:


Maximum off-diagonal element of DM in NLMO basis:  0.99147D-10


Hybridization/Polarization Analysis of NLMOs in NAO Basis:
NLMO/Occupancy/Percent from Parent NBO/ Atomic Hybrid Contributions
----------------------------------------------------------------

  1. (2.00000)  99.8363%  BD ( 1) N  1- C  2
                     62.004%  N  1 s( 38.56%)p 1.59( 61.34%)d 0.00( 0.10%)
                     37.864%  C  2 s( 33.01%)p 2.03( 66.87%)d 0.00( 0.12%)
                      0.071%  O  3 s(  6.27%)p14.77( 92.69%)d 0.17( 1.04%)
                      0.013%  H  4 s( 75.65%)p 0.32( 24.35%)
                      0.012%  H  5 s( 73.57%)p 0.36( 26.43%)
                      0.036%  H  6 s( 98.35%)p 0.02(  1.65%)
  .
  .
 12. (2.00000)  92.5780%  LP ( 2) O  3
                      1.292%  N  1 s( 66.44%)p 0.50( 33.05%)d 0.01( 0.50%)
                      3.787%  C  2 s(  1.46%)p64.45( 94.17%)d 2.99( 4.37%)
                     92.578%  O  3 s(  0.01%)p 1.00( 99.91%)d 0.00( 0.09%)
                      0.086%  H  4 s( 98.90%)p 0.01(  1.10%)
                      0.294%  H  5 s( 99.85%)p 0.00(  0.15%)
                      1.962%  H  6 s( 99.58%)p 0.00(  0.42%)
```

As shown in I/O-5.7, NLMOs 1–12 each have exact double occupancy, with dominant contribution from the parent NBO (99.8% for NLMO1, ..., 92.6% for NLMO 12) and small residual tails whose percentage contributions are given in terms of atomic hybrid composition at each contributing atomic center. For example, NLMO 12 is primarily the expected large contribution (92.58%) from $(p_y)_O$ on O_3, with smaller contributions from $(sp^{64})_C$ (3.79%) on C_2, $(s)_{H(6)}$ (1.96%) on H_6, and so forth. [Note that such NLMO output is generated for other keywords that involve implicit use of NLMOs (such as PLOT, DIPOLE, etc.), and therefore may appear even if no NLMO keyword request was included in the $NBO keylist.]

The standard NLMO printout includes qualitative information about the percentage contributions (but not signs) from various centers, but does not give full details of the $\{t_{ij}\}$ coefficients in (5.37) and (5.38) that express each NLMO in terms of contributing NBOs. To see printout of the full transformation (IN $\rightarrow$ OUT) from starting NBO ("IN") to final NLMO ("OUT") orbitals, one can simply invoke the corresponding NBONLMO ("INOUT") keyword (see Appendix C for further details), namely:

$$\text{\$NBO NBONLMO \$END} \tag{5.40}$$

The NBONLMO keyword leads to printout of the full table of NBO $\rightarrow$ NLMO transformation coefficients, with the coefficients of each NBO (rows) listed under each NLMO (columns), as shown (in severely abridged form, neglecting CR, RY* contributions) for NLMOs 1–8 in I/O-5.8:

```
─────────────────────────── I/O-5.8 ───────────────────────────

 NLMOs in the NBO basis:
                                                                  .

           NBO        1        2        3        4        5        6        7        8
      ---------- -------  -------  -------  -------  -------  -------  -------  -------
   1.  N 1- C 2   0.9992  -0.0003 -0.0003  0.0000   0.0000  -0.0002  0.0000   0.0001
   2.  N 1- H 4  -0.0011   0.9978  0.0005  0.0000   0.0000  -0.0001  0.0003   0.0001
   3.  N 1- H 5   0.0011   0.0016  0.9980  0.0000  -0.0001   0.0003  0.0004   0.0000
   4.  C 2- O 3   0.0000   0.0000  0.0000  0.9995   0.0000   0.0000  0.0000   0.0000
   5.  C 2- O 3  -0.0002  -0.0001 -0.0001  0.0000   0.9992   0.0003  0.0000  -0.0001
   6.  C 2- H 6   0.0008   0.0003  0.0006  0.0000   0.0015   0.9959 -0.0001  -0.0001
   :
  10.  N 1(lp)    0.0000   0.0000  0.0000  0.0001   0.0000   0.0000  0.0000   0.0000
  11.  O 3(lp)    0.0005  -0.0014 -0.0002  0.0000  -0.0003   0.0007  0.0002  -0.0004
  12.  O 3(lp)    0.0032   0.0114 -0.0015  0.0000   0.0005  -0.0038 -0.0008  -0.0005
   :
  82.  N 1- C 2*  0.0000  -0.0066 -0.0052  0.0000  -0.0150   0.0238 -0.0010  -0.0005
  83.  N 1- H 4* -0.0031   0.0000  0.0007  0.0000   0.0108  -0.0722 -0.0005  -0.0021
  84.  N 1- H 5* -0.0065   0.0006  0.0000  0.0000  -0.0260   0.0106 -0.0003  -0.0038
  85.  C 2- O 3*  0.0000   0.0000  0.0000  0.0000   0.0000   0.0000  0.0000   0.0000
  86.  C 2- O 3* -0.0165  -0.0267  0.0474  0.0000   0.0000  -0.0075 -0.0023   0.0013
  87.  C 2- H 6* -0.0153  -0.0450  0.0165  0.0000   0.0064   0.0004  0.0010   0.0015
```

From the first two columns, for example, one can express the compositions of NLMOs 1, 2 as

$$\text{NLMO 1: } {}^{\text{sl}}\sigma_{\text{CN}} = 0.9992\sigma_{\text{CN}} - 0.0165\sigma^*_{\text{CO}} - 0.0153\sigma^*_{\text{CH}} - \ldots \tag{5.41}$$

$$\text{NLMO 2: } {}^{\text{sl}}\sigma_{\text{NH(s)}} = 0.9978\sigma_{\text{NH(s)}} - 0.0450\sigma^*_{\text{CH}} - 0.0267\sigma^*_{\text{CO}} - \ldots \tag{5.42}$$

As expected, the largest NLMO tail coefficients are usually associated with the valence antibonds (if any such exist) at *vicinal anti*-positions relative to the parent NBO, such as the σ^*_{CH} tail of ${}^{\text{sl}}\sigma_{\text{NH(s)}}$ in NLMO 2 (5.42). (Note that the order of tail contributions generally parallels the listed "principal delocalizations" in the NBO summary table, I/O-4.5.)

Although they are obtained quite independently, the NLMO delocalization tails reflect the relative importance of $\Omega_i^{(\text{L})}$-$\Omega_j^{(\text{NL})}$ donor–acceptor interactions in a manner that correlates with corresponding perturbative (Section 5.2) or \$DEL-variational (Section 5.3) estimates. This overall correlation is exhibited in Fig. 5.8, which plots the perturbative estimates $\Delta E_{ij}^{(2)}$ versus squared tail amplitude $|t_{ij}|^2$ for all the interactions of I/O-5.1. Thus, the relative strengths of $\Omega_i^{(\text{L})}$-$\Omega_j^{(\text{NL})}$ interactions can be judged from NLMO t_{ij} tail amplitudes even if energetic estimates $[\Delta E_{ij}^{(2)}, \Delta E(\$\text{DEL})]$ of donor–acceptor interaction strength are unavailable (e.g., for correlation methods lacking an effective 1e Hamiltonian).

The signs of the t_{ij} delocalization tails can also be used to verify that the stabilizing $\Omega_i^{(\text{L})}$-$\Omega_j^{(\text{NL})}$ interaction is indeed of "in-phase" character. Although the phase (sign) of an individual $\Omega_i^{(\text{L})}$ or $\Omega_j^{(\text{NL})}$ could be chosen rather arbitrarily (i.e., through

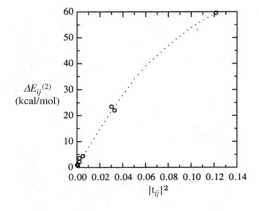

Figure 5.8. Second-order perturbative stabilization ($\Delta E_{ij}^{(2)}$) versus squared NLMO tail amplitude $|t_{ij}|^2$ for the formamide $\Omega_i^{(L)} - \Omega_j^{(NL)}$ donor–acceptor interactions of I/O-5.1 (with dotted qualitative trend curve to aid visualization).

different choices of Cartesian axes) to give apparent "negative overlap" S_{ij} of $\Omega_i^{(L)}$ and $\Omega_j^{(NL)}$, the sign of t_{ij} will generally be such as to restore the favorable in-phase orbital mixing pattern that is dictated on physical grounds (i.e., $t_{ij} < 0$ if $S_{ij} < 0$). [For this reason, we have generally chosen orbital phases to make the in-phase mixing apparent in orbital overlap diagrams such as Fig. 5.4, reversing (if necessary) the phase choice of the ESS coordinate system.]

The most severely distorted NLMO is that for the amine lone pair $^{sl}n_N$ (NLMO 10), which can be expressed as

$$\text{NLMO 10:}\ ^{sl}n_N = 0.9357 n_N + 0.3484 \pi_{CO}^* - \dots \tag{5.43}$$

corresponding to strong allylic-type $n_N \rightarrow \pi_{CO}^*$ conjugative delocalization. But even such extreme conjugative delocalization features preserve the essentially recognizable form of the parent NBO in contour and surface plots of the NLMO, as shown in Fig. 5.9. (Corresponding comparisons for other formamide NLMOs would reveal much smaller visual differences.)

Compared to NBOs, the NLMOs are naturally somewhat less transferable, due to their inclusion of interaction features (delocalization tails) with the specific molecular environment. Nevertheless, the NLMOs provide the natural starting point for reexpressing many properties of SCF-MO wavefunctions in more transparent NBO-based "semi-localized" form, as illustrated, for example, in Chapters 7 and 8.

5.5 HOW TO $CHOOSE ALTERNATIVE LEWIS STRUCTURES

By default, the NBO program is instructed to find the "best possible" Lewis structure diagram, and the subsequent determination of NLMOs and localized description of chemical properties will be based on this default NBO assignment. But what if you wish to carry out the analysis in terms of some *alternative* Lewis structure of your own choosing—perhaps because it is more consistent with other systems you are analyzing, or because you would like to check whether it is really inferior to the

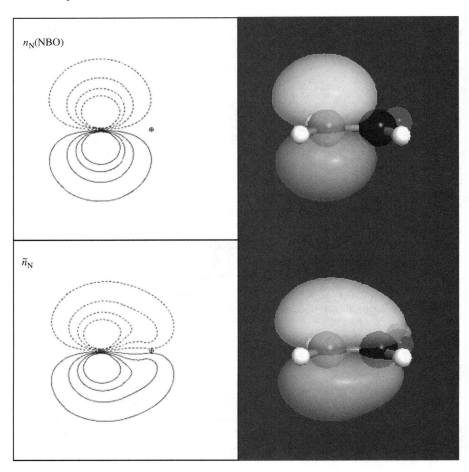

Figure 5.9. Contour and surface plots comparing NBO (upper panels) and NLMO (lower panels) for conjugatively delocalized amine lone pair n_N [Eq. (5.43)] of formamide. (See the color version of this figure in Color Plates section.)

default NBO structure, or just for general curiosity? In this case, the NBO program provides $CHOOSE keylist options that allow exploration and usage of such alternative Lewis (resonance) structural representations in a very general manner.

To $CHOOSE an alternative Lewis structure, you should first draw the desired Lewis diagram, identifying the valence LONE, BOND, or 3CBOND pairs [whether of single (S), double (D), triple (T), quadruple (Q), pentuple (P), or hextuple (H) type] for all valence electrons. (For an open-shell species, corresponding LONE, BOND, 3CBOND specifications are used for individual electrons, typically with different Lewis structural patterns for ALPHA and BETA spin.) When the Lewis structure is input (as detailed below) in the $CHOOSE ... $END keylist, the NBO program uses this keylist to direct the search for NBO hybrids and polarization coefficients that are optimal for this bonding pattern, reporting (as usual) the associated non-Lewis error of the $CHOOSE structure. The $CHOOSE-based NBOs are subsequently employed to construct NLMOs and perform other tasks of NBO analysis, in complete analogy to the NLS-based NBOs of default analysis.

For closed-shell systems the $CHOOSE . . . $END keylist (which usually follows the main $NBO . . . $END keylist) has the following schematic layout:

```
$CHOOSE
   LONE    (1c list)   END
   BOND    (2c list)   END
   3CBOND (3c list)    END
$END
```

[Note that each LONE, BOND, or 3CBOND keyword list (if included at all) must be closed by a corresponding "END" keyword.] The "1c list" (LONE. . .END) entries are integer pairs (AC_i, N_i), each giving the atomic center number (AC_i) and multiplicity of valence lone pairs (N_i) on a lone pair bearing atom; for example, the alternative formamide Lewis structure (5.24), with two valence lone pairs on N_1 and one on O_3, is specified by

```
LONE 1 2  3 1 END
```

whereas (5.25), with one lone pair each on N_1, O_3, and H_6, is specified by

```
LONE 1 1  3 1  6 1 END
```

The "2c list" (BOND. . .END) entries consist of bond-multiplicity descriptors (S = single, D = double, . . .) followed by atom center numbers of the two bonded atoms (such as "S 1 2" for a single bond between atoms 1 and 2, etc.); for example, the BOND list for formamide structure (5.24) (with triple bond between N_2 and O_3 and single bonds N_1-H_4, N_1-H_5, and C_2-H_6) can be specified as

```
BOND S 1 4  S 1 5  T 2 3  S 2 6 END
```

whereas that for structure (5.25) is

```
BOND S 1 4  S 1 5  S 1 2  T 2 3 END
```

The "3c list" (3CBOND. . .END) entries are analogous, with each S/D/T. . . bond-multiplicity descriptor followed by *three* atom center numbers.

[The $CHOOSE input format is rather flexible: The order of LONE, BOND (or 3CBOND) lists is immaterial, as is the order of bond types or bonded atoms within each BOND list, and even a condensed single-line form (not recommended) is valid, for example,

```
$CHOOSE LONE (1c list) END BOND (2c list) END $END
```

As usual, all input is case-insensitive and keyword entries may be separated by commas or (any number of) spaces. Note that only *valence* lone pairs (not CR-type) are to be included in the LONE. . .END list. See the *NBO Manual*, p. B-14ff for further details.]

For open-shell systems, the schematic $CHOOSE layout is similar, but with separate ALPHA . . . END and BETA . . . END sections for each spin set, namely,

```
$CHOOSE
   ALPHA
      LONE   (α 1c list)   END
      BOND   (α 2c list)   END
      3CBOND (α 3c list)   END
   END
   BETA
      LONE   (β 1c list)   END
      BOND   (β 2c list)   END
      3CBOND (β 3c list)   END
   END
$END
```

to specify different Lewis structures for different spins (Section 4.5). (In this case, it is even more challenging to include all "END" keywords that close the various keyword sublists, as well as the terminal "$END" that closes the $CHOOSE keylist itself.)

As a simple example of closed-shell $CHOOSE keylist input, let us consider the principal alternative resonance structure of formamide [the second structure in (5.21)], which can be specified as shown in I/O-5.9:

```
──────────────────────────── I/O-5.9 ────────────────────────────
#N b3lyp/6-311++g**  POP=NBORead

H2NCHO

  0  1
N
C  1  1.361
O  2  1.212   1    124.90
H  1  1.009   2    119.42   3    0.00
H  1  1.007   2    121.45   3  180.00
H  2  1.107   1    112.44   4  180.00

$NBO file=h2ncho $END
$CHOOSE
  LONE 3 3 END
  BOND S 1 4  S 1 5  D 1 2  S 2 3  S 2 6 END
$END
```

The NBO search report for this $CHOOSE job is shown in I/O-5.10:

```
───────────────────────── I/O-5.10 ─────────────────────────

NATURAL BOND ORBITAL ANALYSIS:
Reading $CHOOSE input list.


                 Occupancies      Lewis Structure   Low   High
          Occ.   -----------------  --------------   occ   occ
 Cycle  Thresh.  Lewis  Non-Lewis   CR  BD  3C  LP   (L)   (NL)   Dev
======================================================================

  1(1)   1.90   23.22622  0.77378    3   6   0   3    2     1    0.73
----------------------------------------------------------------------


Structure accepted: NBOs selected via the $CHOOSE keylist


----------------------------------------------------------

   Core                5.99886 ( 99.981% of    6)
   Valence Lewis      17.22735 ( 95.708% of   18)
 ==================   ============================

   Total Lewis        23.22622 ( 96.776% of   24)
----------------------------------------------------------

   Valence non-Lewis   0.72036 (  3.002% of   24)
   Rydberg non-Lewis   0.05342 (  0.223% of   24)
 ==================   ============================

   Total non-Lewis     0.77378 (  3.224% of   24)
----------------------------------------------------------
```

As expected, the directed NBO search now requires only one cycle, leading to the $CHOOSE-selected structure, which is seen to be inferior to the default NBO structure (cf. I/O-4.4), with non-Lewis error of 0.7738e (3.22%) versus 0.4552e (1.90%) for the default NLS.

Some details of the $CHOOSE-based NBOs are shown in I/O-5.11, for comparison with corresponding NLS default NBOs in I/O-4.9. Bonding features that are common to both structures (e.g., σ_{NH}, σ_{CN},...) are seen to have NBOs that are fairly similar but not identical. The new "π_{CN}" (NBO 1) of the $CHOOSE structure is highly occupied (1.998e), but its extreme polarization (83.3% on N) betrays its essential parentage in the nitrogen lone pair NBO of the dominant NLS resonance form. The new out-of-plane $n_O^{(\pi)}$ lone pair (NBO 12) of the $CHOOSE structure has very low occupancy (1.475e), corresponding to a large non-Lewis error ($\sim$0.525e, mostly contained in π^*_{CN}) that already exceeds the 0.455e non-Lewis error for the entire NLS. Thus, as advertised, the NLS description is indeed significantly "better" than the alternative $CHOOSE structure description, but the fairly comparable non-Lewis errors (1.90% versus 3.22%) suggest (correctly) that the $CHOOSE structure gains significant weighting in a formal resonance-hybrid description, as will be demonstrated in the following section.

```
───────────────────────── I/O-5.11 ─────────────────────────

   (Occupancy)   Bond orbital/ Coefficients/ Hybrids
 ---------------------------------------------------------------

  1. (1.99810) BD ( 1) N  1- C  2
          ( 83.32%)   0.9128* N  1 s(  0.00%)p 1.00( 99.99%)d 0.00(  0.01%)
          ( 16.68%)   0.4084* C  2 s(  0.00%)p 1.00( 99.67%)d 0.00(  0.33%)
```

```
 2. (1.99418) BD ( 2) N  1- C  2
    ( 62.20%)   0.7886* N  1 s( 39.03%)p 1.56( 60.87%)d 0.00(  0.10%)
    ( 37.80%)   0.6148* C  2 s( 36.81%)p 1.71( 63.06%)d 0.00(  0.13%)
 3. (1.99130) BD ( 1) N  1- H  4
    ( 70.18%)   0.8377* N  1 s( 30.40%)p 2.29( 69.50%)d 0.00(  0.10%)
    ( 29.82%)   0.5461* H  4 s( 99.96%)p 0.00(  0.04%)
 4. (1.99200) BD ( 1) N  1- H  5
    ( 69.73%)   0.8350* N  1 s( 30.49%)p 2.28( 69.41%)d 0.00(  0.10%)
    ( 30.27%)   0.5502* H  5 s( 99.96%)p 0.00(  0.04%)
 5. (1.99606) BD ( 1) C  2- O  3
    ( 34.51%)   0.5874* C  2 s( 35.85%)p 1.78( 63.97%)d 0.00(  0.18%)
    ( 65.49%)   0.8093* O  3 s( 37.08%)p 1.69( 62.72%)d 0.01(  0.20%)
 6. (1.98211) BD ( 1) C  2- H  6
    ( 56.85%)   0.7540* C  2 s( 28.24%)p 2.54( 71.62%)d 0.01(  0.14%)
    ( 43.15%)   0.6569* H  6 s( 99.96%)p 0.00(  0.04%)
   :
10. (1.98437) LP ( 1) O  3           s( 60.38%)p 0.66( 39.59%)d 0.00(  0.03%)
11. (1.81459) LP ( 2) O  3           s( 2.67%)p36.42( 97.23%)d 0.04(  0.10%)
12. (1.47466) LP ( 3) O  3           s( 0.00%)p 1.00( 99.89%)d 0.00(  0.11%)
```

Other possible Lewis structures may be similarly tested. For example, the structures (5.24), (5.25) associated with the strong hyperconjugative delocalizations of the $n_O^{(y)}$ NBO (Fig. 5.4) could be specified with LONE/BOND lists given previously as

```
$CHOOSE
  LONE 1 2  3 1 END
  BOND S 1 4  S 1 5  T 2 3  S 2 6 END
$END
```

for (5.24), or

```
$CHOOSE
  LONE 1 1  3 1  6 1  END
  BOND S 1 4  S 1 5  S 1 2  T 2 3 END
$END
```

for (5.25). Table 5.2 compares the non-Lewis error (ρ_{NL}) of these subsidiary structures (RS 3, 4) with the leading structures (RS 1, 2) described above. As seen

Table 5.2 Comparison of non-Lewis errors (ρ_{NL}) for alternative $CHOOSE structures RS 1–4 of formamide (see text), with associated NBO donor–acceptor delocalization.

RS	NBO deloc.	ρ_{NL} (e)	%-ρ_{NL}
1	[NLS]	0.4552	1.90
2	$n_N \to \pi^*_{CO}$	0.7738	3.22
3	$n_O^{(y)} \to \sigma^*_{CN}$	1.2044	5.02
4	$n_O^{(y)} \to \sigma^*_{CH}$	1.4628	6.10

in the table, the ρ_{NL} values are appreciably greater for RS 3, 4, suggesting their lesser weighting in the formal resonance-hybrid description (Section 5.6).

Finally, we show an illustrative open-shell $CHOOSE keylist for the alternative "spin-flipped" structure of ozone [Section 4.5; cf. (4.29)]:

```
$CHOOSE
  ALPHA
     LONE  1 3   2 1   3 2   END
     BOND  S 1 2   D 2 3  END
  END
  BETA
     LONE  1 2   2 1   3 3   END
     BOND  D 1 2   S 2 3  END
  END
$END
```

Of course, this merely returns an "identical" structure to that shown in Section 4.5 (but with α and β spin output sections interchanged), corresponding to the equivalent weightings that such spin-flipped resonance structures are expected to have in a resonance-hybrid picture.

Once you have gained some practice, the $CHOOSE format will be found to offer convenient expression for Lewis structures of quite general form, including the exotic high-order (quadruple and higher) metal–metal bonds of transition metal species (see *V&B*, p. 413ff).

5.6   NATURAL RESONANCE THEORY

Many topics in this chapter have portended close relationship to the "resonance" picture of molecular electronic structure. This simple generalization of the Lewis structure concept envisions a molecular species as a weighted-average "hybrid" of two or more contributing Lewis structures, harking back to Kekulé's famous imagery for benzene. The essentials of the chemical resonance concept (known earlier as "electromerism theory") were developed extensively by Robinson, Ingold, and other physical organic chemists, long before the discovery of quantum mechanics and Pauling's famous 1931–1933 wavefunctional reformulation of the theory (see L. Pauling, *Nature of the Chemical Bond*, 3rd ed., Cornell U. Press, Ithaca NY, 1960). Today, Pauling's resonance concepts pervade practically every elementary chemistry textbook, and students of chemistry will naturally be interested to explore how (if at all) such concepts are manifested in modern molecular wavefunctions.

Strictly speaking, Pauling's mathematical formulation of resonance theory did not behave as its author intended (Sidebar 5.3). However, the theory was initially applied only in a qualitative empirical fashion that obscured these difficulties. Nearly a half-century elapsed before reliable polyatomic calculations allowed a rigorous test of Pauling's approximations, by which time the qualitative concepts of Pauling's resonance theory had become firmly entrenched in chemistry textbooks in more or less present form. Although some theorists continue to believe that resonance concepts should *only* be defined in terms of the specific type of multiconfigurational valence bond wavefunctions proposed by Pauling, such wavefunctions (except for H_2) are of limited accuracy and practicality and have played little real role in modern computational investigations that successfully account for all the phenomena traditionally attributed to resonance.

Fortunately, a simple extension of the NBO method allows easy evaluation of resonance weights and bond orders from wavefunctions of quite general form. Instead of searching for the *single* Lewis-like density (or strictly, the "first-order density operator" for a Lewis-like wavefunction; see *V&B,* p. 21ff) that best matches a target wavefunction density, the "natural resonance theory" method searches for a *manifold* of localized Lewis densities and associated weighting factors (positive numbers summing to unity) that satisfies this criterion. The NRT manifold of candidate Lewis structures and associated densities is generated in an orderly manner from the starting NBO structure and its principal delocalizations, each mapped onto a formal resonance structure (cf. Fig. 5.6) whose mathematical details are evaluated by the $CHOOSE procedure (Section 5.5). As usual, the accuracy of the NRT description is quantified as the mean square difference between the target density and its resonance hybrid approximation (Sidebar 5.4). The NRT formulation of resonance theory conforms closely (but not exactly) to Pauling's original precepts, and leads, as Pauling's formulation could not, to an *exact* "resonance averaging" relationship for *every* formal one-electron property, including electron density, kinetic energy, nuclear–electron attraction, dipole moment, and molecular geometry, consistent with the earlier tenets of electromerism theory. Even if not tied to Pauling's original wavefunction assumptions, the NRT resonance weights and bond orders map easily onto the qualitative empirical concepts of Pauling's resonance theory, and can be considered a more accurate and practical computational implementation of those concepts.

Performing NRT analysis of a modern wavefunction is typically as simple as inserting the "NRT" keyword into the $NBO keylist. For formamide, for example, the NRT keyword leads first to summary search diagnostics, as shown in I/O-5.12:

─────────── *I/O-5.12* ───────────

```
NATURAL RESONANCE THEORY ANALYSIS:                        .


Maximum reference structures :    20
Maximum resonance structures : 1439
Memory requirements :    9917757 words of    9927226 available


  5 candidate reference structure(s) calculated by SR LEWIS
Total of   5 candidate reference structures.
Initial loops searched    7 bonding pattern(s); all but   3 were discarded
Reference   1:  rho*=0.45516, f(w)=0.93270 converged after  26 iterations
Reference   2:  rho*=0.73148, f(w)=0.95934 converged after  21 iterations
Reference   3:  rho*=0.97867, ionic; deleted
Multi-ref( 2):  D(W)=0.07563, F(W)=0.11941 converged after 214 iterations
```

		non-Lewis		fractional accuracy f(w)		
Ref	Wgt	density	d(0)	all NBOs	val+core	valence
1	0.66505	0.45516	0.04199	0.93270	0.94988	0.94991
2	0.33495	0.73148	0.08125	0.95934	0.96773	0.96774

Full explanation of this output involves algorithmic details that are beyond the scope of the present work (see *NBO Manual*, p. B-72ff and the original NRT papers referenced therein). However, one can see in a general way that the NRT search involves initial identification of "reference structures" (each with secondary resonance corrections) to maximize the "fractional accuracy" $f(\underline{w})$ factors, followed by "Multiref" optimization to maximize the corresponding multireference fractional accuracy $F(\underline{W})$. [Cf. the *NBO website* sample NRT output, www.chem.wisc.edu/~nbo5/nrt.pdf, for further details of the $f(\underline{w})$, $F(\underline{W})$ variational criteria.] In this case, two reference structures were identified among five candidate structures initially selected from the large number of secondary structures generated internally, and both single- and multireference optimizations converged successfully. (Unless something went wrong, the diagnostics of this section can usually be safely ignored.)

Next follows the "TOPO matrix" (bond order table) for the lead resonance structure and full listing of NRT structures and percentage weightings, as shown in I/O-5.13.

```
───────────────────── I/O-5.13 ─────────────────────
TOPO matrix for the leading resonance structure:


    Atom  1   2   3   4   5   6

    ----  --- --- --- --- --- ---

 1.  N    1   1   0   1   1   0
 2.  C    1   0   2   0   0   1
 3.  O    0   2   2   0   0   0
 4.  H    1   0   0   0   0   0
 5.  H    1   0   0   0   0   0
 6.  H    0   1   0   0   0   0


         Resonance
  RS     Weight(%)                  Added(Removed)
 ------------------------------------------------------------------------

  1*(2)   62.82
  2*      29.42    N  1- C  2, ( C  2- O  3), ( N  1),  O  3
  3 (2)    2.62    C  2- O  3, ( C  2- H  6), ( O  3),  H  6
  4 (2)    2.39    ( N  1- C  2),  C  2- O  3,  N  1, ( O  3)
  5        1.30    N  1- C  2, ( C  2- H  6), ( N  1),  H  6
  6        0.47    N  1- C  2, ( N  1- H  4), ( C  2- H  6),  H  4
  7        0.24    N  1- C  2, ( N  1- H  5), ( C  2- O  3),  O  3
  8        0.24    N  1- C  2,  N  1- C  2, ( N  1- H  4), ( C  2- O  3),
                   ( C  2- H  6), ( N  1),  O  3,  H  4
  9        0.20    N  1- C  2, ( N  1- H  5), ( C  2- O  3),  H  5
 10        0.12    N  1- C  2,  N  1- C  2, ( N  1- H  5), ( C  2- O  3),
                   ( C  2- O  3), ( N  1),  O  3,  O  3
 11        0.10    N  1- C  2,  N  1- C  2, ( N  1- H  5), ( C  2- O  3),
                   ( C  2- O  3), ( N  1),  O  3,  H  5
 12-13     0.08
 ------------------------------------------------------------------------

         100.00    * Total *              [* = reference structure]
```

To reconstruct the leading resonance structure from the TOPO matrix, use the off-diagonal (i,j)-entry (in row i, column j) to find the number of bonds between atoms i, j (e.g., two bonds between atoms C_2 and O_3). Similarly, use the diagonal (i,i)-entry to find the number of lone pairs on atom i (e.g., two lone pairs on O_3). In this case, the leading resonance structure coincides as usual with the default NBO structure shown at the left in (5.23). This leading structure (RS 1, with 62.82% weighting) is the starting point for reconstructing each remaining structure of the full NRT listing, by adding (or removing) the listed 1c (lone pair) and 2c (bond) structural elements. For example, to form RS 2 (with 29.42% weighting), one should *add* a bond to N_1-C_2 and lone pair to O_3 and *subtract* a bond from C_2-O_3 and lone pair from N_1, thereby obtaining the expected principal amide resonance structure correction shown at the right in (5.23). These two leading structures (RS 1, 2) are identified by asterisks as dominant "reference structures" for this species.

Similarly, for "secondary" RS 3 (2.62%) and RS 4 (2.39%) one obtains the resonance structures shown in (5.25) and (5.24), respectively. Qualitatively, the NRT weightings are seen to be in reasonable accord with the rankings previously anticipated from perturbative $\Delta E_{ij}^{(2)}$ or $DEL-variational $\Delta E($DEL) estimates (cf. Table 5.1) or corresponding NLMO t_{ij} tail amplitudes (cf. Fig. 5.8), consistent with the visualizations provided by NBO overlap diagrams (cf. Figs. 5.2 and 5.4) for donor–acceptor delocalizations. As shown in I/O-5.13, only structures contributing at least 0.1% are identified explicitly in the NRT listing, while the remaining structures (RS 12,13 in this case) are simply grouped together under their combined weighting. (Additional details for such low-weighted structures can be obtained by including the NRTDTL keyword; cf. *NBO Manual*, p. B-75.)

The 13 listed resonance structures of the final NRT expansion are seen to have assigned total weightings of 100% (as would also be the case if greater or lesser number of possible NRT structures were included, based on other numerical thresholds). However, it should be recalled [from the $f(\underline{w})$, $F(\underline{W})$ values of I/O-5.12] that this NRT expansion is incomplete, and the structures included in the NRT listing do not represent a "complete set" of resonance structures as assumed in conventional Pauling-type resonance expressions (Sidebar 5.4). For structures 1, 3, and 4, the parenthesized "(2)" indicates that two *distinct* NBO configurations map onto the same Lewis structural diagram (as can be further investigated with the NRTDTL keyword), but only the "expected" valence-type configuration typically accounts for the overwhelming majority of the assigned total weighting.

The NRT weightings of each resonance structure are then used to obtain the resonance-averaged *NRT bond order* b_{ij} between each $i–j$ atom pair, as shown in I/O-5.14 and depicted graphically in (5.44):

──────────────── *I/O-5.14* ────────────────

```
Natural Bond Order:  (total/covalent/ionic)


   Atom       1       2       3       4       5       6
   ----    ------  ------  ------  ------  ------  ------

1.  N   t 0.7117  1.3028  0.0000  0.9922  0.9934  0.0000
        c   ---   0.8207  0.0000  0.5917  0.6015  0.0000
        i   ---   0.4821  0.0000  0.4005  0.3919  0.0000


2.  C   t 1.3028  0.0000  1.7443  0.0000  0.0000  0.9529
        c 0.8207    ---   1.0781  0.0000  0.0000  0.8228
        i 0.4821    ---   0.6662  0.0000  0.0000  0.1301


3.  O   t 0.0000  1.7443  2.2527  0.0000  0.0000  0.0000
        c 0.0000  1.0781    ---   0.0000  0.0000  0.0000
        i 0.0000  0.6662    ---   0.0000  0.0000  0.0000
```

```
4.  H  t 0.9922 0.0000 0.0000 0.0070 0.0000 0.0000
       c 0.5917 0.0000 0.0000  ---   0.0000 0.0000
       i 0.4005 0.0000 0.0000  ---   0.0000 0.0000


5.  H  t 0.9934 0.0000 0.0000 0.0000 0.0030 0.0000
       c 0.6015 0.0000 0.0000 0.0000  ---   0.0000
       i 0.3919 0.0000 0.0000 0.0000  ---   0.0000


6.  H  t 0.0000 0.9529 0.0000 0.0000 0.0000 0.0401
       c 0.0000 0.8228 0.0000 0.0000 0.0000  ---
       i 0.0000 0.1301 0.0000 0.0000 0.0000  ---
```

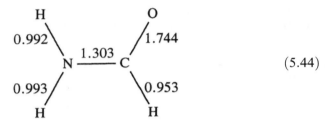

$$(5.44)$$

(The NRT bond order b_{ij} supplants the assortment of alternative "bond orders" produced by the obsolete BNDIDX keyword.) As shown in I/O-5.14, each total (t) bond order is divided into covalent (c) and ionic (i) components, based on the resonance-averaged bond ionicity for each contributing resonance structure. The covalent contribution often dominates for organic compounds, and the ionic component for inorganic compounds, but only the total b_{ij} is expected to exhibit bond-order–bond-length correlations of the type assumed in elementary bonding theory. As expected from the approximate 30% weighting of the alternative dipolar amide structure in (5.21), the b_{CN} bond order is about 30% greater than that for an ordinary C–N single bond, while b_{CO} is correspondingly reduced from ordinary C=O double-bond order, all consistent with the unique structural and torsional properties of amide groups. The qualitative Pauling-type picture of amide resonance is therefore recovered in the NRT description, even if underlying mathematical details differ from those originally assumed by Pauling.

The NRT bond orders around each atom are next summed to give the total *valency* of the atom, as shown in I/O-5.15:

---------------------------------- *I/O-5.15* ----------------------------------

Natural Atomic Valencies:

Atom	Valency	Co-Valency	Electro-Valency
1. N	3.2883	2.0139	1.2745
2. C	4.0000	2.7216	1.2784
3. O	1.7443	1.0781	0.6662
4. H	0.9922	0.5917	0.4005
5. H	0.9934	0.6015	0.3919
6. H	0.9529	0.8228	0.1301

For example, one can see from the sum of bond orders around C_2 (1.303 + 1.744 + 0.953 = 4.000) that the carbon atom is aptly described as "tetravalent," and each hydrogen atom is similarly "monovalent," consistent with the usual assignment of atomic valency in the periodic table. (Due to the strong effects of resonance, "trivalent" nitrogen and "divalent" oxygen are seen to exhibit somewhat anomalous departures from integer valency values in formamide, but this may be considered an "exception that proves the rule" with respect to common organic species that the student may encounter.) As for the contributing bond orders, each total atomic valency can be divided into covalent ("covalency") and ionic ("electrovalency") contributions, but only the total NRT valency exhibits the expected near-integer association with idealized periodicity assignments.

The final section of NRT output (I/O-5.16) provides a sample $NRTSTR ... $END keylist describing the two primary reference structures that were found in this case:

```
───────────────── I/O-5.16 ─────────────────
$NRTSTR
  STR            ! Wgt = 62.82%
    LONE 1 1 3 2 END
    BOND S 1 2 S 1 4 S 1 5 D 2 3 S 2 6 END
  END
  STR            ! Wgt = 29.42%
    LONE 3 3 END
    BOND D 1 2 S 1 4 S 1 5 S 2 3 S 2 6 END
  END
$END
```

As seen in this example, the syntax of each resonance structure specification (STR ... END) is very similar to that of a $CHOOSE keylist (see *NBO Manual*, p. B-77ff). Such $NRTSTR keylist may be included in the input file (after the usual $NBO ... $END keylist) to *dictate* the choice of reference structures for NRT analysis.

For open-shell species, NRT analysis proceeds similarly with corresponding single-spin resonance weightings, bond order, and valencies in each spin set, followed by a composite spin average that yields the final overall bond orders and valencies of the species. For ozone (Section 4.5), for example, the α-spin weightings are shown in I/O-5.17, leading to α-bond orders $b_{12} = 0.9278$, $b_{23} = 0.5722$, with corresponding β-spin output differing only by interchange of O_1 and O_3:

```
───────────────── I/O-5.17 ─────────────────
          Resonance
  RS      Weight (%)                    Added (Removed)
  ------------------------------------------------------------------------
  1*(2)    80.78
  2*(2)    16.13    ( O 1- O 2),   O 2- O 3,   O 1, ( O 3)
  3         2.62    O 1- O 2, ( O 2- O 3), ( O 1),   O 3
  4         0.47    ( O 1- O 2), ( O 1- O 2),   O 2- O 3,   O 2- O 3,   O 1,
                    O 1, ( O 3), ( O 3)
  ------------------------------------------------------------------------
           100.00   * Total *              [* = reference structure]
```

The composite NRT bond orders and valencies then follow as shown in I/O-5.18, leading to the expected symmetries ($b_{12} = b_{23} = 1.5$; $V_1 = V_3 = 1.5$, $V_2 = 3$) of the composite $\alpha + \beta$ resonance hybrid:

─── *I/O-5.18* ───

```
Natural Bond Order (total):


   Atom    1       2       3
   ----  ------  ------  ------
 1.  O   2.5000  1.5000  0.0000
 2.  O   1.5000  1.0000  1.5000
 3.  O   0.0000  1.5000  2.5000


Natural Atomic Valencies (total):

                     Co-     Electro-
   Atom  Valency  Valency  Valency
   ----  -------  -------  -------
 1.  O   1.5000   1.0864   0.4136
 2.  O   3.0000   2.1728   0.8272
 3.  O   1.5000   1.0864   0.4136
```

Further examples of open-shell $NRTSTR keylists and other special NRT job control keywords for difficult cases (see *NBO Manual*, p. B-75ff) will be illustrated in the following chapters.

SIDEBAR 5.3   *ELECTROMERISM CONCEPTS AND PAULING'S RESONANCE THEORY*

The concept of "electromers" (electronic isomers) has roots tracing back to Kekulé's dream-like visualization of benzene. In the ensuing evolution of Lewis's theory of electron-pair bonding, it became clear that the two distinct electromeric formulations of benzene are "hybridized" into a single observed high-symmetry species having structural properties (P) "averaged" between those (P_α) of the idealized electromeric forms $\alpha = 1,2,\ldots,RS$. Mathematically, such averaging is expressed most generally as

$$P = \sum_\alpha w_\alpha P_\alpha \tag{5.45}$$

where $\{w_\alpha\}$ are nonnegative weighting factors summing to unity,

$$\sum_\alpha w_\alpha = 1, \text{ all } w_\alpha \geq 0 \tag{5.46}$$

(e.g., $w_1 = w_2 = 1/2$ for the two electromeric forms of benzene). The general hybrid-averaging concepts of electromerism theory were extensively developed by Robinson, Ingold, and other physical organic chemists to explain the unusual structural and reactive properties of "conjugated" and "aromatic" species. Such electromerism concepts served as a useful empirical extension of Lewis's original electron-pair bonding concept (1916) until given more formal quantum mechanical "resonance" expression by Pauling (1931–1933). Pauling himself repeatedly expressed indebtedness to Lewis and the established empirical facts of physical organic chemistry as basis for his formulation of resonance theory.

Pauling was deeply impressed by Heisenberg's "resonance" language for expressing the quantum mechanical origin of chemical bonding in H_2. Mathematically, Heisenberg's bonding concept was formulated in the Heitler–London wavefunction for H_2,

$$\Psi_{AB}(1,2) = \mathscr{N}[s_A(1)s_B(2) + s_B(1)s_A(2)](\text{singlet}) \tag{5.47}$$

where s_A, s_B are 1s-type atomic orbitals on nuclei H_A, H_B, $\mathcal{N}$ is a normalization factor, and "(singlet)" denotes spin pairing. This Heitler–London "valence bond" *Ansatz* incorporates the superposition between alternative electron configurations, $s_A(1) s_B(2)$ versus $s_B(1) s_A(2)$, that is the essence of Heisenberg's resonance concept.

Pauling (and Slater) envisioned the extension of the Heitler–London *Ansatz* to each localized g_{AB} electron-pair bond of a Lewis structural representation by replacing hydrogenic orbitals with directed hybrids h_A, h_B,

$$g_{AB}(1,2) = \mathcal{N}[h_A(1)h_B(2) + h_B(1)h_A(2)](\text{singlet}) \tag{5.48}$$

but otherwise preserving the "perfect pairing" (localized singlet spin coupling) in each localized A-B bond. This leads to the resonance-structure wavefunction $\Psi_\alpha(1, 2, \ldots, N)$ [the so-called "Heitler–London–Slater–Pauling Perfect-Pairing Valence Bond" (HLSP-PP-VB) approximation], which can be expressed as

$$\Psi_\alpha(1, 2, \ldots, N) = A_{op}\{g_{AB}(1,2)g_{CD}(3,4)\ldots g_{XY}(N-1, N)\} \tag{5.49}$$

where A_{op} is the antisymmetrizer operator that imposes the proper electron indistinguishability (exchange antisymmetry) required by the Pauli exclusion principle. Pauling considered the HLSP-PP-VB wavefunctions (5.49) as the quantum mechanical epitomization of the Lewis structural concept.

To incorporate the empirical electromeric extensions (5.45) of Lewis-structure concepts, Pauling envisioned additional configurational superposition ("resonance hybridization") of the form

$$\Psi = \sum_\alpha c_\alpha \Psi_\alpha \tag{5.50}$$

where, in some sense, the squares of the wavefunction coefficients c_α give the resonance weightings w_α, namely,

$$c_\alpha = (w_\alpha)^{1/2} \tag{5.51}$$

This appears to be consistent with (5.46) *if* the $\{\Psi_\alpha\}$ are a complete orthonormal set satisfying

$$\int \Psi_\alpha{}^* \Psi_\beta d\tau = \delta_{\alpha\beta}(1 \text{ if } \alpha = \beta, 0 \text{ otherwise}) \tag{5.52}$$

However, evaluation of the proper quantum mechanical expectation value $\int \Psi^* P_{op} \Psi d\tau$ for property P then implies that

$$P = \sum_\alpha w_\alpha P_\alpha + \sum_{\alpha,\beta} (w_\alpha w_\beta)^{1/2} \int \Psi_\alpha{}^* P_{op} \Psi_\beta d\tau \tag{5.53}$$

where the signs of $(w_\alpha w_\beta)^{1/2}$ may be presumed positive (in-phase) for the favored ground-state superposition. Compared to (5.45), the Pauling *Ansatz* (5.53) only makes sense if *all* the cross-terms in the summations are somehow vanishing, for example,

$$\int \Psi_\alpha{}^* P_{op} \Psi_\beta d\tau = 0 \text{ (all } \alpha \neq \beta) \tag{5.54}$$

Even if (5.52) were accepted, it seems difficult to imagine how the cross-terms in (5.54) could be safely neglected for all properties P_{op} to which Pauling's resonance theoretic concepts were blithely applied in the subsequent decades. However, except for the original Heitler–London wavefunction for H_2, practically none of the HLSP-PP-VB-type approximations (5.48–5.54) could be evaluated for polyatomic molecules without wholesale empirical substitutions, and questions of the reliability of these assumptions remained largely unaddressed during the heyday of Pauling's resonance theory.

In truth, many errors were uncovered in Pauling's trail of assumptions as rigorous polyatomic calculations finally became feasible in the 1970s. Even before Norbeck and Gallup carried out decisive numerical tests of the Pauling-type resonance assumptions for benzene (*J. Am. Chem. Soc.* **96**, 3386, 1974), it was recognized that the success of the Heitler–London approximation (5.47) for H_2 was a rather fortuitous special case, and that this mathematical function usually leads to severely "overcorrelated" (excessively diradical-type) description for more typical electron pair bonds. Compared to a corresponding bond orbital description (e.g., of NLS type), in which the homopolar bond function can be written as a doubly occupied σ_{AB} orbital, namely,

$$g_{AB}^{(BO)}(1,2) = \mathscr{N}[\sigma_{AB}]^2(\text{singlet})$$
$$= \mathscr{N}[(s_A(1) + s_B(1))][(s_A(2) + s_B(2))](\text{singlet}) \qquad (5.55)$$

the Heitler–London approximation (5.47) is formally equivalent to an *equal-weighted* mixture of σ_{AB} (bonding) and σ^*_{AB} (antibonding) terms, namely,

$$g_{AB}^{(HL)}(1,2) = \mathscr{N}\{[\sigma_{AB}]^2 - [\sigma^*_{AB}]^2\}(\text{singlet}) \qquad (5.56)$$

that is grossly inaccurate (unless far from equilibrium) for all bonds except H_2. For H_2, the s_A and s_B orbitals are so highly overlapping that each atomic orbital numerically approximates the equilibrium σ_{AB} orbital, and the unphysically weighted σ^*_{AB} contribution is largely self-cancelling. However, for more general polyatomic molecules, the starting HLSP-PP-VB building-block $g_{AB}^{(HL)}$ is found to be a qualitatively *un*reasonable representation of a localized A–B chemical bond. The unphysical character of its building blocks leads to a cascade of chemically unreasonable consequences in the multiconfigurational extensions (5.50–5.54) to benzene and other molecules (see the Norbeck– Gallup paper referenced above).

For heteronuclear diatomics or other cases involving polar covalent bonding, the failures of HL-type approximations become even more acute. Whereas bond orbital approximations are easily generalized to deal with polarity variations [cf. Eq. (4.17)], the inflexible HL approximation cannot accommodate such variations except by including "covalent-ionic resonance," i.e., additional terms in (5.50), an artificial (and numerically impractical) complication of localized Lewis structural concepts. The apparent necessity for covalent-ionic resonance created an unfortunate impression (all too common in elementary textbook expositions) of dichotomous chemical bonding "types." The student should reject all such suggestions, focusing instead on the *continuously* variable polarities of localized bond orbitals (generally requiring no special multiconfigurational wavefunction character).

Still another aspect of Pauling-type approximations (5.50–5.54) is inconsistent with empirical resonance concepts. For benzene and related molecules, it was recognized empirically that resonance hybridization is associated with unusual stability, such that the energy is *lower* than (rather than "an average of") its localized constituents. However, for $P_{op} = H_{op}$, where (5.54) could be expected to follow from (5.51), the Pauling formulation apparently predicts the same type of resonance-averaging for total energy as for structural properties, contrary to observation. (This point was lost in the general empirical confusion, for as shown by Norbeck and Gallup, the resonance wavefunction usually assumed in empirical HLSP-PP-VB treatments is not even the lowest-energy root of the secular determinant.)

Nowadays, multiconfigurational VB-type wavefunctions are usually employed (if at all) in "generalized" form in which each VB pair includes "self-consistent mixing corrections" from basis functions *throughout* the molecule, thus sacrificing the original conceptual association with localized A, B orbitals. The legacy of Pauling's resonance theory is thus only weakly preserved in the modern quantum chemical research literature, but its grip on elementary textbook expositions remains formidable.

SIDEBAR 5.4 *SOME BASICS OF NATURAL RESONANCE THEORY VERSUS PAULING RESONANCE THEORY*

A fundamental mathematical feature of chemical systems is that the relevant Hamiltonian (total energy) operator H_{op} can be written in terms of one-electron $(P_{op}^{(1e)})$ and two-electron $(P_{op}^{(2e)})$ operators *only*. This implies that all relevant chemical information can be obtained from reduced "density operators" $(\Gamma^{(1e)}, \Gamma^{(2e)})$ that condense and simplify the N-electron wavefunction information (see *V&B*, p. 21ff). Instead of a conventional wavefunction expectation value, namely,

$$P^{(1e)} = \int \Psi^* P_{op}^{(1e)} \Psi d\tau \tag{5.57}$$

the equivalent density-operator evaluation of any one-electron property can be written as

$$P^{(1e)} = \text{Tr}\{\Gamma^{(1e)} P_{op}^{(1e)}\} \tag{5.58}$$

where Tr ("trace") denotes a certain one-electron (density-like) integral whose details need not concern us here.

Given this simplification (which is *exact*, not an approximation), one can see that a resonance-type assumption for the density operator $\Gamma^{(1e)}$, namely,

$$\Gamma^{(1e)} = \sum_\alpha w_\alpha \Gamma_\alpha^{(1e)} \tag{5.59}$$

would be *necessary and sufficient* to insure the basic resonance averaging assumption (5.45)

$$P^{(1e)} = \text{Tr}\{\Gamma^{(1e)} P_{op}^{(1e)}\} = \sum_\alpha w_\alpha \text{Tr}\{\Gamma_\alpha^{(1e)} P_{op}^{(1e)}\} = \sum_\alpha w_\alpha P_\alpha^{(1e)} \tag{5.60}$$

for *all* one-electron properties. In this expression, $\Gamma_\alpha^{(1e)}$ is the density operator for an idealized localized resonance structure α (for example, for a corresponding NLS-type wavefunction, obtained by the $CHOOSE procedure). Although it is unrealistic to assume that (5.59) and (5.60) hold exactly, we may *seek* the weightings and resonance structures that satisfy (5.59) and (5.60) as nearly as possible in mean-squared sense, namely,

$$||\Gamma^{(1e)} - \sum_\alpha w_\alpha \Gamma_\alpha^{(1e)}||^2 = \text{minimum} \tag{5.61}$$

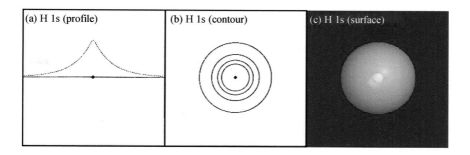

Figure 2.1 Hydrogen atom 1s orbital in (a) 1D profile, (b) 2D contour, and (c) 3D surface plot.

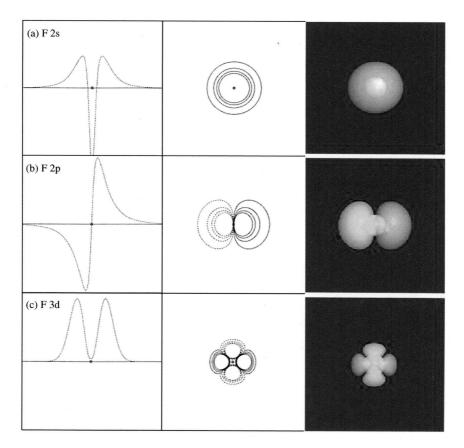

Figure 2.2 Fluorine atom (a) 2s, (b) 2p, and (c) 3d orbitals in 1D profile (left), 2D contour (middle), and 3D surface plot (right). The depicted orbitals have respective occupancies of 2, 1, and 0 in the F atom ground state. (Note that the four outermost contour lines of default *NBOView* contour output do not include the negative 2s "inner spike" near the nucleus, which is better seen in the 1D profile plot.)

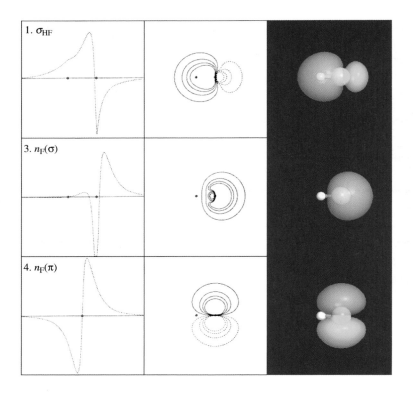

Figure 4.1 Distinct valence (P)NBOs of HF of bonding (BD:σ_{HF}) and nonbonding type (LP: on-axis $n_F^{(\sigma)}$ and off-axis $n_F^{(\pi)}$), shown in profile, contour, and surface plots. The profile of the p_y-type LP (NBO 4) is along a vertical line through the F nucleus, perpendicular to the equivalent p_x-type LP (NBO 5, not shown) that points out of the page.

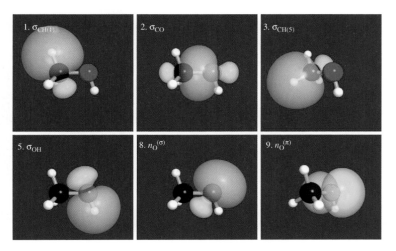

Figure 4.2 Lewis-type valence NBOs of CH_3OH (cf. I/O-4.3).

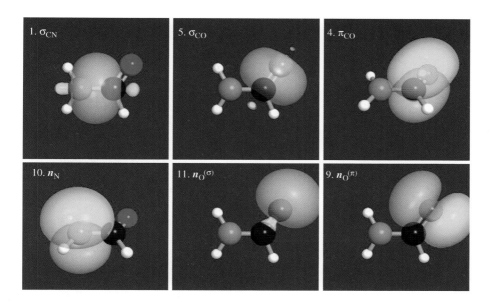

Figure 4.3 Leading Lewis-type valence NBOs of formamide (cf. I/O-4.6).

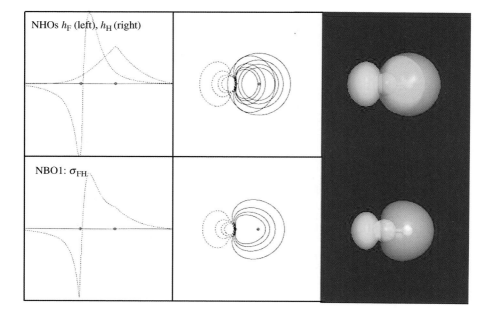

Figure 4.5 σ_{FH} bond of hydrogen fluoride, shown as overlapping NHOs (upper) or as final NBO (lower); (cf. I/O-4.7).

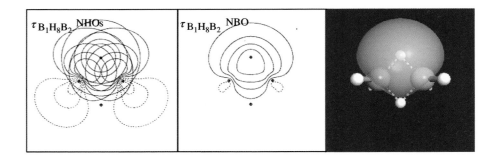

Figure 4.7 Lewis-type three-center τ_{BHB} bond of B_2H_6, showing contour plots for overlapping NHOs (left) and final NBO (center), and corresponding surface plot (right).

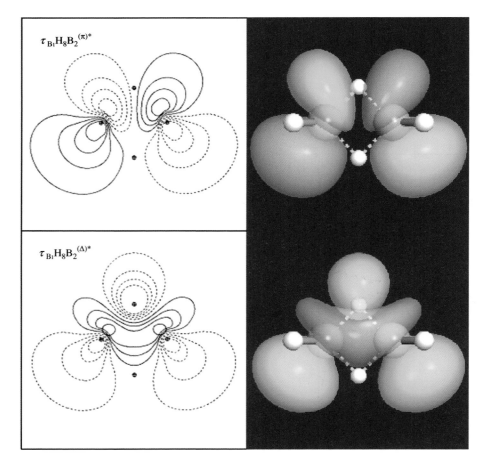

Figure 4.8 Non-Lewis-type three-center antibonds $\tau_{BHB}^{(\pi)*}$ (upper) and $\tau_{BHB}^{(\Delta)*}$ (lower) of B_2H_6, shown in contour and surface plots.

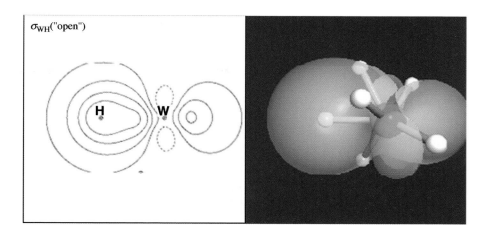

Figure 4.9 "Open" σ_{WH} NBO (4.51) of WH_6, shown in contour and surface plots.

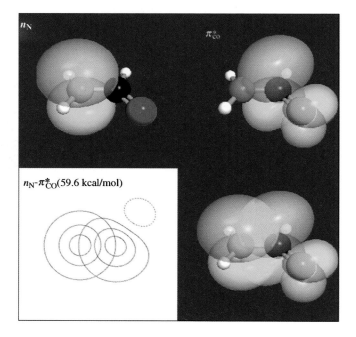

Figure 5.2 Formamide n_N and π^*_{CO} NBOs, shown individually (upper panels) and in interaction (lower panels) as contour and surface plots. (The contour plot is a top–down view of the π system, with chosen contour plane slicing through the π orbitals 1 Å above the molecular plane.)

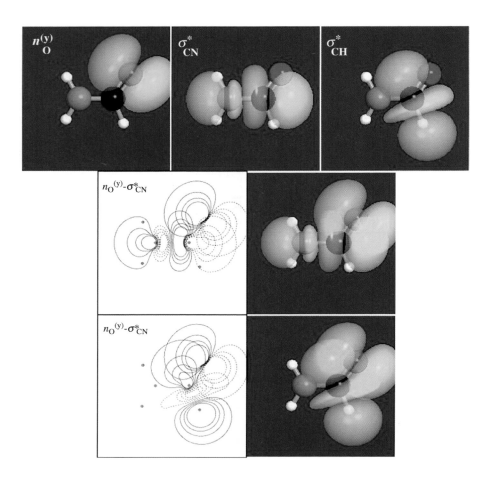

Figure 5.4 Principal in-plane lone-pair → antibond delocalizations of formamide, showing individual donor $\left(n_O^{(y)}\right)$ and acceptor $\left(\sigma^*_{CN}, \sigma^*_{CH}\right)$ NBOs (upper panels), and overlapping donor–acceptor pairs (lower panels) in contour and surface plots.

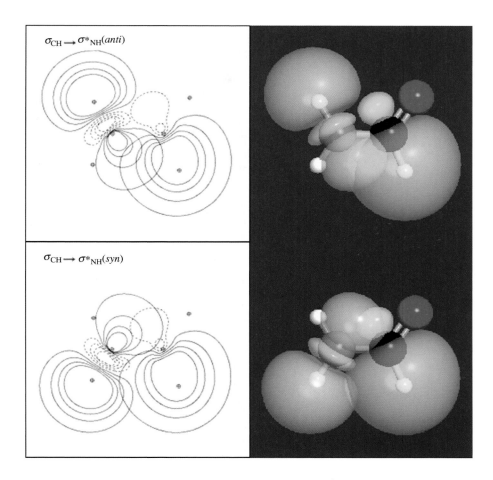

Figure 5.5 Comparison contour and surface plots of vicinal $\sigma_{CH} - \sigma^*_{NH}$ interactions in *anti* (upper) versus *syn* (lower) orientations, showing the far more favorable NBO overlap in *anti*periplanar arrangement [consistent with the stronger hyperconjugative stabilization evaluated in the E(2) table].

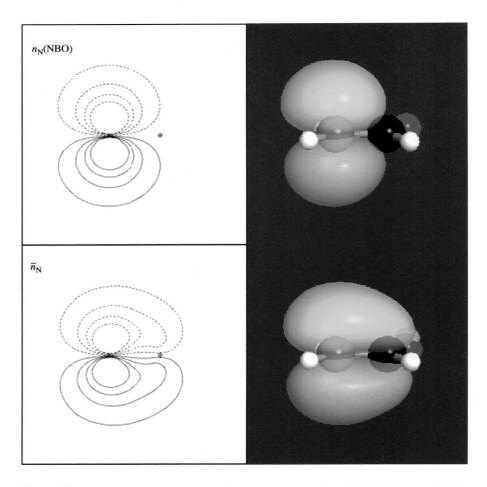

Figure 5.9 Contour and surface plots comparing NBO (upper panels) and NLMO (lower panels) for conjugatively delocalized amine lone pair n_N [Eq. (5.42)] of formamide.

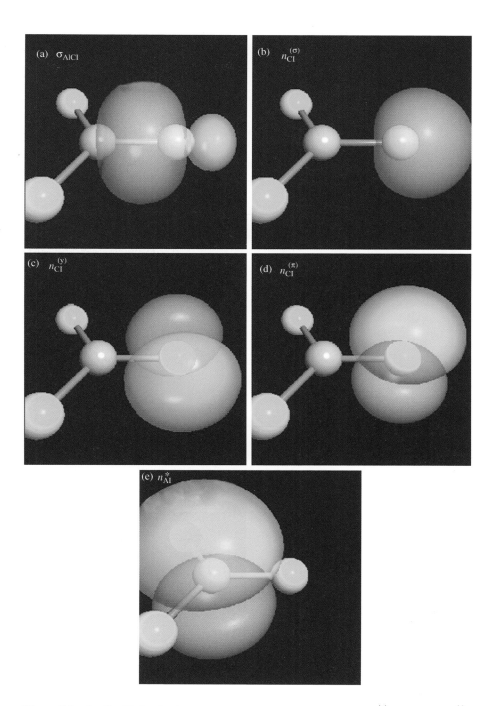

Figure 8.1 Leading NBOs of $AlCl_3$, showing Lewis-type (a) σ_{AlCl} (NBO 1), (b) $n_{Cl}^{(\sigma)}$ (NBO 24), (c) $n_{Cl}^{(y)}$ (NBO 25), (d) $n_{Cl}^{(\pi)}$ (NBO 26), and non-Lewis-type (e) n_{Al}^* (NBO 33), the characteristic "LP*" acceptor of a strong Lewis acid (cf. I/O-8.1).

Figure 8.2 Endocyclic coordinative $\sigma_{Al:Cl}$ bond of Al_2Cl_6 (NBO 1; cf. I/O-8.1). Except for the "missing" $n_{Cl}^{(\pi)}$ that is the coordinative "parent" of $\sigma_{Al:Cl}$, other NBOs of Al_2Cl_6 closely resemble those shown in Fig. 81 for $AlCl_3$.

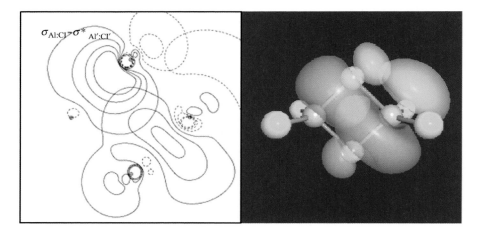

Figure 8.3 Coordinative $\sigma_{Al:Cl}$-$\sigma^*_{Al':Cl'}$ interaction of Al_2Cl_6 in contour and surface plots, showing nonvanishing hyperconjugative overlap ($\Delta E^{(2)} = 2.32$ kcal/mol) despite the unfavorable (cyclobutadiene-like) vicinal bond–antibond alignment.

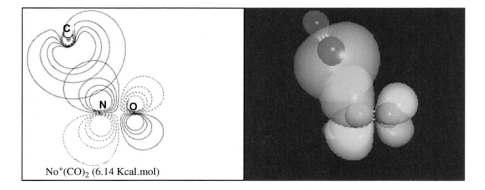

Figure 9.9 NBO contour diagram (left) and surface plot (right) of n_C-π^*_{NO} donor–acceptor interaction in $NO^+(CO)_2$ (with estimated $\Delta E^{(2)}_{n \to \pi^*}$ stabilization). The O atom of CO lies slightly out of the contour plane in the left panel.

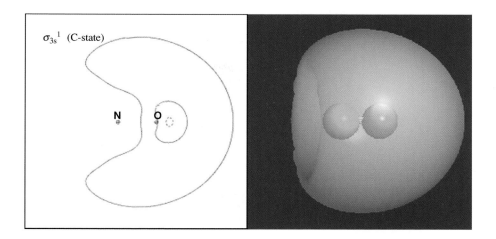

Figure 11.6 Contour and surface plots for σ_{3s} Rydberg-type NBO in quadruply bonded C-state inner well (1.025 Å; cf. Table 11.2) of NO (CIS/6-311++G** level).

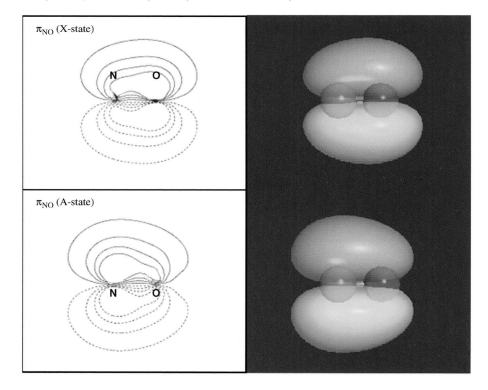

Figure 11.7 Contour and surface plots comparing π_{NO} NBO for normal-polarized ground-state (X; upper panels) versus reversed-polarized excited state (A; lower panels) π-bonds of NO (CIS/6-311++G** level).

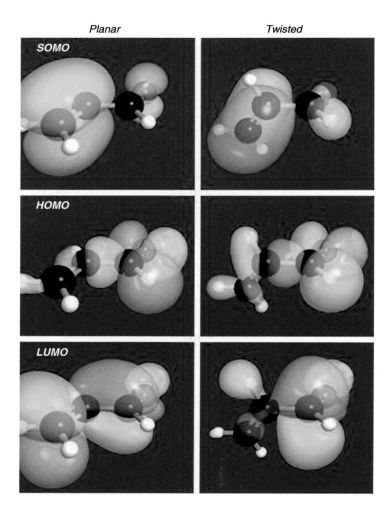

Figure 11.8 Frontier MOs of ground-state acrolein (SOMO = second occupied MO; HOMO = highest occupied MO; LUMO = lowest unoccupied MO), showing qualitative variations of form with torsions from planar (left) to twisted (right) geometry.

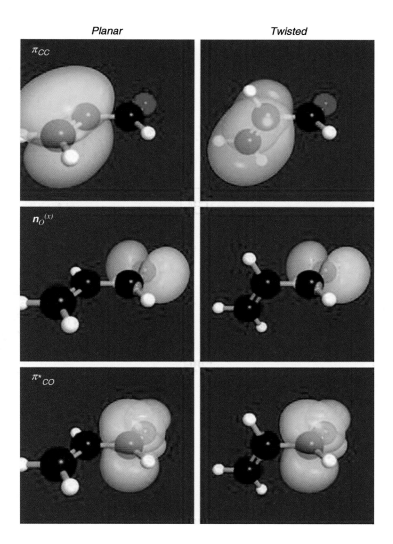

Figure 11.9 Similar to Fig. 11.8, for π_{CC}, $n_O^{(\pi)}$, π^*_{CO} NBOs that are leading contributors to MOs of Fig. 11.8, showing near-transferable NBO forms in planar and twisted geometry.

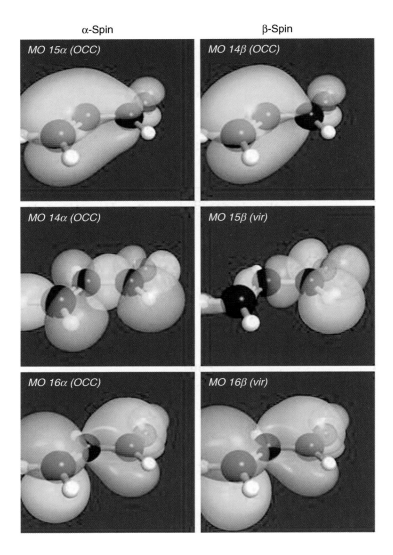

Figure 11.10 Frontier MOs of lowest triplet excited state of acrolein in vertical ground-state geometry (cf. left panels of Fig. 11.8 for ground singlet state), showing significant variations in state-to-state MO forms.

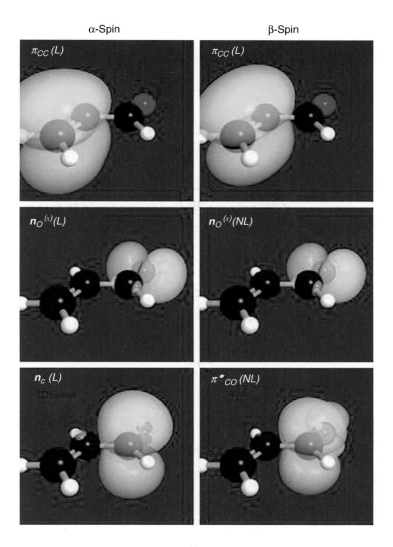

Figure 11.11 Similar to Fig. 11.10, for π_{CC}, $n_O^{(\pi)}$, π^*_{CO} NBOs (cf. left panels of Fig. 11.9 for ground singlet state), showing high state-to-state NBO transferability.

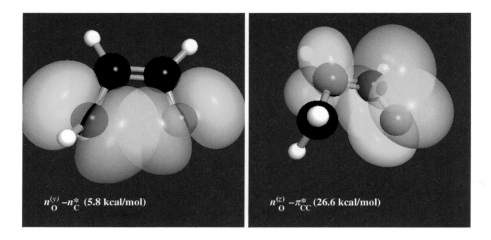

$n_O^{(y)} - n_C^*$ (5.8 kcal/mol) $n_O^{(z)} - \pi_{CC}^*$ (26.6 kcal/mol)

Figure 11.17 Leading (a) in-plane and (b) out-of-plane α NBO interactions (and estimated second-order stabilization energies) for CI-S_0 conical intersection of acrolein (UHF/6-311++G** level in geometry of I/O-11.2).

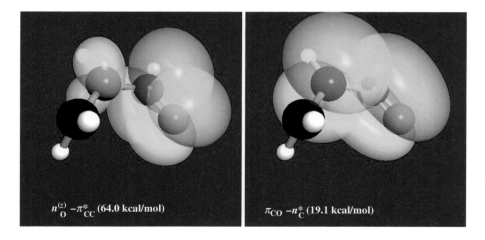

$n_O^{(z)} - \pi_{CC}^*$ (64.0 kcal/mol) $\pi_{CO} - n_C^*$ (19.1 kcal/mol)

Figure 11.18 Leading (a) α spin and (b) β spin π-type NBO donor–acceptor interactions (and estimated second-order stabilization energies) for ISC-S_0 intersystem crossing of acrolein (UHF/6-311++G** level in geometry of I/O-11.3). [In the ISC-S_0 species, the in-plane α spin $n_O^{(y)}$-n_C^* interaction (cf. Fig. 11.17a for the CI-S_0 species) is negligibly weak (0.27 kcal/mol).]

The variational minimization (Eq. 5.61) of mean-squared difference ("error") between target $\Gamma^{(1e)}$ (actual density operator for the given wavefunction) and its best possible resonance-type representation is the essence of the NRT algorithm (see *NBO Manual*, p. B-72ff, and references therein).

Algorithmic details of the NRT variational minimization (5.61) differ somewhat for dominant "reference" structures (where all elements of the density operators are considered) and weaker "secondary" structures (which are treated by a simpler perturbative-type approximation involving diagonal density operator elements only). In each case, the variational minimization of (5.61) can be equivalently expressed as the *max*imization of a corresponding fractional improvement $f(\underline{w})$ (for secondary structures) or $F(\underline{W})$ (for reference structures) that expresses the percentage reduction of the multiresonance "error" in (5.61) from its initial single-resonance value. Of course, there is no assurance that this error can be reduced to zero [i.e., that $f(\underline{w})$ or $F(\underline{W})$ can achieve 100% accuracy], because the included resonance structures have no necessary connection to a "complete set." The typical high accuracy of NRT expansions is therefore a computational *result*, not an initial assumption based on supposed multiconfigurational wavefunction completeness (as in Pauling resonance theory).

Many other differences from conventional Pauling resonance theory can be cited. The automatic inclusion of bond-polarity effects in the individual $CHOOSE resonance structures makes Pauling-type "covalent-ionic resonance" wholly superfluous. Whereas the Pauling-type resonance assumptions (5.50–5.54) could hardly justify general resonance averaging for *any* property P (other than total energy, where it is contrary to empirical evidence; cf. Sidebar 5.3), the NRT variational criterion (5.61) assures a comparable resonance averaging property for *all formal one-electron properties* (including kinetic energy, nuclear–electron attraction, dipole moment, electron and spin density, and all geometrical bond lengths and angles), consistent with usual resonance theoretic empiricism. Of course, the success of the NRT resonance averaging assumption for the one-electron density operator $\Gamma^{(1e)}$ carries no implication for the two-electron density operator $\Gamma^{(2e)}$, which is needed to evaluate total energy. The success of NRT theory is therefore fully consistent with the empirical concept of "resonance stabilization" (not "averaging" of total energy), and the close relationship between resonance structures and donor–acceptor stabilizations (cf. Fig. 5.6) makes it obvious why such resonance *lowering* of energy must be a general chemical phenomenon.

PROBLEMS AND EXERCISES

5.1. For the five isomeric COH_2 species considered in Problem 4.1:

   (a) Evaluate $E(L)$ and $E(NL)$, and rank the species according to NL delocalization from idealized Lewis structure representation.

   (b) Find the leading second-order $\Delta E_{ij}^{(2)}$ delocalization energy for each species, and use $DEL-deletion to evaluate the corresponding $\Delta E($DEL$)$ variational estimate of the associated donor–acceptor interaction. Are these two estimates qualitatively consistent with one another?

   (c) Similarly, obtain the NBO Fock matrix element (F_{ij}) and the PNBO overlap matrix element (S_{ij}) for the leading NBO donor–acceptor interaction. If *NBOView* is available, obtain orbital overlap diagrams for each such interaction in contour and surface forms. Do the graphical visualization impressions correspond qualitatively to the various numerical measures (e.g., F_{ij}, S_{ij}, $\Delta E_{ij}^{(2)}$) of donor–acceptor strength?

(d) For each of these leading i–j donor–acceptor interactions, find the leading contribution to the "tail" of NLMO i and verify that it originates from the same acceptor NBO j as implicated in (b), (c) above.

(e) If possible for your ESS, change the method from "B3LYP" to "MP2" and attempt to carry out (a)–(d) for Isomer-1. Which NBO descriptors are still available in MP2 calculation, and which could not be obtained (because no 1e Hamiltonian is defined for this level)?

5.2. In each isomer of Problem 4.1 (described respectively by optimal Lewis structures R_1, R_2, R_3, R_4, R_5), use the \$CHOOSE keylist to evaluate the ρ^* "error" for choosing the *wrong* Lewis structure for each isomer (i.e., the error of choosing R_2, R_3, ... for Isomer-1, etc.). For each isomer, order the structures R_1–R_5 according to their apparent \$CHOOSE errors (from least to worst error), and rank the isomers according to greatest difference between "best" and "second-best" structural representation. See if you can relate (in some qualitative manner) the ranking of isomers and "second-best" structures to the descriptors $E(\text{NL})$, $\Delta E_{ij}^{(2)}$, ... studied in (a)–(d).

5.3. For the five isomeric COH_2 species considered in Problem 4.1, determine the NRT bond orders b_{ij} and the weighting of the principal resonance structure for default NRT analysis of each species. Do the relative weightings correspond (at least qualitatively) to your \$CHOOSE rankings in Problem 5.2?

5.4. Construct the \$NRTSTR keylist that specifies inclusion of the five principal resonance structure (Problem 5.3) for isomeric forms 1–5 as reference structures. Repeat NRT analysis of each isomer with inclusion of the \$NRTSTR keylist. Report the significant changes (if any) in calculated weightings or bond orders for any species.

5.5. Repeat the default and \$NRTSTR-directed NRT analysis of each isomeric species 1–5 with the same changes of method or basis set suggested in Problem 4.1f. Report the significant changes (if any) for any species. Can you see evidence of systematic NRT shifts with improved treatment of electron correlation?

5.6. Consider again (cf. Problem 4.2) the formal intramolecular hydride shift $1 \rightarrow 3$ reaction and use the LST geometries between these two isomers to find the *continuous* variations of NRT bond orders and weighting factors along this pathway. (Include a \$NRTSTR keylist that includes at least structures R_1, R_3, in order to insure balanced treatment of these two reference structures along the entire pathway.) Plot the changes of $b_{CH}(\lambda)$ and $b_{OH}(\lambda)$ for the transferred hydride bond, and check for satisfaction of the "natural transition state" (NTS) half-bond criterion, $b_{CH}(\lambda_{TS}) \cong b_{OH}(\lambda_{TS}) \cong 1/2$, near the energetic transition state.

5.7. Repeat Problem 5.6 for the $3 \rightarrow 4$ bond-dissociation reaction to find the λ_{TS} at which the NTS criterion, $w_3 = w_4$, is satisfied. Use this example to describe why the NTS criterion is a more flexible and general "transition state" characterization than the usual energetic saddle point of an IRC coordinate.

Chapter 6

Steric and Electrostatic Effects

The Lewis-type $E(L)$ contribution is considered the "easy" part of chemical wavefunction analysis, because it corresponds closely to the elementary Lewis structure model of freshman chemistry. Nevertheless, controversy often arises over the magnitude of "steric" or "electrostatic" effects that are associated with the Lewis model itself [i.e., distinct from the resonance-type effects contained in $E(NL)$]. The NBO program offers useful tools for quantifying both steric and electrostatic interactions in terms of the space-filling (size and shape) and dielectric properties (charge, dipole moment, etc.) of the electron pair bonds and lone pairs that comprise the Lewis structure model. This chapter discusses the physical nature and numerical quantitation of these important chemical effects, which are often invoked in a "hand-waving" manner that reflects (and promotes) significant misconceptions.

In principle, we are attempting to dissect classical-like steric exchange ($E^{(sx)}$) and electrostatic ($E^{(es)}$) contributions to $E(L)$ from an idealized uncrowded and electroneutral starting point (E_{ideal}), namely,

$$E(L) = E_{ideal}(L) + E^{(sx)}(L) + E^{(es)}(L) \tag{6.1}$$

Such dissection assumes, somewhat superficially, that steric and electrostatic (polarity) contributions belong exclusively to the dominant Lewis structure component $E(L)$ in (5.8) [i.e., with no coupling terms to $E(NL)$], consistent with the hoped-for interpretation for each contribution as a classical-like correction to an apolar ball and stick image of the starting Lewis structure.

More precisely, we may write

$$E^{(sx)} = E^{(sx)}(L) + E^{(sx)}(NL) \tag{6.2}$$

$$E^{(es)} = E^{(es)}(L) + E^{(es)}(NL) \tag{6.3}$$

where $E^{(sx)}(NL)$ and $E^{(es)}(NL)$ are "doubly small" corrections that couple classical-like crowding and polarity effects with resonance delocalization. We can

Discovering Chemistry With Natural Bond Orbitals, First Edition. Frank Weinhold and Clark R. Landis.
© 2012 John Wiley & Sons, Inc. Published 2012 by John Wiley & Sons, Inc.

conveniently "neglect" such coupling by simply adopting an NLMO-based (rather than NBO-based) conception of Lewis structural constituents for present qualitative purposes. (By expressing these NLMOs in terms of NBOs, we can subsequently dissect $E^{(sx)}$, $E^{(es)}$ into pure L versus NL components, if desired.) Accordingly, our mathematical Lewis structural building blocks in this chapter are primarily the NLMOs (and associated PNLMOs; cf. Section 5.4), consistent with the manner in which STERIC and DIPOLE analysis are implemented in the current *NBO 5* program.

6.1   NATURE AND EVALUATION OF STERIC INTERACTIONS

What is "steric repulsion"? The generic term refers to the *space-filling* property of atoms and molecules, as manifested in crystal packing densities, molecular collision cross-sections, and other lines of experimental evidence. Indeed, space-filling molecular models are among the most useful tools of the chemistry student, and "atomic radii" are among the first properties called to the student's attention to illustrate atomic periodicity trends.

In the atomic theories of antiquity, atoms were considered the indivisible, incompressible, and indestructible constituent units of material substances, with no further internal structure. It was therefore quite surprising when Rutherford discovered (from the scattering patterns of α-particles on thin gold foils) that atoms were mostly empty space, having virtually all mass concentrated in an infinitesimally small "nucleus" (something like a small marble on the 50-yard line of a football stadium), with the surrounding "atomic volume" composed only of an "electron cloud" of near-zero mass density. Although we continue to envision atoms as classical-like billiard balls in our molecular models, the origins of the apparent repulsions of the wispy electronic clouds must be deeply nonclassical.

We now recognize that steric repulsions arise fundamentally from the Pauli exclusion principle, the "exchange antisymmetry" that guarantees electron indistinguishability and resists electrons being crowded into small regions of space. [Some books mistakenly confuse this with the Coulombic repulsion between nuclear charges, but the (inverse square) Coulombic forces are generally negligible compared to the *exponential* repulsions due to electronic exchange forces.] In essence, the occupancy limit and orthogonality requirement for each electronic orbital require increasing oscillatory nodal features (high-frequency Fourier components, corresponding to increased kinetic energy "curvature" of the wavefunction) as electrons are forced into reduced volume. The NBO algorithm for STERIC evaluations connects directly to this "kinetic energy pressure" picture of electronic exchange repulsions (Sidebar 6.1).

[As described in Sidebar 6.1, the steric-free (non-antisymmetrized) electron density that serves as starting point for STERIC energetics puts unusual stresses on current-generation DFT energetic functionals. Depending on details of exchange approximations, empirical DFT methods can lead to unphysical artifacts in STERIC evaluations, making them unreliable unless checked closely against comparison *ab initio* HF evaluations. For that reason, we employ HF/6-311++G** evaluation in this section, with comments on some artifacts of B3LYP evaluation in Fig. 6.6.]

To perform NBO steric analysis, simply include the STERIC keyword in the $NBO keylist. The first output section produces a tabulation of "steric exchange energy" contributions for each occupied NLMO [$\Delta E_i^{(sx)}$, "d$E(i)$"] and the summed total $E^{(sx)}$ by unit and total species, as shown for formamide (Section 4.1.3) in I/O-6.1.

As seen in the table, outer valence NLMOs tend to give negative $\Delta E_i^{(sx)}$ contributions, but the energy of inner core orbitals is strongly *increased* by anti-symmetrization, so the net steric exchange energy $E^{(sx)}$ is *positive* (by 35.44 kcal/mol) as expected on physical grounds. However, neither the individual $\Delta E_i^{(sx)}$'s nor total $E^{(sx)}$ values of a single-point calculation are particularly informative, until differenced with respect to a suitable reference state to determine *changes* of chemical interest (as described below).

Next follow the pairwise steric exchange energies [$\Delta E_{ij}^{(sx)}$, "d$E(i,j)$"] for occupied NLMOs i, j and the "total disjoint NLMO steric exchange energy from pairwise sum" (sum of pairwise $\Delta E_{ij}^{(sx)}$ values) as shown in I/O-6.2.

I/O-6.1

```
NBO/NLMO STERIC ANALYSIS

  Occupied NLMO contributions dE(i) (kcal/mol) to total steric exchange energy

  NLMOs (i) in unit  1                 dE(i)
     1. BD ( 1) N 1- C 2             -37.71
     2. BD ( 1) N 1- H 4             -39.70
     3. BD ( 1) N 1- H 5             -39.25
     4. BD ( 1) C 2- O 3             -13.30
     5. BD ( 2) C 2- O 3             -70.67
     6. BD ( 1) C 2- H 6             -28.04
     7. CR ( 1) N 1                  105.66
     8. CR ( 1) C 2                   77.61
     9. CR ( 1) O 3                  164.66
    10. LP ( 1) N 1                   -8.89
    11. LP ( 1) O 3                  -88.07
    12. LP ( 2) O 3                   13.15


  Steric exchange energy, unit 1:    35.44 kcal/mol
  -----------------------------------------------------
  Total steric exchange energy:      35.44 kcal/mol
  -----------------------------------------------------
```

─────────────── *I/O-6.2* ───────────────

```
Pairwise steric exchange energies dE(i,j) (kcal/mol) and associated
pre-NLMO overlaps S(i,j) for disjoint (no common atoms) interactions
between NLMOs i,j:


        Threshold for printing:      0.50 kcal/mol

                                                    PNLMO     dE(i,j)
        NLMO (i)                     NLMO (j)        S(i,j)    kcal/mol
================================================================================


within unit  1
    1. BD ( 1) N 1- C 2       11. LP ( 1) O 3        -0.0524     1.77
    1. BD ( 1) N 1- C 2       12. LP ( 2) O 3         0.1986    14.88
    2. BD ( 1) N 1- H 4        6. BD ( 1) C 2- H 6    0.0863     5.23
    2. BD ( 1) N 1- H 4        8. CR ( 1) C 2         0.0069     0.53
    2. BD ( 1) N 1- H 4       12. LP ( 2) O 3        -0.1135     1.04
    3. BD ( 1) N 1- H 5        5. BD ( 2) C 2- O 3    0.0993     4.32
    3. BD ( 1) N 1- H 5        6. BD ( 1) C 2- H 6   -0.0968     2.86
    3. BD ( 1) N 1- H 5        8. CR ( 1) C 2         0.0078     0.63
    4. BD ( 1) C 2- O 3       10. LP ( 1) N 1         0.2145    19.31
    6. BD ( 1) C 2- H 6       11. LP ( 1) O 3        -0.0378     1.38
    6. BD ( 1) C 2- H 6       12. LP ( 2) O 3        -0.1517    15.70
    8. CR ( 1) C 2            11. LP ( 1) O 3        -0.0229     4.20


                               sum within unit 1:              71.54


--------------------------------------------------------------------
Total disjoint NLMO steric exchange energy from pairwise sum:    71.54
--------------------------------------------------------------------
```

As before, the pairwise sum is *positive* (71.54 kcal/mol) and requires comparison to a reference state before inferences can be drawn. (The pairwise-sum values are intrinsically referenced to a different "zero" than $E^{(sx)}$, so no particular significance should be attributed to the disparity between $E^{(sx)}$ and $\sum \Delta E_{ij}^{(sx)}$ values.)

The individual $\Delta E_{ij}^{(sx)}$ entries are often most useful, because they allow interesting steric inferences to be drawn from a single-point molecular geometry. From the final column of $\Delta E_{ij}^{(sx)}$ values one can see that the three largest steric repulsions involve filled NLMOs $n_N - \pi_{CO}$ (19.3 kcal/mol), $n_O^{(y)} - \sigma_{CH}$ (15.7 kcal/mol), and $n_O^{(y)} - \sigma_{CN}$ (14.9 kcal/mol). The next-to-last column shows corresponding PNLMO overlap integrals ["S(i,j)"], whose significant magnitudes (0.15–0.21) and contour plot depictions (Fig. 6.1a–c) aptly suggest the dominance of these three donor–donor contacts. [Of course, these donor–donor repulsions are compensated by the powerful $n_N - \pi^*_{CO}$, $n_O^{(y)} - \sigma^*_{CH}$, and $n_O^{(y)} - \sigma^*_{CN}$ donor–acceptor stabilizations (cf. Figs. 5.2 and 5.4) that adequately "reward" the apparently crowded equilibrium geometry]. Just as for analogous donor–acceptor interactions of Chapter 5, judicious use of graphical plots of the overlapping PNLMOs can give a powerful visual image of donor–donor overlap and the associated steric "clash" between filled orbitals.

Closer examination of I/O-6.2 reveals an important paradox: Whereas the hydride bonds ($C_2 - H_6$ and $N_1 - H_5$) in *cis*-like arrangement superficially appear "closer" than those in *trans*-like arrangement ($C_2 - H_6$ and $N_1 - H_6$), the steric

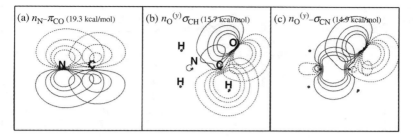

Figure 6.1 PNLMO overlap diagrams for leading donor–donor interactions in formamide (with associated $\Delta E_{ij}^{(sx)}$ steric repulsion values). The contour plane in (a) is perpendicular to the molecular plane, passing through the CN axis.

repulsions $\Delta E_{ij}^{(sx)}$ for the former pair are actually *weaker* than for the latter (2.9 versus. 5.2 kcal/mol). Figure 6.2 exhibits contour plots of PNLMO overlap for these donor–donor pairs that suggest the origin of this paradox. As shown in Fig. 6.2, the NH–CH overlap is actually somewhat higher for *cis* (left) than *trans* arrangement (0.10 versus 0.09), but the overlapping regions lie *nearer* the heavy-atom nuclei in the latter case (always involving a tight "backside" orbital lobe), and is therefore associated with *deeper* energy values and *stronger* steric repulsions, as shown by the $\Delta E^{(sx)}$ values. The moral is that one must look carefully at the *orbital shapes and energy content* (not just a ball and stick model) to envision "proximity" and "steric contact" between donor groups (particularly, hydride bonds in vicinal arrangements).

Let us now discuss possible choices of "reference system" for particular types of steric questions. As one simple example, we may consider evaluation of an atomic "van der Waals radius" R_{VdW} by bringing up a probe species (e.g., He) to the atom until the steric exchange energy increment $\Delta E^{(sx)}$ (value at R_{VdW} minus value at infinite separation) matches a preset threshold, such as the energy (kT) of ambient thermal collisions, namely,

$$\Delta E^{(sx)}(R_{VdW}) = kT \cong 0.6 \, \text{kcal/mol (at 298 K)} \tag{6.4}$$

Such He-probe calculations are the basis for evaluating "natural van der Waals radii" [J. K. Badenhoop and F. Weinhold, *J. Chem. Phys.* **107**, 5422, 1997], as tabulated for elements 1–18 in Table 6.1. The NBO-based R_{VdW} values are generally in sensible agreement with empirical values inferred from X-ray data, and they allow one to see

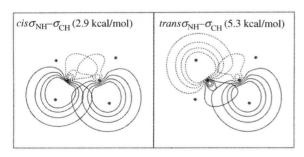

Figure 6.2 Similar to Fig. 6.1, for *cis* (left) versus *trans* (right) NH-CH hydride bond interactions, showing the primary involvement of "backside" lobes and spatial regions close to the heavy-atom nuclei in the *trans* case.

Table 6.1   Natural atomic Van der Waals radii R_{vdw} (Å) of elements 1–18.

Z	Atom	R_{vdw}	Z	Atom	R_{vdw}	Z	Atom	R_{vdw}
1	H	1.42	7	N	1.63	13	Al	2.30
2	He	1.07	8	O	1.46	14	Si	2.21
3	Li	2.76	9	F	1.27	15	P	2.44
4	Be	2.22	10	Ne	1.22	16	S	2.16
5	B	1.78	11	Na	3.07	17	Cl	1.89
6	C	1.62	12	Mg	2.75	18	Ar	1.78

steric subtleties such as anisotropic differences between longitudinal and transverse bonding directions, atomic charge variations, or derivative ("hardness") properties that are beyond empirical reach.

As a further example, let us consider the *cis* versus *trans* configurational preference in *N*-methylformamide (NMF), which might be thought to have a steric basis. For simplicity, we merely replace H_4 or H_5 of formamide (Section 4.1.3) by an idealized CH_3 group (tetrahedral angles, 1.09 Å R_{CH}, 1.49 Å R_{CN}), retaining all other details of the parent geometry in order to isolate the *cis* versus *trans* configurational difference without reoptimization of either structure (Fig. 6.3). From STERIC evaluation of each isomer, we find a difference in steric exchange energy

$$\Delta E^{(sx)} = E^{(sx)}(cis) - E^{(sx)}(trans) = 3.18 \text{ kcal/mol} \tag{6.5}$$

or in the sum of pairwise contributions

$$\Delta \sum\nolimits_{i,j} E_{ij}{}^{(sx)} = 1.47 \text{ kcal/mol} \tag{6.6}$$

both indicating that steric repulsions are favoring the *trans* isomer. However, the total energy calculation indicates that the *cis* isomer is actually slightly favored (by 1.00 kcal/mol). Hence, we must look elsewhere for an explanation of the observed *cis*-NMF preference.

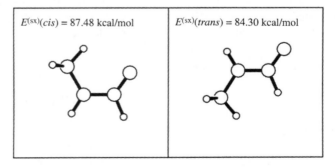

| $E^{(sx)}(cis) = 87.48$ kcal/mol | $E^{(sx)}(trans) = 84.30$ kcal/mol |

Figure 6.3   *Cis* (left) and *trans* (right) isomers of *N*-methylformamide (NMF), comparing steric exchange energies $E^{(sx)}$ for each isomer.

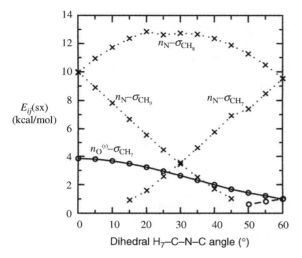

Figure 6.4 Leading pairwise $E_{ij}^{(sx)}$ steric exchange interactions for methyl torsional variations in *cis*-NMF (cf. Fig. 6.3), showing localized methyl σ_{CH} steric interactions with n_N (triangles, dotted curves) and $n_O^{(y)}$ (circles, solid curve). Whereas the three strong $\sigma_{CH}-n_N$ steric interactions (summing to a constant value at each dihedral angle, as required by symmetry) can be ignored, the proximal $\sigma_{CH}-n_O^{(y)}$ interaction (for example, $\sigma_{CH_7}-n_O^{(y)}$ in the dihedral range shown) dominates the overall steric dependence of methyl torsions.

Let us finally examine some details of methyl torsion in the more sterically crowded *cis* isomer of *N*-methylformamide, focusing on steric interactions involving the methyl protons (numbered H_7, H_8, and H_9). If we rigidly rotate the methyl group about the connecting C—N single bond (with H_7 the initial in-plane methyl proton, as shown at the left in Fig. 6.3) we see the variations of pairwise $E_{ij}^{(sx)}$ repulsions shown in Fig. 6.4 for a symmetry-unique 0–60° dihedral range.

Most conspicuous in Fig. 6.4 are the steric variations (x's, dotted lines) of the three methyl protons as they successively twist into coplanarity with the adjacent n_N lone pair (with H_8 achieving such coplanarity at 30° for the dihedral range shown in Fig. 6.4). However, the *sum* of the three $n_N-\sigma_{CH}$ interactions is *constant* (as the rigid C_{3v} symmetry of the methyl group demands), so these repulsions make no net contribution to the overall methyl torsional dependence. Instead, the most important CH methyl repulsion is expected to be that with the $n_O^{(y)}$ lone pair (circles, solid line), which varies by about 4 kcal/mol between the proximal in-plane maximum (shown for H_7 at 0° in Fig. 6.4) to the distal minimum (at 180°). (A small portion of the symmetrically related repulsion with CH_8 is shown as the dashed curve near 60°.) Thus, we can anticipate that the overall $E^{(sx)}$ dihedral variation (of the order of 4 kcal/mol) is dominated by the pairwise $n_O^{(y)}-\sigma_{CH}$ repulsion with the "nearby" methyl proton, as physical intuition would suggest. The significant steric contact between CH_7 and O_3 is also suggested by the relatively short $H_7\cdots O_3$ distance of 2.34 Å, well inside the expected van der Waals contact distance of 2.88 Å (Table 6.1). Figure 6.5 displays the PNLMO overlap diagram for the $n_O^{(y)}-\sigma_{CH_7}$ interaction, confirming the appreciable overlap (and $E_{ij}^{(sx)}$ value) that is achieved in the proximal $CH_7\cdots O$ geometry.

Figure 6.6 shows the total $E^{(sx)}$ variation for rigid methyl torsions of *cis*-NMF ("HF": circles, solid line), which exhibits reasonable agreement with the ca. 4 kcal/mol steric barrier expected from the pairwise $E_{ij}^{(sx)}$ values (Fig. 6.4). [Small numerical "glitches" (ca. 0.3 kcal/mol) are seen near 20°, 45°, probably resulting

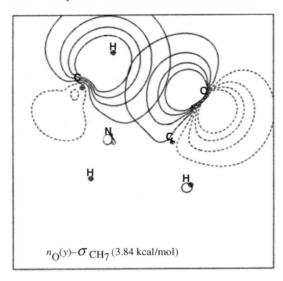

Figure 6.5   PNLMO overlap contour diagram (and $E_{ij}^{(sx)}$ steric exchange value) for leading methyl $n_O^{(y)} - \sigma_{CH_7}$ steric interaction in *cis*-NMF (cf. Fig. 6.4).

$n_O(y) - \sigma_{CH_7}$ (3.84 kcal/mol)

from linear-dependence instabilities in the augmented 6-311++G** basis.] Of course, the total $E^{(sx)}$ value includes the fully coupled effect of *all possible* pairwise interactions (including many not shown in Fig. 6.4), and is therefore considered to be the more accurate measure of overall torsional sterics. Similar connections between overall $E^{(sx)}$ variations and those for the "closest" few pairwise $E_{ij}^{(sx)}$ interactions could be demonstrated for many other geometry alterations, confirming that the STERIC descriptors are usually in excellent accord with physical intuition (but with notable exceptions such as the vicinal hydride interactions shown in Fig. 6.2).

We have shown angular details of the *ab initio* HF $E^{(sx)}$ values in Fig. 6.6 in order to make an instructive numerical comparison with B3LYP evaluations (cf. Sidebar 6.1). As shown in Fig. 6.6, the B3LYP method ("DFT": triangles, dotted

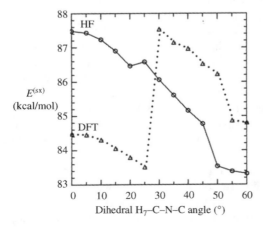

Figure 6.6   Dihedral variation of overall $E^{(sx)}$ for methyl torsions in *cis*-NMF (Fig. 6.3), comparing *ab initio* HF values (circles, solid line) with hybrid DFT result (triangles, dotted line) to illustrate unphysical artifacts of the latter STERIC evaluation (see Sidebar 6.1). (Both HF/6-311++G** and B3LYP/6-311++G** calculations were performed with identical geometries and SCF=TIGHT convergence thresholds.) The overall $\Delta E^{(sx)}$ variation of ~4 kcal/mol in this angular range agrees sensibly with the cruder single-term $\sigma_{CH} - n_O^{(y)}$ variation shown in Fig. 6.4.

line) exhibits an absurd numerical discontinuity near $25°$ ($>4\,\text{kcal/mol}$) that renders DFT evaluation of STERIC descriptors quite useless in this case. Used with care, the NBO descriptors of STERIC interactions provide powerful tools for exploring numerical details of steric exchange interactions, based on computational algorithms that are deeply related to the underlying "kinetic energy pressure" concept (Sidebar 6.1). However, careful thought is required to select an appropriate reference state for the specific steric question, and proper caution should be exercised in using current-generation DFT methods for this purpose. Superficial usage of STERIC keyword output can be a two-edged sword.

SIDEBAR 6.1	*NBO EVALUATION OF STERIC "KINETIC ENERGY PRESSURE"*

Theoretical physicist Victor Weisskopf first expressed the quantum mechanical essence of electronic steric repulsions in terms of "kinetic energy pressure" and employed this concept as one of four basic principles governing the qualitative physics of our universe [including fundamental limits on the heights of mountains or size of stars; see V.W. Weisskopf, "*Of Atoms, Mountains, and Stars: A Study in Qualitative Physics*," Science **187**, 605, 1975].

Weisskopf's concept rests on the fundamental Pauli exchange antisymmetry of the N-electron wavefunction, which forces each electron of given spin to be accommodated by a *distinct* spin-orbital, orthogonal to those occupied by other electrons. (The mutual orthogonality of electronic orbitals follows rigorously from the Hermitian property of physical Hamiltonian or density operators from which these orbitals originate.) When a given number of electrons are forced into reduced spatial volume, such orbital orthogonality can only be maintained by an increased density of oscillatory and nodal features, corresponding to shorter wavelength, or higher-frequency components, in the orbital waveform. Such high-frequency oscillations correspond to increasing second-derivative "curvature" of the wavefunction, as sampled by the quantum mechanical kinetic energy (Laplacian) operator, and kinetic energy therefore *rises* in "repulsive" response to the electronic volume reduction.

The kinetic energy response to volume decrease is naturally described as a "pressure" due to "overcrowding" of electrons. However, unlike the more familiar Boyle-type ($P \propto 1/V$) relationship for gases, the wavefunction amplitude is typically growing *exponentially* near its outer extremities. The Laplacian kinetic energy pressure grows accordingly, and the abrupt onset of steric repulsive forces (interatomic orbital overlap of filled orbitals) is much more "brick wall"-like than ordinary power-law forces, for example, of Coulombic type. The freshman-level picture of atomic billiard-balls of definite van der Waals radius "colliding" when they come into steric contact is therefore essentially correct.

To evaluate the energy change associated with interatomic orthogonalization, we note that the eigenorbitals of the one-electron density operator allow *exact* evaluation of the formal one-electron kinetic energy operator (Sidebar 5.4). For HF or DFT wavefunctions, where a one-electron effective Hamiltonian operator (Fock or Kohn–Sham operator F_{op}) is available, these eigenorbitals of the density operator are the NLMOs

$\{\varphi_i^{(\text{NLMO})}\}$ (Section 5.4), and their orbital energies can be evaluated as expectation values of the one-electron Hamiltonian:

$$\varepsilon_i^{(\text{NLMO})} = \int \varphi_i^{(\text{NLMO})} * F_{\text{op}} \varphi_i^{(\text{NLMO})} d\tau = F_{ii}^{(\text{NLMO})} \tag{6.7}$$

The Pauli-free starting point in which such interatomic orthogonalization effects are *absent* can be taken as that described by the corresponding PNLMOs $\{\varphi_i^{(\text{PNLMO})}\}$, with orbital energies:

$$\varepsilon_i^{(\text{PNLMO})} = \int \varphi_i^{(\text{PNLMO})} * F_{\text{op}} \varphi_i^{(\text{PNLMO})} d\tau = F_{ii}^{(\text{PNLMO})} \tag{6.8}$$

because these PNLMOs are constructed to resemble the NLMOs as closely as possible *except* for omission of the final interatomic orthogonalization step. The total NBO "steric exchange energy" $E^{(\text{st})}$ is therefore evaluated from the sum of doubly occupied NLMO versus PNLMO orbital energy changes, namely,

$$E^{(\text{st})} = 2\sum_i \left(\varepsilon_i^{(\text{NLMO})} - \varepsilon_i^{(\text{PNLMO})} \right) \tag{6.9}$$

(Numerical values of $\varepsilon_i^{(\text{NLMO})}$ and $\varepsilon_i^{(\text{PNLMO})}$ orbital energies are easily obtained from FNLMO and FPNLMO keywords; see Appendix C.) The simple estimate (6.9) is known to well approximate the effect of neglecting wavefunction antisymmetrization (e.g., difference between Hartree and Hartree–Fock energies). It also provides an excellent approximation for the total interaction energy of closed-shell rare gas atoms, which are generally regarded as the prototype system for steric repulsive forces (see *V&B*, p. 36ff and references therein).

The direct estimate (6.9) incorporates the effects of full N-electron antisymmetrization (and implicit orthogonalization), but does not allow a direct estimate of the *local* steric repulsion between distinct NLMOs i and j. To this end, the NBO program employs a "partial" deorthogonalization procedure to obtain "PNLMO/2 orbitals" that form the reference system for the local i–j contribution $(\Delta E_{ij}^{(\text{st})})$ to steric exchange. The sum of such pairwise steric exchange energies $(\Delta E_{ij}^{(\text{st})})$ (each corresponding to partial antisymmetrization of only the two electron pairs occupying NLMOs i and j) generally shows R-dependent variations that reasonably approximate the full $E^{(\text{st})}(R)$ dependence. The pairwise $(\Delta E_{ij}^{(\text{st})})$ values therefore allow one to follow details of how *individual* electron pairs come into "steric collision" with changes of intra- or intermolecular geometry. However, such uncoupled ("disjoint") pairwise contributions, although chemically informative, can only provide a rough approximation to the more accurate $E^{(\text{st})}$ value that properly incorporates the fully coupled effects of N-electron antisymmetrization. Further details of STERIC evaluation (see *NBO Manual*, p. B-100ff and references therein) are beyond the scope of this discussion.

One *caveat*: Evaluation of PNLMO energies (6.8) with DFT methods involves evaluation of the chosen functional $°[\rho_{\text{PNLMO}}]$ with a highly unusual electron density ρ_{PNLMO} that is inconsistent with the Pauli exclusion principle. Because this density differs appreciably from those originally used to parameterize ("train") the density functional, the numerical DFT results may be unreliable compared to Hartree–Fock results (cf. text Fig. 6.6), where the integral evaluations are under full *ab initio* control. When in doubt, use Hartree–Fock evaluation of STERIC effects as a check on the vagaries of empirical local versus nonlocal exchange approximations in hybrid and nonhybrid DFT methods.

6.2 ELECTROSTATIC AND DIPOLAR ANALYSIS

What is an "electrostatic effect"? At the most superficial level, the term might be applied to practically *any* quantum mechanical quantity, because the potential energy V_{op} of the molecular Hamiltonian is merely Coulomb's law for the charged nuclei (positions $\mathbf{R}_A$) and electrons (positions $\mathbf{r}_i$), namely,

$$V_{op} = -\sum_A \sum_i Z_A / |\mathbf{R}_A - \mathbf{r}_i| + \sum_{i<j} 1 / / |\mathbf{r}_i - \mathbf{r}_j| + \sum_{A<B} Z_A Z_B / |\mathbf{R}_A - \mathbf{R}_B|$$

(6.10)

Indeed, electrostatics enthusiasts have labeled a remarkable variety of quantum mechanical integrals as "electrostatic" (or "inductive") in nature. In this limit, the term becomes meaninglessly vague.

Even if we acknowledge that *all* molecular quantum mechanics originates in the Coulombic potential (6.10), and thus shares the essential weirdness of quantum phenomena, we might hope that limited aspects of molecular behavior could be understood in more intuitive classical terms. If so, the weirdness of quantal electrostatics might be (partially) replaced by the multipole-type formulas of classical electrostatics, along the lines of London's long-range perturbation theory (see *V&B*, p. 585ff). London showed how to separate the *exchange-free* long-range limit (where exponential "overlap" contributions disappear and the benign power-law behavior of classical electrostatics is restored) from the short-range domain of intrinsic covalency and exchange effects. We therefore wish to focus on this more familiar and intuitive limit, making contact with classical polarity concepts as understood by experimental chemists seeking theoretical guidance to successful chemical modifications.

As will be discussed in the subsequent Sections 6.2.1 and 6.2.2, the NBO program offers useful tools for (1) evaluating the classical electrostatic potential energy associated with the quantal charge distribution, or (2) dissecting molecular dipole moment or polarizability into localized bond dipole and resonance-type contributions of recognizable chemical origin. (Related decomposition of intermolecular interaction energy into terms of distinctive electrostatic and steric character will be discussed in Chapter 9.) Our emphasis throughout is on analyzing electrostatic-type descriptors in localized NBO terms, rather than exploiting the variety of multipole shapes and functional forms to create "electrostatic models" of general molecular properties. Thus, the NBO program makes no provision for obtaining numerically fitted "atomic charges," "distributed dipoles," etc., to reproduce selected features of the *ab initio* potential energy surface.

6.2.1 Natural Coulombic Energy

Although formation of atoms from electrons and nuclei is inherently of short-range quantal nature, one may nevertheless suppose that certain aspects of classical long-range behavior are emerging at the level of *atom–atom* interactions. If so, a simple electrostatic descriptor can be formulated in terms of the effective net atomic charges (q_A) and associated Coulombic potential energy function (V_{NCE})

$$V_{NCE} = \sum\nolimits_{A<B} q_A q_B / |\mathbf{R}_A - \mathbf{R}_B| \qquad (6.11)$$

For the NPA-based atomic charges (Section 3.2) at the chosen geometry, Equation (6.11) expresses the *Natural Coulombic Energy* (NCE) $V_{NCE}(\mathbf{R}_A, \mathbf{R}_B, \ldots)$ as a qualitative measure of overall atom–atom electrostatics.

Given two distinct isomeric geometries and the associated natural charges, one can evaluate the potential energy difference (ΔV_{NCE}) that might reasonably be attributed to electrostatic-type forces. Because the natural charges $\{q_A\}$ include effects of NL-type resonance delocalization as well as L-type covalency and bonding interactions, V_{NCE} is only superficially a "classical electrostatic" potential energy. [For example, Equation (6.11) includes the charge shifts due to alternative dipolar amide resonance structure in (5.20), which is certainly not "classical electrostatic" in nature. See *V&B*, p. 602ff, for other examples of resonance-enhanced polarity patterns and equilibrium geometries that resonance-free classical forces could never achieve.] Nevertheless, Equation (6.11) appears correct for the long-range limit of negligible atom–atom exchange interactions, and might be considered (if not taken too seriously) as some type of "continuation" of classical Coulombic potential energy into the strongly nonclassical domain of exchange-type and chemical bonding forces. (A more satisfactory dissection of intermolecular interaction energy into components of classical electrostatic origin is described in Section 9.3.)

To briefly illustrate the NCE concept, let us reconsider the *cis* and *trans* isomers of *N*-methylformamide (Fig. 6.3). From the calculated NPA charges and interatomic distances in each isomer (further details not given), we find from Equation (6.11):

$$\begin{aligned} \Delta E^{(es)} &= V_{NCE}(cis) - V_{NCE}(trans) = (-0.381483) - (-0.376354) \\ &= -0.005129 \text{ a.u.} = -3.22 \text{ kcal/mol} \end{aligned} \qquad (6.12)$$

The favorable $\Delta E^{(es)}$ value (-3.22 kcal/mol) is evidently just sufficient to overcome the unfavorable $\Delta E^{(sx)}$ value [$+3.18$ kcal/mol, Eq. (6.5)], leading to a slight net advantage for the *cis* isomer. This is qualitatively consistent with the observed slight L-type difference favoring the *cis* isomer:

$$\Delta E(L) = -0.000568 \text{ a.u.} = -0.36 \text{ kcal/mol} \qquad (6.13)$$

to which NL-type resonance stabilization

$$\Delta E(NL) = -0.64 \text{ kcal/mol} \qquad (6.14)$$

contributes a slight additional net *cis* advantage. The results are consistent with the suggestion that Coulomb electrostatic interactions are principally responsible for overcoming the evident steric disadvantages of the *cis* isomer in this idealized model.

Equation (6.11) shows that one can readily dissect NCE into individual atom–atom contributions. By decomposing each q_A into contributions from L- versus NL-type NBOs, one can further separate each NCE contribution into L versus NL components, allowing estimates of classical-type versus resonance-enhanced electrostatic effects, and so forth. However, details of such quasi-classical electrostatic interactions are seldom of principal chemical interest, and are not considered further here.

6.2.2 Natural DIPOLE Analysis

The "dipole moment" ($\boldsymbol{\mu}$) of a diatomic molecule is an easily visualized polarity descriptor that is introduced to all students of chemical bonding. In order to explore the ramifications of this concept in polyatomic species, we need to recall the three-dimensional vectorial character of $\boldsymbol{\mu}$ and its mathematical representation as the resultant (vector sum) of constituent "bond dipoles" and other localized components of the molecular electron distribution.

Quantum mechanically, the total electric dipole moment $\boldsymbol{\mu}$ is evaluated as a vector sum of electronic ($\boldsymbol{\mu}^{(e)}$) and nuclear ($\boldsymbol{\mu}^{(n)}$) contributions,

$$\boldsymbol{\mu} = (\mu_x, \mu_y, \mu_z) = \boldsymbol{\mu}^{(e)} + \boldsymbol{\mu}^{(n)} \tag{6.15}$$

Evaluation of $\boldsymbol{\mu}$ requires the quantum mechanical integral ("first moment" of the charge distribution)

$$\boldsymbol{\mu}^{(e)} = \int \Psi^* \boldsymbol{\mu}_{op} \Psi d\tau \tag{6.16}$$

over the electric dipole operator (in a.u.)

$$\boldsymbol{\mu}_{op} = -\sum_i \mathbf{r}_i \tag{6.17}$$

[The nuclear contribution is merely the classical-like expression for the positions of the nuclear point charges

$$\boldsymbol{\mu}^{(n)} = +\sum_A Z_A \mathbf{R}_A \tag{6.18}$$

and is easily combined with the electronic integral (6.16).] The electric dipole moment provides the simplest and most important descriptor of overall electronic charge asymmetry, the first member of the classical multipole series describing successively longer-range details (dipole, quadrupole, octupole, . . .) of the electrostatic charge distribution. Dipole-related quantities are usually expressed in terms of

"Debye units" (1 D $\cong 3.3356 \times 10^{-30}$ C m; Appendix E), such that charges of $\pm e$ separated by 1 Å correspond to $|\boldsymbol{\mu}| = 1$ D.

As a formal one-electron property, $\boldsymbol{\mu}$ can be evaluated exactly from the one-electron density operator (see *V&B*, p. 21ff). For single-configuration SCF-MO or DFT description, this implies in turn that $\boldsymbol{\mu}$ can be simply evaluated (and visualized) as a sum of localized "NLMO bond dipoles," namely, for NLMO $\varphi_i^{(NLMO)}$ (cf. Section 5.4),

$$(\boldsymbol{\mu}^{(NLMO)})_{ii} = \int \varphi_i^{(NLMO)*} \boldsymbol{\mu}_{op} \varphi_i^{(NLMO)} d\tau \tag{6.19}$$

(A multiconfigurational wavefunction leads to an additional "NLMO coupling" contribution that appears near the end of DIPOLE output.) Each occupied $\varphi_i^{(NLMO)}$ can in turn be expanded in terms of its parent Lewis-type NBO $\varphi_i^{(L)}$ and weak delocalization tails on surrounding non-Lewis NBOs [with respective coefficients t_{ii} and t_{ij}; cf. Equation (5.24)] to give the following:

$$(\boldsymbol{\mu}^{(NLMO)})_{ii} = t_{ii}^2 (\boldsymbol{\mu}^{(L)})_{ii} + \sum_{j,k} t_{ij} t_{ik} (\boldsymbol{\mu}^{(NL)})_{jk} \tag{6.20}$$

A mathematical identity (see *NBO Manual*, p. B-25) then allows total $\boldsymbol{\mu}$ to be expressed equivalently as a sum of localized "NBO bond dipoles":

$$(\boldsymbol{\mu}^{(L)})_{ii} = \int \varphi_i^{(L)*} \boldsymbol{\mu}_{op} \varphi_i^{(L)} d\tau \tag{6.21}$$

together with associated resonance-type delocalization corrections into surrounding NL-type NBOs $\varphi_j^{(NL)}$.

Of course, the total electronic dipole moment integral (6.16) must include contributions from core (CR) and lone pair (LP) as well as bond (BD) NBOs of the Lewis structure. The near-spherical core orbitals normally make insignificant contributions to the dipole integral, but the contributions of valence lone pairs usually cannot be ignored at any reasonable level of approximation. Thus, the superficial freshman-level "sum of bond dipoles" picture (even more superficially, with bond dipoles envisioned in terms of isolated point charges at each atomic nucleus) cannot give a realistic description of the molecular dipole moment of most chemical species.

NBO analysis of the molecular dipole moment is requested by inclusion of the DIPOLE keyword in the $NBO keylist. As a simple example, we consider the formamide molecule (Section 4.13), whose nonvanishing dipole components all lie in the *x–y* molecular plane. Results of DIPOLE analysis for formamide are shown in I/O-6.3, slightly truncated (by inclusion of the "DIPOLE = 0.05" keyword) to include only delocalization corrections exceeding 0.05 D (rather than default 0.02 D):

```
────────────────────────── I/O-6.3 ──────────────────────────
DIPOLE MOMENT ANALYSIS:

[Print threshold: Net dipole >0.050 Debye]


                        NLMO bond dipole         NBO bond dipole
                   ------------------------   ------------------------
          Orbital     x     y     z  Total      x     y     z  Total
==============================================================================
 1. BD ( 1) N 1- C 2  0.603 0.602 0.000 0.853   0.560 0.663 0.000 0.868

 2. BD ( 1) N 1- H 4 -0.167 0.807 0.000 0.824  -0.151 0.827 0.000 0.841

 3. BD ( 1) N 1- H 5  0.859-0.178 0.000 0.878   0.880-0.156 0.000 0.894
                         deloc  86:-0.048-0.024 0.000 0.054

 4. BD ( 1) C 2- O 3 -2.373 0.261 0.000 2.387  -2.382 0.290 0.000 2.399

 5. BD ( 2) C 2- O 3 -0.805 0.124 0.000 0.815  -0.801 0.135 0.000 0.812

 6. BD ( 1) C 2- H 6  0.839-1.607 0.000 1.813   0.733-1.787 0.000 1.932
                         deloc  83: 0.033 0.090 0.000 0.096
                         deloc  82: 0.062 0.054 0.000 0.082

 7. CR ( 1) N 1       0.000-0.001 0.000 0.001  -0.001-0.002 0.000 0.002

 8. CR ( 1) C 2       0.001 0.004 0.000 0.004  -0.001 0.005 0.000 0.005

 9. CR ( 1) O 3       0.005-0.001 0.000 0.005   0.004-0.001 0.000 0.004

10. LP ( 1) N 1      -1.276-1.047 0.000 1.651   0.239 0.226 0.000 0.329
                         deloc  85:-1.510-1.185 0.000 1.919
                         deloc  32:-0.024-0.078 0.000 0.081

11. LP ( 1) O 3      -2.842 0.444 0.000 2.877  -2.887 0.444 0.000 2.921

12. LP ( 2) O 3       1.165-0.066 0.000 1.167  -0.183 0.056 0.000 0.191
                         deloc  87: 0.509-0.122 0.000 0.523
                         deloc  82: 0.430-0.038 0.000 0.432
                         deloc  31: 0.171 0.064 0.000 0.183
                         deloc  84: 0.097 0.037 0.000 0.104
                         deloc  30: 0.019-0.096 0.000 0.098
                     -------------------------------------------------
    Net dipole moment -3.991-0.657 0.000 4.045  -3.989 0.700 0.000 4.050
Delocalization correction                       -0.002-1.356 0.000 1.356
                     -------------------------------------------------
  Total dipole moment -3.991-0.657 0.000 4.045  -3.991-0.657 0.000 4.045
```

As shown in I/O-6.3, the vector (x,y,z) components and total length are given for each NLMO and NBO bond dipole of the formal Lewis structure. The NLMO bond dipoles sum directly to the total molecular dipole moment, with components

$$\boldsymbol{\mu} = (-3.991, -0.657, 0.000) \tag{6.22}$$

and total length

$$\mu = |\boldsymbol{\mu}| = (\mu_x^2 + \mu_y^2 + \mu_z^2)^{1/2} = 4.045 \text{ D} \tag{6.23}$$

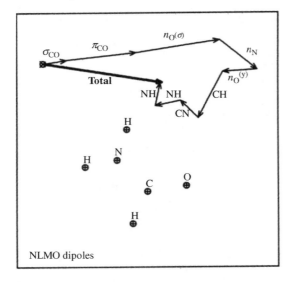

Figure 6.7   Vector addition of NLMO bond dipoles (light arrows) to give the total molecular dipole moment (heavy arrow) in the x–y plane of formamide, with nuclear positions shown in the same (principal) axis system for comparison. Note the large contribution of lone pairs (particularly $n_O^{(\sigma)}$ to the total dipole, which lies roughly parallel to the N$\cdots$O axis in this species.

as shown in "net dipole moment" and "total dipole moment" entries of the NLMO column. The corresponding NBO bond-dipole sum [$\mathbf{\mu}_{NBO} = (-3.989, 0.700, 0.000)$] must be added to the "delocalization correction" [$\mathbf{\mu}_{deloc} = (-0.002, -1.356, 0.000)$] to obtain the correct total dipole moment, as shown at the bottom of the NBO column.

Figure 6.7 depicts the vector geometry of NLMO bond dipoles in formamide. The labeled arrows for each NLMO are arranged head-to-tail (negative head, positive tail) to give the resultant total dipole (heavy arrow), plotted in the x–y molecular plane (principal axes) of the nuclei, which are shown in correct relative orientation for comparison. [The center of charge (crossed circle) from which the dipole vector emanates (upper left) is located near C.]

As expected, both σ_{CO} and π_{CO} dipoles (parallel to the C=O bond axis) contribute significantly to the total moment. However, the largest single contributor in Fig. 6.7 is actually the $n_O^{(\sigma)}$ lone pair (cf. Fig. 4.3), which also aligns along this axis. The n_N and $n_O^{(y)}$ lone pair NLMOs are also seen to make highly significant contributions (greater, e.g., than σ_{CN} or σ_{NH} bond NLMOs). As expected, each σ-bond dipole is roughly parallel to the corresponding bond axis, and the $n_O^{(y)}$, n_N dipoles also align approximately parallel to CO and CN axes, respectively. Figure 6.7 shows that attempted description of dipole geometry without adequate account of lone pair contributions is fundamentally erroneous.

Figure 6.8 shows a closely related vector diagram of NBO dipole geometry. In this case, the NBO bond dipoles (light arrows) are shown with corresponding delocalization corrections (light dotted lines) that sum to the same resultant dipole moment (heavy solid arrow). The vector resultant of NBO dipoles (heavy dashed arrow) and delocalization corrections (heavy dotted line) are also shown for comparison. Comparison of Figs. 6.7 and 6.8 shows that NLMO and NBO bond dipoles are fairly similar, with NL-type delocalization "gaps" barely visible between the L-type arrows. However, the resonance-induced dipole shifts are quite conspicuous for n_N

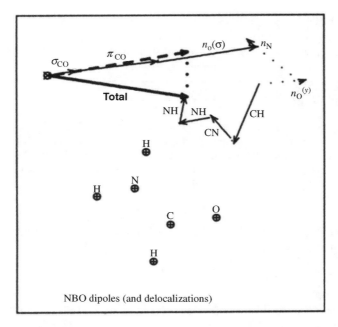

Figure 6.8   Similar to Fig. 6.7, for the NBO bond dipole geometry of formamide. Each NBO bond dipole (light arrows) is shown with its delocalization correction (light dotted lines), resulting in the same total dipole moment (heavy solid arrow) as in Fig. 6.7. The resultant sum of NBO dipoles is shown as the heavy dashed arrow and the resultant delocalization correction as the heavy dotted line. Note the large dipole reorientation due to resonance-type delocalizations, which twist the final molecular dipole significantly out of parallelism with the C=O double bond.

and $n_O^{(y)}$ lone pairs, resulting in significant overall reorientation of the dipole vector, relative to an elementary localized Lewis-like picture (heavy dashed arrow).

As particular examples from I/O-6.3 and Fig. 6.8, we may observe that the delocalization "correction" for the n_N lone pair (1.92 D) is more than five *times* larger than that (0.33 D) of the parent NBO dipole (as well as oppositely oriented). Similarly large resonance effects are found for the $n_O^{(y)}$ lone pair. From such examples we can conclude that attempted classical-like descriptions of dipolar charge distributions, without adequate account of resonance-type delocalization, are generally superficial and misleading.

The large resonance-induced dipole shifts found in I/O-6.3 are precisely those due to NBO donor–acceptor delocalizations studied previously (cf. Figs. 5.2 and 5.4). Although the quantity of charge transferred in such delocalizations appears modest, the transfer *distance* is appreciable, and such distance-dependence is directly sampled by the dipole operator (6.17) and integral (6.16). Thus, some of the most striking consequences of nonclassical resonance-type delocalizations are to be found in dipole moments and transition values (spectral intensities) that are often portrayed superficially as "classical electrostatic" in nature.

The DIPOLE keyword also allows localized analysis of electric polarizability, hyperpolarizability, and other dielectric response properties, as sketched in Sidebar 6.2. However, further discussion of such higher-order electrical properties is beyond the scope of present treatment.

SIDEBAR 6.2 *POLARIZABILITY ANALYSIS*

Electric dipole "polarizability" $\boldsymbol{\alpha}$ is a second-order tensorial (3×3 matrix) quantity whose α_{ij} element describes the change of dipole moment component μ_i induced by a change of electric field in direction j, where $i, j = \mathrm{x,y,z}$ denote arbitrarily chosen Cartesian directions. Formally, if $\boldsymbol{\mu}$ is the field-free dipole moment, and $\Delta\boldsymbol{\mu}_j$ is the change in dipole moment induced by a static electric field ΔF_j in the j direction, then α_{ij} can be evaluated as the limiting ratio

$$\alpha_{ij} = \lim_{\Delta Fj \to 0} (\Delta\boldsymbol{\mu}_j)_i / \Delta F_j \qquad (6.24)$$

[Equivalently, α_{ij} could be expressed as a second derivative of energy with respect to crossed field variations, namely,

$$\alpha_{ij} = -\partial^2 E / \partial F_i \partial F_j \qquad (6.25)$$

but Equation (6.24) is more useful for present purposes.]

Although somewhat tedious, Equation (6.24) provides a blueprint for using standard DIPOLE analysis to obtain a localized dissection of each α_{ij} component, assuming that the host ESS is capable of finite-field calculations with variable external electric field. One merely performs DIPOLE analysis for two separate calculations, differing by a small field change (say, ΔF_z in the z-direction), then evaluates the numerical ratio (6.24) from *differences* in the two DIPOLE analysis outputs. In principle, only four judiciously chosen finite-field calculations (with fields of $0, F_x, F_y, F_z$) are required to obtain the six independent α_{ij} polarizability components.

A similar DIPOLE-based procedure could evidently be extended to higher-order "hyperpolarizability" components. However, the number of tensorial components rises steeply with tensor order, and the numerical differencing problems associated with accurate evaluation of limiting ratios such as (6.24) become increasingly challenging. Practical DIPOLE-based analysis of such higher-order polarizability properties may therefore be limited to the leading few components.

From a chemical viewpoint, a more informative alternative analysis of polarizability-type properties may be based on the concept of *freezing* the form of Lewis-type NBOs, in order to prevent the orbital deformations ("repolarization") associated with external perturbations. In effect, by recalculating the L-type electronic response in the *absence* of the usual rehybridization and repolarization (orbital-distortion) effects, one can partially isolate the classical-like orbital-polarizability deformations from more exotic resonance-type donor–acceptor effects of wholly nonclassical character. Such a frozen-NBO recalculation can be carried out by "importing" the NBOs (using the NAONBO=R keyword: "read the stored NAO → NBO transformation") from a disk file to which NBOs of a field-free calculation were previously saved (using the NAONBO=W keyword; cf. Appendix C and *NBO Manual*, p. B-8). Comparison of the frozen-NBO E(L) with the default E(L) for optimally repolarized NBOs then gives an intuitive measure of classical-type induction or polarization effects, distinct, for example, from resonance-type "intermolecular charge transfer" or other effects of distinctively nonclassical origin (see Chapter 9).

PROBLEMS AND EXERCISES

6.1. For the five isomeric COH_2 species considered in Problem 4.1, choose formaldehyde (isomer 1) as the "zero" for comparing differences with other isomers in the problems below:

(a) Use the NBO program to evaluate the steric-exchange difference $[\Delta E^{(sx)}]$ for each isomer 2–5 compared to isomer 1. Do the rankings appear qualitatively sensible? Why or why not?

(b) Similarly, evaluate the electrostatic difference $[\Delta E^{(es)}]$ for each isomer 2–5 compared to isomer 1, and discuss the reasonability of the resulting rankings.

(c) To what extent can classical-like steric or electrostatic differences explain the overall energetic rankings and energy differences among these species? To what extent would characteristic chemical bonding effects (starting, e.g., from empirical bond energies for the NBO Lewis structure diagram) be necessary to successfully rationalize the isomeric energetics? Discuss briefly.

6.2. *Cis* and *trans* isomers of difluoroethylene

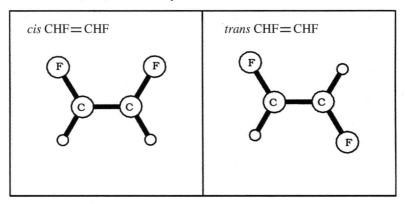

exhibit a counterintuitive energetic preference for the *cis* species, despite its apparent steric and electrostatic disadvantages. [For this problem, consider each species in idealized geometry ($R_{CC} = 1.34$ Å, $R_{CH} = 1.08$ Å, $R_{CF} = 1.33$ Å, $120°$ bond angles) to make the comparisons as "fair" as possible.]

(a) Evaluate the overall steric difference between isomers $[\Delta E^{(sx)} = E^{(sx)}(trans) - E^{(sx)}(cis)]$ and the specific pairwise interactions most responsible for this difference. Are these results consistent with your physical intuition?

(b) Similarly, evaluate the overall electrostatic difference between isomers $[\Delta E^{(es)} = E^{(es)}(trans) - E^{(es)}(cis)]$ and the specific pairwise interactions most responsible for this difference. Are these results consistent with your physical intuition?

(c) Using the methods of Chapter 5, evaluate the $\Delta E(L)$ and $\Delta E(NL)$ differences in Lewis and non-Lewis energy for each isomer. Are your results in Problems (a) and (b) qualitatively consistent with the L-type difference $\Delta E(L)$? Can you intuit (or work out) the main donor–acceptor interactions responsible for the large $\Delta E(NL)$ difference favoring the *cis* isomer? [If in doubt, consult the *NBO Tutorials* ($DEL) section of the *NBO Website*.]

(d) Evaluate the total dipole moment of *cis* CHF=CHF and the local bond dipole of each CH and CF bond. For comparison, evaluate the bond ionicities [Equation (4.18)] of σ_{CH}, σ_{CF} NBOs. Can you see evidence of hyperconjugative delocalization effects on the total μ_{cis} dipole moment? Explain briefly.

(e) Similarly, evaluate the total dipole, bond dipoles, and bond ionicities in the *trans* isomer. Do the bond dipoles in the two isomers vary with bond ionicities in a chemically reasonable way?

(f) Compare the strength of individual bond-dipole delocalization corrections in the two isomers. What NBO donor–acceptor interactions contribute principally to these corrections? (If NBOView capabilities are available, plot the PNBO overlap diagrams for the interactions most responsible for the *cis–trans* differences.) Can you rationalize how these donor–acceptor interactions are further enhanced by the slight ionicity variations noted in (e)?

(g) Summarize your results briefly, by quantifying and describing the steric and electrostatic advantages of the *trans* isomer versus the hyperconjugative advantages of the *cis* isomer.

[Similar problems can be set for CH_2FCH_2F, NHF=NHF, and many related species.]

6.3. [For this problem, it is assumed that your ESS program can perform finite-field calculations for chosen electric fields along chosen directions. In the Gaussian program, use the NOSYMM keyword to obtain the desired molecular orientation with respect to Cartesian x–y axes.]

Many polyatomic molecules possess the interesting ability to induce a dipole moment in a direction (say, *y*) *perpendicular* to the applied electric field (say, *x*), corresponding to nonvanishing α_{xy} polarizability component. Consider the simple triatomic species HN=O, with N=O along the horizontal *x* axis, and with a finite electric field F_x of strength 0.001 a.u. in the *x*-direction, as shown in the Gaussian input file below:

```
#B3LYP/6-311++G** nosymm pop=nboread field=x+10

HN=O with finite field (10*0.000110*0.0001 = 0.001 a.u.) in x-direction

0 1
O
N  1  1.33
H  2  1.01  1  118.

$nbo file=hno_001 dipole $end
```

(a) For the field-free problem (omit the FIELD keyword), evaluate the dipole moment of HN=O and sketch the vector components for NLMO and NBO bond dipoles in a vector diagram that includes the nuclei (cf. Figs. 6.7 and 6.8). What are the relative magnitudes of L-type versus NL-type (delocalization) components of the overall dipole vector?

(b) From the finite-field DIPOLE analysis (as sketched in Sidebar 6.2), evaluate the α_{xx} and α_{xy} polarizability components. From the results of your analysis, can you rationalize the physical origin of the nonvanishing α_{xy} value? Would you predict α_{xy} to be larger or smaller in FN=O? Explain briefly.

Chapter 7

Nuclear and Electronic Spin Effects

Nuclear magnetic resonance (NMR) and electron spin resonance (ESR) provide some of the most powerful and versatile spectroscopic tools of the modern chemist. Each spectroscopy depends on the intrinsic "spin" angular momentum and associated magnetic dipole moment that is exhibited by nuclei with odd numbers of neutrons or protons, as well as by all electrons.

In one sense, spin appears to be a chemically *negligible* property of nuclei and electrons. Indeed, the usual chemical Hamiltonian of standard ESS packages ignores nuclear and electronic spin variables entirely. (Although electron spin is commonly introduced as a wavefunction variable, it serves merely for convenient "bookkeeping" of the important Pauli exchange antisymmetry, and a completely spin-free formulation of nonrelativistic quantum chemistry would be equally valid.) Nevertheless, the weakness of its chemical effects makes spin an ideal "spectator" of the chemical environment. Magnetic resonance spectroscopists employ clever manipulations with magnetic fields and radio frequency pulses to interrogate particular spins and read out a bonanza of chemically useful information, particularly from the nuclear spins. Indeed, experienced chemists can often infer key structural features of an organic molecule from a mere glance at the ^{1}H- or ^{13}C-NMR spectrum of its principal nuclei.

Recent theoretical progress in first principles calculation of NMR properties has greatly enhanced the accuracy and specificity of chemical information that can be inferred from the NMR spectrum of common organic species. Assuming that the correct theoretical structure is employed (if necessary, with proper account of vibrational averaging), the GIAO-based ("gauge-including atomic orbital") methods can now be expected to routinely reproduce experimental ^{1}H-NMR shieldings within the measurement uncertainties. Theoretical calculations that correctly reproduce the measured NMR spectrum can therefore be employed with considerable confidence to interpret the electronic origins of even the most subtle spectral features within current experimental resolution.

In this chapter, we wish to briefly illustrate NBO-based tools for analyzing the principal features of a calculated magnetic resonance spectrum, with primary

Discovering Chemistry With Natural Bond Orbitals, First Edition. Frank Weinhold and Clark R. Landis.
© 2012 John Wiley & Sons, Inc. Published 2012 by John Wiley & Sons, Inc.

emphasis on NMR. NBO analysis reveals the detailed chemical origins of the characteristic positions ("chemical shifts") and multiplet splittings ("J-couplings") of NMR spectral lines. These spectral parameters are described in terms of elementary Lewis structural and resonance concepts, similar to those employed throughout this book, allowing standard empirical correlations to be put on a firmer theoretical basis or significantly enhanced in accuracy and specificity. Section 7.1 first describes analysis of the chemical shielding effects, leading to the characteristic resonance frequency (chemical shift) of each NMR-active nucleus in response to its unique chemical environment. Section 7.2 similarly describes analysis of the scalar spin–spin couplings (J-splittings) that survive molecular tumbling in the external magnetic field, giving rise to the characteristic multiplet splittings for each shifted nucleus. Finally, Section 7.3 briefly describes NBO analysis of the unpaired spin density distribution in open-shell systems, the fundamental property probed by ESR spectroscopy. [Further background information on magnetic resonance spectroscopy and details of GIAO-based methods for calculating and analyzing NMR properties are cited in the original NCS paper: J. A. Bohmann, F. Weinhold, and T. C. Farrar, *J. Chem. Phys.* **107**, 1173, 1997; *NBO Manual*, p. B-138ff).]

7.1   NMR CHEMICAL SHIELDING ANALYSIS

"Chemical shielding" (σ, measured in parts per million—ppm) refers to the fractional change by which the resonance frequency of a bare nucleus is shifted due to its electronic environment. In effect, the external magnetic field (strength B_0) induces electronic circulation patterns that slightly alter the effective magnetic field "seen" by the nucleus, thus shifting its resonance frequency ($v = \gamma_A B_0$, γ_A = nuclear gyromagnetic ratio) relative to the bare-nucleus value.

Because the electronic environment is generally anisotropic, the associated induced fields have components transverse to the external field direction. As a result, chemical shielding has the mathematical character of a second-rank tensor (3×3 matrix), with components $(\boldsymbol{\sigma})_{ij}$ for any chosen pair of Cartesian axes ($i, j = $ x,y,z). However, under the usual liquid-state ^{1}H-NMR conditions of rapid sample spinning and molecular tumbling, the shielding tensor is effectively averaged over all orientations. The isotropic shielding average

$$\sigma = \sigma_{iso} = \tfrac{1}{3}[(\boldsymbol{\sigma})_{xx} + (\boldsymbol{\sigma})_{yy} + (\boldsymbol{\sigma})_{zz}] \tag{7.1}$$

then becomes the principal quantity of experimental interest.

Experimentally, the isotropic chemical shielding σ_A of a given nucleus A is generally expressed as the "chemical shift" δ_A with respect to a chosen reference signal σ_{ref}, such as tetramethylsilane (TMS) for ^{1}H spectra:

$$\delta_A = \sigma_{TMS} - \sigma_A \tag{7.2}$$

Increasing chemical shift δ_A is generally plotted leftward ("downfield") of the TMS signal, so that chemical shielding σ_A increases toward the right ("upfield"

I/O-7.1

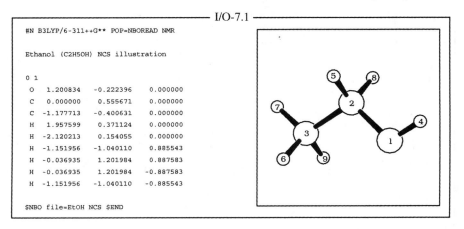

```
#N B3LYP/6-311++G** POP=NBOREAD NMR

Ethanol (C2H5OH) NCS illustration

0 1
    O    1.200834    -0.222396    0.000000
    C    0.000000     0.555671    0.000000
    C   -1.177713    -0.400631    0.000000
    H    1.957599     0.371124    0.000000
    H   -2.120213     0.154055    0.000000
    H   -1.151956    -1.040110    0.885543
    H   -0.036935     1.201984    0.887583
    H   -0.036935     1.201984   -0.887583
    H   -1.151956    -1.040110   -0.885543

$NBO file=EtOH NCS $END
```

from the bare nucleus) in this convention. For theoretical purposes, we must therefore calculate the reference TMS shielding at a consistent theoretical level (namely, $\sigma_{TMS} = 31.98$ ppm for B3LYP/6-311++G** level) in order to make direct contact with experimentally determined chemical shift data.

Localized NBO/NLMO-based analysis of chemical shielding tensors requires interactive (linked) cooperation with the host ESS program, and is currently implemented only in NBO5-linked versions of the Gaussian program (not in the binary version distributed by Gaussian Inc.). Default "natural chemical shielding" (NCS) analysis of the calculated GIAO shielding (Gaussian keyword "NMR") is invoked simply by including the NCS keyword in the $NBO keylist, as illustrated in I/O-7.1 for ethanol.

Although we focus on the default localized NCS analysis of isotropic shielding in this simple illustration, other keyword options (see *NBO Manual*, p. B-142) allow one to readily analyze individual shielding tensor components, chemical shielding anisotropy (CSA), or field-free ("diamagnetic") versus field-induced ("paramagnetic") shielding contributions. [The NCS=MO keyword also allows one to see the corresponding analysis in terms of delocalized MOs, which offers an informative contrast to the localized focus of NBO/NLMO results.]

The default NCS analysis output for ethanol is shown in I/O-7.2 for nuclei 1–7. The rows display localized contributions due to L-type NBOs and their NL-type corrections (from the corresponding NLMO delocalization tails), which sum to the total isotropic shielding at the bottom of the column.

The formal theory of NMR shielding is beyond the scope of the present discussion. However, one can expect in a general way that shielding depends on the size and shape of occupied orbitals in proximity to the nucleus. The size-dependence (r^{-3} radial weighting) emphasizes the contribution of electrons closest to the nucleus, while the angular momentum shape-dependence (s, p, d ... type) emphasizes the contribution of s-character (peaked at the nucleus) versus p-character (noded at the nucleus) for orbitals centered on the nucleus of interest.

─────────────── I/O-7.2 ───────────────

```
NATURAL CHEMICAL SHIELDING ANALYSIS

NCS Tensor analysis is using        120960 words


Summary of isotropic NMR chemical shielding

Total Lewis (L) and non-Lewis (NL) contributions: (ppm)
```

NBO	O(1)	C(2)	C(3)	H(4)	H(5)	H(6)	H(7)
1. O 1- C 2 L	-13.37	-9.08	0.08	1.77	-0.04	0.01	-0.25
NL	-0.63	0.59	-0.25	-0.01	0.01	0.01	-0.07
2. O 1- H 4 L	-0.38	0.03	-0.75	19.32	0.26	0.18	0.11
NL	1.31	-0.80	0.18	-0.32	0.03	0.01	-0.01
3. C 2- C 3 L	5.67	-27.61	-12.06	-0.37	-0.06	1.33	1.10
NL	-4.09	-1.18	-0.88	-0.15	0.10	0.08	0.17
4. C 2- H 5 L	2.70	-22.74	0.09	-0.06	27.25	-0.07	0.21
NL	-3.26	0.97	-0.98	-0.02	-0.03	-0.05	-0.11
5. C 2- H 8 L	2.70	-22.74	0.09	-0.06	0.87	-0.22	0.21
NL	-3.26	0.97	-0.98	-0.02	0.03	-0.05	-0.11
6. C 3- H 6 L	-0.22	-0.11	-9.60	0.01	0.13	25.73	1.93
NL	-0.46	0.23	0.71	0.03	-0.01	0.02	0.04
7. C 3- H 7 L	-0.84	0.29	-5.38	0.15	0.27	1.68	25.67
NL	-1.55	-0.47	1.59	-0.06	0.01	0.24	-0.04
8. C 3- H 9 L	-0.22	-0.11	-9.60	0.01	0.01	1.76	1.93
NL	-0.46	0.23	0.71	0.03	-0.24	0.09	0.04
9. O 1(cr) L	276.00	-0.08	-0.03	0.08	-0.04	-0.01	-0.01
NL	0.09	0.00	0.00	-0.01	0.00	0.00	0.00
10. C 2(cr) L	0.04	203.59	-0.12	0.05	0.23	-0.06	-0.05
NL	0.01	-0.03	-0.02	-0.01	-0.06	0.01	0.00
11. C 3(cr) L	-0.04	-0.15	203.62	-0.03	-0.04	0.09	0.36
NL	-0.01	-0.03	0.01	0.00	0.01	-0.01	-0.05
12. O 1(lp) L	13.09	-2.63	-1.05	4.96	0.31	-0.43	-0.08
NL	1.49	0.77	-0.16	0.05	-0.01	0.22	0.11
13. O 1(lp) L	11.06	-0.73	-0.06	6.03	-0.24	0.00	-0.07
NL	1.31	0.96	0.33	0.46	-0.62	0.04	-0.01
Lewis	296.19	117.92	165.22	31.86	28.91	29.98	31.06
non-Lewis	-9.50	2.23	0.25	-0.03	-0.77	0.61	-0.03
Total	286.69	120.15	165.48	31.83	28.14	30.60	31.03

How can we begin to understand the chemical shielding patterns in terms of localized bonding concepts? For ^{1}H nuclei, the 1s-orbital amplitude at the nucleus is controlled principally by the bond ionicity i_{AH} (Eq. 4.18), which in turn is sensitive to the electronegativity and hybrid p-character of the directly bonded A nucleus (through Bent's rule; cf. *V&B*, p. 138ff). Qualitative empirical relationships often stress the correlation of NMR shielding with hybridization and electronegativity of the bonded nucleus, and such correlation is clearly evident in the NCS contributions of the primary σ_{AH} NBOs that dominate total shielding [i.e., NBOs 2, 4, 6, and 7 for protons H(4)–H(7), respectively]. Figure 7.1 illustrates how the direct-bonded NCS contribution correlates with bond ionicity of the associated σ_{AH} NBO for the four distinct hydride bonds of ethanol. Bond ionicity can in turn be estimated from atomic electronegativity differences [Equation (4.35), further corrected, if necessary, for

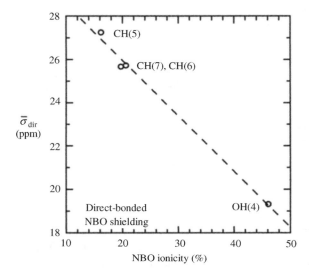

Figure 7.1 "Direct" isotropic chemical shielding (σ_{dir}, ppm) for hydride NBOs 2, 4, 6, 7 of ethanol (cf. I/O-7.2), shown as a function of NBO bond ionicity (%). An approximate linear regression [$\sigma_{dir} = -0.26(\%-i_{AH}) + 31$] is shown as a dashed line to aid visualization.

hybridization variations] to give qualitative predictions of the direct-bonded NBO contribution for many hydride species.

Although the direct-bonded NBO interaction clearly dominates the overall shielding magnitude in I/O-7.2, appreciable nonbonded contributions are seen to arise from other nearby NBOs of the bonding skeleton. Geminal ("1–3") NBOs are typically far more important in this respect than those at vicinal ("1–4") or more remote positions (but see the counterexample discussed below). For hydroxyl H(4) shielding, for example, the contributions of geminal NBOs 1 (1.77 ppm), 12 (4.96 ppm), or 13 (6.03 ppm) far exceed those of vicinal NBOs 3 (−0.37 ppm) or 4, 5 (−0.06 each), and similar geminal versus vicinal disparities are seen for other protons.

How can a proton of interest be partially shielded by neighboring geminal or vicinal NBOs? Quantum mechanical orbitals are known to extend continuously in all directions from the atomic nucleus, with long-range "tails" of exponentially decaying amplitude. Nuclear spins on nearby atoms may therefore "see" such long-range orbital tails as weak perturbations of the local electronic environment. Note that contributions from stray fringes of other-atom orbitals are not subject to the usual angular momentum hierarchy of importance (s ≫ p > d, etc.) that governs same-atom orbitals.

As examples, Fig. 7.2 illustrates the long-range tails of neighboring geminal (long-dash) and vicinal (short-dash) NBOs in the vicinity of hydroxyl proton H(4) (left panel) and in-plane methyl proton H(7) (right panel) for ethanol. Consistent with intuition, the geminal neighbors contribute significantly stronger fringe amplitudes than those from vicinal or more remote locations. The relative NCS shielding contributions in I/O-7.2 are seen to agree qualitatively with plotted NBO amplitude variations at the proton of interest. (Only qualitative agreement is expected, because the calculated shielding involves r^{-3} radial convolution with the amplitude profiles

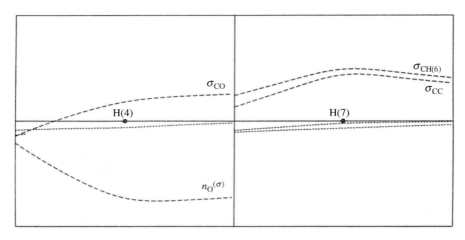

Figure 7.2 NBO amplitude profiles for long-range tails of neighboring geminal (long-dash) and vicinal (short-dash) NBOs in the vicinity of hydroxyl H(4) (left) and methyl H(7) (right) protons of ethanol (cf. I/O-7.1), showing the significantly stronger tail amplitudes (and shielding) from geminal NBOs. [In each case, the profile is along the hydride bonding axis, oriented in the molecular symmetry plane (so that $n_O^{(\pi)}$ does not appear in the left panel). In each panel, the horizontal range is ±0.2 Å around each nucleus, and the vertical amplitude range is ±0.05 a.u.]

plotted in Fig. 7.2.) Thus, a fairly simple picture of long-range NBO tails seems to satisfactorily rationalize the leading L-type NCS contributions.

Table 7.1 compares the calculated shieldings with experimental liquid ethanol values, both expressed as chemical shifts (δ_A) with respect to the TMS resonance position for each NMR-distinguishable proton type. [Here, "type" refers to hydroxyl ("OH"), methylene ("CH$_2$"), or methyl ("CH$_3$") features of the spectrum, with integrated intensities summing to 1, 2, or 3 protons, respectively, with δ_{exp} taken as the high-resolution centroid or low-resolution peak-maximum of the associated spectral feature(s), and with δ_{theor} a corresponding orientational average (see below).] Theoretical δ_{theor} values are expressed to the full precision of NCS output, whereas experimental δ_{exp} values are quoted to 0.1 ppm precision, approximating the systematic measurement uncertainty.

Table 7.1 Comparison of calculated (δ_{theor}) versus experimental (δ_{exp}) chemical shift values (ppm) for hydroxyl, methylene, and methyl protons of ethanol [see, e.g., I. Weinberg and J. R. Zimmerman, *J. Chem. Phys.* **23**, 748, 1955]. Conformationally averaged theoretical methyl-proton values are given for comparison with experiment.

^{1}H-type	δ_{theor}	δ_{exp}
OH	0.15	5.3
CH$_2$	3.84	3.6
CH$_3$	1.24	1.1

Compared to the experimental ^{1}H-NMR ethanol spectrum for the neat liquid as summarized in Table 7.1, the calculated shieldings for molecular ethanol raise additional interesting questions, such as follows:

(1) Why do the methyl protons H(6), H(7) show slightly *in*equivalent theoretical shieldings (30.60 versus 31.03 ppm)? On the experimental side, such differences are erased by CH_3 torsional averaging that is rapid on the NMR time scale, whereas the theoretical spectrum retains the asymmetries of each instantaneous spatial configuration. As suggested by symmetry, the direct $\sigma_{CH(6)}$, $\sigma_{CH(7)}$ shielding contributions (25.78 versus 25.67 ppm) are nearly identical, as are the combined shielding contributions of geminal NBOs. The combined shielding differences of the weaker vicinal contributions are also rather negligible. However, I/O-7.2 indicates a surprising -0.43 ppm *de*shielding of H(6) by the $n_O^{(\sigma)}$ lone pair, which accounts rather well for the overall shielding difference. Evidently, the difference in through-space proximity (despite the common through-bond connectivity) leads to a surprising stray-fringe difference at the two nuclei.

(2) Why does the hydroxyl proton H(4) lie so far *up*field (31.83 ppm, near the TMS resonance), compared to its usual position in the experimental ethanol spectrum? The answer is that the theoretical spectrum describes a free gas-phase ethanol molecule, whereas the experimental liquid spectrum reflects the ubiquitous influence of *hydrogen bonding* interactions. The dramatic effect of H-bonding might be anticipated from the large contributions (totaling ca. 11 ppm) of oxygen lone pair NBOs 12, 13, because these lone pairs participate as Lewis-base donors in the H-bonding donor–acceptor phenomenon (Chapter 9). To see the effect more directly, we may consider the simple model ethanol dimer as shown in I/O-7.3.

The model geometry has a single strong O_1-$H_3 \cdots O_2$ interaction that distinguishes the H-bonded OH(3) and "free" OH(4) hydroxyl protons, which resonate at 28.23 and 31.41 ppm, respectively. Thus, the pronounced downfield-shifting effect of H-bonding is demonstrated even in this simplest dimer model, and the shift is found to provide a very general and distinctive measure of H-bond strength in higher cluster species. NCS analysis gives a clear picture of oxygen lone pair involvement, but further discussion of the H-bonding phenomenon is postponed to Chapter 9.

The examples above only hint at the richness of chemical structural information that is available from NMR shielding values. Indeed, the usual liquid-state emphasis on isotropic shielding (7.1) masks the still more intimate bonding details that are available from the individual shielding tensor components $(\sigma)_{ij}$, as measured, for example, in solid-state NMR. Theoretical progress in first-principles calculations and NBO/NLMO analysis of shieldings for heavy-atom species (see, e.g., J. Autschbach, *J. Chem. Phys.* **127**, 124106, 2007, **128**, 164112, 2008) promises exciting prospects for NMR structural investigations in many areas of modern biophysical, inorganic, and materials research.

—————————————— I/O-7.3 ——————————————

```
#N B3LYP/6-311++G** POP=NBOREAD NMR

(EtOH)2 nmr properties

0 1
  H    -1.5428   -0.7480    1.0342
  H    -2.1534   -0.2532    3.3590
  H    -0.0711    0.4313   -0.3278
  H    -1.3513   -2.4242    3.7440
  H    -1.6775    1.1768   -0.3025
  H    -0.5082   -1.1905    4.7058
  H    -0.8644    0.9119   -2.6437
  H     1.1211   -2.6973    3.5558
  H    -2.2999   -0.0952   -2.3700
  H     1.2702   -1.0368    2.9523
  H    -0.6932   -0.8374   -2.3965
  H     0.4277   -2.2603    1.9837
  O    -1.6450   -0.8707    0.0771
  O    -1.3346   -0.5083    2.9221
  C    -1.1257    0.2714   -0.5947
  C    -0.6942   -1.5484    3.6864
  C    -1.2530    0.0506   -2.0923
  C     0.6102   -1.9060    3.0008

$nbo file=EtOH_2 ncs $end
```

7.2 NMR J-COUPLING ANALYSIS

At higher resolution, the chemically shifted proton resonance frequencies are seen to be split into multiplets by "spin–spin couplings." Because these couplings are of scalar character (carrying no angular dependence), they survive rotational averaging. The resulting "J-splittings" (measured in Hz) provide valuable structural clues to the number and location of other nuclear spins in the chemical bonding environment. Recent theoretical progress now allows the J-couplings to be calculated with reasonable accuracy by finite-field techniques (Sidebar 7.1). The accompanying development of NBO/NLMO-based "natural J-coupling" (NJC) analysis [S. Wilkens, W. M. Westler, J. L. Markley, and F. Weinhold, *J. Am. Chem. Soc.* **123**, 12026, 2001] provides powerful structural insights into the localized chemical origin of specific J values, complementing the structural information provided by chemical shielding. Thus, combined NCS and NJC analysis yields a remarkably detailed picture of major NMR spectral features, exploiting the complementary structural information that is provided by electronically distinct shielding and J-splitting mechanisms.

Although several distinct electronic effects can contribute to scalar J-coupling, the most important is that due to the "Fermi contact" (FC) mechanism. This involves the subtle manner in which remote nuclear spins A, B can communicate by means of exchange-type interactions between electrons having close contact ("collisions") with the two nuclei, thereby relaying nuclear spin information through the weak perturbations of spin pairing in the chemical bonding network. As expected, this

SIDEBAR 7.1 *FINITE-FIELD PERTURBATION THEORY OF FERMI-CONTACT INTERACTIONS*

Similar to polarizability (response to an external electric field, Sidebar 6.2), the FC electronic response to a nuclear spin may be calculated by finite-field techniques. In effect, a "point" (Dirac delta function) spin source of small magnitude (say, 0.02 a.u.) is positioned at the nucleus of interest, where it slightly spin polarizes the surrounding electronic orbitals (particularly of s-type), producing slight imbalance of $1s^\uparrow$ and $1s^\downarrow$ occupancy. Occupied orbitals having significant s-type contributions from both nuclei are thus effective in transmitting the FC coupling between nuclei, whether through direct NBO bonding ($^1J_{AB}$) or hyperconjugative NLMO delocalization tails (through-bond $^2J_{AB}$, $^3J_{AB}$, etc.). The $^nJ_{AB}$ values therefore provide a virtual blueprint of bonding and hyperconjugative delocalization pathways that link A, B.

The J-coupling can be evaluated from the small energy lowering and UHF-type orbital distortions that accompany response to the FC spin perturbation. The contribution of a given NBO or NLMO to J_{AB} coupling requires calculation of the spin polarization response to perturbations at each nucleus, and J-couplings among N nuclei therefore require N separate UHF-type finite-field calculations.

The *Gaussian* syntax for FC ("F") perturbation of magnitude M (in multiples of 0.0001 a.u.) at nucleus n is of the form

```
FIELD = F(n)M
```

For example,

```
FIELD = F(4)200
```

for a perturbation with coefficient 0.02 a.u. on nucleus 4. I/O-7.4 illustrates the form of *Gaussian* input file for such perturbations. As shown in the route card, the "SCF=(QC,VERYTIGHT)" keyword can be set to insure best possible wavefunction convergence for the challenging numerical differencing required by the finite-field technique.

Note that accurate evaluation of J-coupling constants requires highly flexible description of core orbitals, so the 6-311++G** basis set provides only a qualitative description for present illustrative purposes.

exchange-type coupling mechanism is particularly effective if A, B are directly bonded (one-bond $^1J_{AB}$ coupling), but alternative two-bond (geminal $^2J_{AB}$), three-bond (vicinal $^3J_{AB}$), or other through-n-bonds pathways may also provide measurable splittings. Most remarkably, such J-coupling has also been found to cross molecular boundaries along "through-H-bond" pathways (e.g., $^{1h}J_{AB}$, $^{2h}J_{AB}$, etc.). Through-H-bond J-couplings provide a quantitative measure of intermolecular electronic *sharing* that is the essence of H-bonding interactions (Chapter 9). The fascinating chemical information provided by intermolecular NJC analysis is beyond the scope of the present discussion. However, even the simple C_2H_5OH application to be described below suggests the rich information content of J-couplings pertaining to specific localized features of the chemical bonding and H-bonding environment.

NJC analysis requires a complex interactive partnership with a host ESS program that is capable of finite-field FC perturbation calculations, and is currently implemented only in NBO5.9-linked versions of the *Gaussian-09* program. Because separate FC perturbations are required for each coupled nucleus of interest, overall NJC analysis involves more *Gaussian*-specific job input detail than is required for other $NBO keylist options (cf. Sidebar 7.1). For further details of NJC algorithms and input syntax, see the *NBO Manual*, p. B-147ff.

As a simple illustration of J-coupling calculations and analysis, let us return to the ethanol example (I/O-7.1) to focus on proton spin–spin couplings that dominate the experimental ^{1}H-NMR spectrum. An input file to compute $^nJ_{HH'}$ couplings and perform NJC analysis for all distinct H, H′ pairs is shown in I/O-7.4. The overall job consists of six chained calculations (with "--Link1--" separators) for successive perturbations of nuclei H(4)–H(9). As shown in the example, z-matrix input can be replaced by "GUESS=READ GEOM=ALLCHECKPOINT" after the first step, and inclusion of $NBO keyword "PRINT=0" avoids repetitive printing of NBO output that differs infinitesimally from that of the first step.

```
──────────────────────── I/O-7.4 ────────────────────────
%chk=EtOH
#UB3LYP/6-311++G** SCF=(QC,VERYTIGHT)  POP=NBORead FIELD=F(4)200


HOCH2CH3

   0  1
   8    1.200834   -0.222396    0.000000
   6    0.000000    0.555671    0.000000
   6   -1.177713   -0.400631    0.000000
   1    1.957599    0.371124    0.000000
   1   -0.036935    1.201984    0.887583
   1   -1.151956   -1.040110    0.885543
   1   -2.120213    0.154055    0.000000
   1   -0.036935    1.201984   -0.887583
   1   -1.151956   -1.040110   -0.885543


$NBO FILE=EtOH_4 NJC $END

--Link1--
%chk=EtOH
#UB3LYP/6-311++g** SCF=(QC,VERYTIGHT)  POP=NBORead FIELD=F(5)200
                  GUESS=READ GEOM=ALLCHECKPOINT


$NBO FILE=EtOH_5 NJC PRINT=0 $END

--Link1--
   :
--Link1--
%chk=EtOH
#UB3LYP/6-311++g** SCF=(QC,VERYTIGHT)  POP=NBORead FIELD=F(9)200
                  GUESS=READ GEOM=ALLCHECKPOINT


$NBO FILE=EtOH_9 NJC PRINT=0 $END
```

Table 7.2 Calculated $^nJ_{HH'}$ scalar couplings (Hz) for protons H(4)-H(9) of ethanol (with parenthesized n of through-n-bond coupling pathway).

	$^nJ_{HH'}$ Spin–Spin Coupling Constants (Hz)				
	H(4)	H(5)	H(6)	H(7)	H(8)
H (5)	0.41(3)				
H (6)	<0.1(4)	4.64(3)			
H (7)	1.78(4)	1.78(3)	−14.51(2)		
H (8)	0.40(3)	−8.24(2)	12.83(3)	1.78(3)	
H (9)	−0.12(4)	12.83(3)	−13.59(2)	−14.51(2)	4.64(3)

The output for the first nucleus contains no NJC output, but each subsequent jobstep contains the J-couplings to all nuclei that were perturbed in previous steps. Table 7.2 summarizes the calculated J values (and through-n-bond connectivities) for protons H(4)–H(9) of ethanol. Note that the theoretical J values include algebraic signs that are difficult to determine experimentally, because the measured splittings depend only on the *magnitude* of scalar coupling.

The values shown in Table 7.2 cannot be directly compared with experimental liquid values, because the latter involve Boltzmann-weighted averaging over torsional motions that are rapid on the NMR time scale. However, for the coupling J[CH₂,CH₃] between methylene and methyl protons one can make an "eyeball estimate" by simply averaging the six vicinal couplings between methylene protons H(5), H(8) and methyl protons H(6), H(7), H(9), leading to the following

$$J_{theor}[CH_2, CH_3] = 6.40 \, Hz \qquad (7.3)$$

which agrees reasonably with the experimentally inferred coupling

$$J_{exp}[CH_2, CH_3] \cong 7 \, Hz \qquad (7.4)$$

{In neat or dilute aqueous solutions the vicinal coupling between hydroxyl and methylene protons is also measurable (J[OH,CH₂] ≅ 5 Hz), but additional H−O−C−H torsional conformers would be needed to obtain the appropriate Boltzmann-weighted theoretical estimate.}

What is the chemical origin of the J-coupling patterns shown in Table 7.2? NJC output for the representative case of $^3J_{H(5)H(9)}$ coupling (12.83 Hz) between vicinal antiperiplanar protons H(5), H(9) is shown in I/O-7.5.

As shown near the top of the printout, the J[H5,H9] coupling of 12.83 Hz is identified as a vicinal $^3J[H(5)−C(2)−C(3)−H(9)]$ (through-three-bond) pathway. Each column details the contributions from a given occupied NBO/NLMO (if above the print threshold of 0.1 Hz), showing the NBO Lewis and repolarization contributions (top), the leading delocalization corrections (middle), and the total NLMO contribution (bottom), all summed to give the values shown in the final column, and

─────────── I/O-7.5 ───────────

```
NATURAL J-COUPLING ANALYSIS
Writing disk file EtOH_9.NJC
  :
  :
J[H5,H9] = 12.83 Hz:  3-J[H5-C2-C3-H9]
    Threshold for printing:   0.10 Hz
                         2.        4.        5.        6.        7.
                      | O 1- H 4   C 2- H 5   C 2- H 8   C 3- H 6   C 3- H 7
         ----------------------------------------------------------------------
            Lewis |    -0.02      4.32      -0.00      0.46       0.03
            Repol.|    -0.00     -0.04       0.00     -0.00       0.00
         ----------------------------------------------------------------------
    14. O 1(ry*) |                -0.17 v    0.02 v    -0.01 r    0.01 r
        C 2(ry*) |     0.00 v                                    -0.04 v   -0.06 v
        C 3(ry*) |    -0.00 r    -0.67 v    -0.06 v
    57. H 5(ry*) |     0.00 r                -0.05 v    -0.05 r   -0.04 r
    61. H 9(ry*) |     0.01 r    -0.37 r    -0.08 r    -0.09 v   -0.04 v
    65. C 2- H 5*|    -0.01 v                0.16 g     0.22 v    0.17 v
    69. C 3- H 9*|    -0.02 r     2.80 v     0.28 v     0.08 g    0.08 g
         Others|       0.01       0.12      -0.00      -0.02     -0.01
  (Total deloc.)|     (-0.00)     (1.72)     (0.26)     (0.09)    (0.11)
         ----------------------------------------------------------------------
            NLMO |    -0.02       5.99       0.26       0.55      0.14

                         8.        12.       13.
                      | C 3- H 9   O 1(lp)   O 1(lp)   Others |    Total
         ----------------------------------------------------------------------
            Lewis |     4.38       0.02      -0.01     -0.05 |     9.12
            Repol.|     0.18      -0.00      -0.00      0.02 |     0.16
         ----------------------------------------------------------------------
    14. O 1(ry*) |     0.01 r                          -0.02 |    -0.16
        C 2(ry*) |    -0.95 v     0.08 v    -0.01 v     0.00 |    -0.97
        C 3(ry*) |               -0.01 r     0.04 r    -0.01 |    -0.73
    57. H 5(ry*) |    -0.35 r     0.01 r    -0.02 r    -0.06 |    -0.55
    61. H 9(ry*) |                0.04 r    -0.06 r    -0.08 |    -0.66
    65. C 2- H 5*|     2.43 v    -0.02 v     0.05 v     0.16 |     3.16
    69. C 3- H 9*|               -0.10 r     0.11 r     0.18 |     3.41
         Others|       0.03      -0.06      -0.03       0.01 |     0.05
  (Total deloc.)|     (1.17)     (-0.06)     (0.08)     (0.18)|    (3.55)
         ----------------------------------------------------------------------
            NLMO |     5.73      -0.04       0.07       0.15 |    12.83
```

totaled to give the final $^nJ_{HH'}$ value at the lower right. Remaining L/NL contributions are simply grouped as "others," but can be displayed in greater or lesser detail by resetting the print threshold.

[The "repolarization" correction refers to the manner in which the parent NBO can respond to the FC perturbation by slightly changing its shape or internal spin polarization, using available orbitals from the *same* parent atom(s). Such "NBO reshaping" correction is usually rather negligible, as in the present case.]

As shown in I/O-7.5, H(5)−H(9) J-coupling originates predominantly from NLMOs 4 and 8, comprising four major contributions: the L-type contributions of "parent" hydride bond NBO 4 (4.32 Hz) and NBO 8 (4.38 Hz), and their mutual NL-type delocalizations into antibond NBO 69 (2.80 Hz) and NBO 65 (2.43 Hz), respectively. Each of the two contributing bond NBOs therefore donates into the

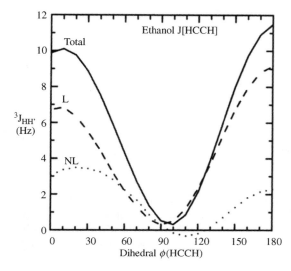

Figure 7.3 Calculated "Karplus curve" for dihedral variations of ^{3}J[HCCH] spin–spin coupling in ethanol (idealized rigid-rotor methyl torsions), showing total $^3J_{HH'}$ (solid) and its Lewis (dashed) and non-Lewis (dotted) contributions (Hz) at each φ(HCCH) dihedral angle.

antibond of the other, leading to strong exchange-type "cross-talk" and J-coupling between the nuclei in each associated NLMO. Both steric-exchange (L-type) and hyperconjugative (NL-type) contributions are expected to exhibit interesting angular dependence that provides additional information concerning stereoelectronic relationships between the coupled nuclei. Although a few other entries of the NJC table warrant secondary attention, the four discussed above provide a particularly simple and satisfying rationale for the dominant electronic origins of strong vicinal antiperiplanar $^3J_{H(5)H(9)}$ coupling.

To exhibit the angular dependence of vicinal $^3J_{HH'}$ couplings, let us consider the methyl torsions in an idealized rigid-rotor model of ethanol. Figure 7.3 displays the vicinal H(5)–H(9) coupling (solid line) and its L-type (dashed line) and NL-type (dotted line) components as a function of dihedral angle φ for 0–180° rotation of the methyl group. As shown in the figure, the J-coupling in *anti* conformation (180°) is significantly stronger than in *gauche* conformation (60°). The plotted dihedral dependence of $^3J_{HH'}$ (often called the "Karplus curve") allows one to directly "read" vicinal angular geometry from measured spin–spin splittings, thus providing one of the principal structural tools of the NMR spectroscopist.

As shown in Fig. 7.3, both L-type and NL-type features of the wavefunction contribute significantly to total $^3J_{HH'}$ (as for other properties), but the "sampling" of these features by the FC spin–spin coupling is distinctive. Initially, it may seem surprising that L-type contributions play any role in $^3J_{HH'}$ coupling, for example, that nucleus H(5) would have direct communication with the C(3)–H(9) NBO on the adjacent carbon. However, such L-type contributions are testimony to the high sensitivity of J-coupling to weak fringes or tails of the NBOs or NLMOs centered on other nuclei. Figure 7.4 displays orbital profile plots of the C(3)–H(9) bond (NLMO 8) near the H(5) nucleus, for *gauche* (60°, left) and *anti* (180°, right) dihedral

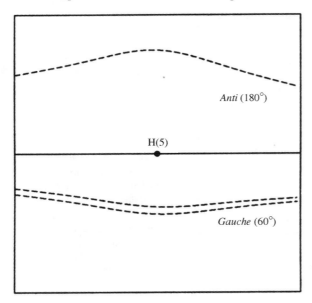

Figure 7.4 NLMO amplitude profiles for long-range tails of vicinal σ_{CH} bonds oriented *anti* or *gauche* to the proton H(5) of interest in ethanol (cf. Fig. 7.3), showing the significantly greater *anti* FC amplitude at the H(5) nucleus. [Profile axis is along the C(2)–H(5) bond, with horizontal range ± 0.2 Å and vertical range ± 0.03 a.u.]

orientation. These plots exhibit the pronounced angular variations of fringe-amplitude at H(5) as "seen" by the FC interaction, showing the dominance of *anti* over *gauche* coupling. Both the parent bond (NBO 8) and the hyperconjugatively coupled antibond tail (NBO 65) contribute to the amplitude of NLMO 8 at H(5). [Of course, antibond NBO 65 has intrinsic large amplitude at H(5), but its contribution as a weak hyperconjugative tail of NLMO 8 is modulated by the angular dependence of hyperconjugative delocalization, giving rise to the angular dependence (dotted line) in Fig. 7.3.]

Although the present discussion has focused primarily on the angular dependence of vicinal $^3J_{HH'}$ coupling, valuable structural information is also available from the distance dependence of direct-bonded $^1J_{AB}$ coupling, as well as other through-bond and through-H-bond coupling pathways. Given the ongoing theoretical progress in calculating NMR spin-coupling properties, NJC analysis promises improved understanding of many such properties and their relationship to the chemical bonding environment.

7.3 ESR SPIN DENSITY DISTRIBUTION

Electron spin resonance [also called electron paramagnetic resonance (EPR)] exhibits both parallels and contrasts with NMR. At the conceptual level, the basic quantum mechanical equation for magnetic resonance is identical for electron and nuclear spins, but the larger magnetic moment of electron spin boosts the ESR resonance frequency into the microwave region. In other respects, the concepts employed to analyze the two spectroscopies are analogous, with the electronic "g-tensor" serving as the ESR analog of the NMR chemical shielding tensor, and nuclear spin-electron spin "hyperfine coupling" as the ESR analog of nuclear spin–spin J-coupling.

However, the nuclear spins are envisioned as point-like spectators of the surrounding electron clouds, with each nucleus signaling its unique chemical bonding environment through a pronounced resonance frequency shift, whereas the electron spins are spatially dispersed over multiple nuclei and exhibit only subtle shifts in resonance frequency. The ESR signals reflect the much more active participation of electron spins in the chemical interactions under study.

The differences between ESR and NMR spectroscopy are most apparent in their practical chemical applications. NMR is conventionally applied to diamagnetic species in their stable liquid form under ordinary laboratory conditions, whereas ESR is applicable only to paramagnetic radical species, often so highly reactive as to require trapping in cryogenic matrices to preserve signal intensity. Radical trapping in turn involves immobilization and undesirable anisotopic broadening of ESR spectral lines, further limiting the resolution and structural information that can be obtained. Superficially, ESR spectra also reflect the experimental convenience of varying magnetic field strength (rather than frequency) to achieve the resonance condition, so that the horizontal axis is measured in field strength (Gauss or Tesla) rather than ppm. In addition, ESR spectroscopists typically prefer to plot the ESR absorption peak *derivative* (rather than the absorption peak itself) along the vertical axis. The ESR spectrum therefore has a distinctive "look and feel" that typically requires careful analysis to extract desired structural information. Nevertheless, ESR spectroscopy provides useful insights into stable or reactive paramagnetic species that are typically outside the purview of conventional NMR techniques. (See, however, experimental NMR studies and comparison theoretical analysis of stable paramagnetic iron–sulfur protein species: S. J. Wilkens, B. Xia, F. Weinhold, J. L. Markley, and W. Westler, *J. Am. Chem. Soc.* **120**, 4806, 1998.)

Because electrons have near-identical g-values, differing only by weak effects of spin-orbit coupling that are difficult to resolve experimentally, the ESR spectral descriptors of principal interest are the isotropic hyperfine coupling parameters a_H to nearby protons, which lead to the characteristic ESR splitting patterns of common organic radicals. The a_H hyperfine coupling parameters [generally expressed in field strength Gauss (G) units] arise primarily from FC-type interactions, analogous to those discussed in Section 7.2. However, theoretical analysis of hyperfine splittings is often confined to empirical fitting of model parameters of a phenomenological spin Hamiltonian, employing, for example, a "McConnell equation"

$$a_H \cong Q_R \rho_R \tag{7.5}$$

where ρ_R denotes spin density at the radical center and Q_R is an empirical proportionality factor (e.g., $Q_C \cong -21$ G for π-type carbon centers).

Given the limitations of the experimental ESR spectrum and theoretical concepts underlying its interpretation, the primary objective of *ab initio* investigation is usually to provide details of the electronic spin-density distribution. At the crudest level, one may merely wish to characterize the radical as "σ-type" or "π-type," and for this purpose the NPA orbital spin density assignments (cf. I/O-3.5) are useful. For the distribution of spin density among atomic centers, the NPA summary table (cf. I/O-3.10) can be employed.

As a simple illustration, let us examine the spin density distribution in Tempone (2,2,6,6-tetramethylpiperidone-*N*-oxyl)

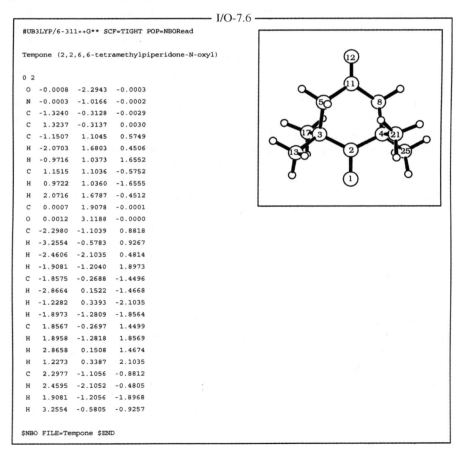

a popular nitroxide "spin label" reagent, which optimizes to the twisted (near-C_2) structure shown in I/O-7.6.

Although Tempone is often depicted as having the unpaired electron localized on the nitroxide oxygen atom, the calculated natural spin density (NSD) distribution presents a more complex picture, as summarized in Table 7.3.

──────────────────── I/O-7.6 ────────────────────

```
#UB3LYP/6-311++G** SCF=TIGHT POP=NBORead

Tempone (2,2,6,6-tetramethylpiperidone-N-oxyl)

0 2
  O  -0.0008  -2.2943  -0.0003
  N  -0.0003  -1.0166  -0.0002
  C  -1.3240  -0.3128  -0.0029
  C   1.3237  -0.3137   0.0030
  C  -1.1507   1.1045   0.5749
  H  -2.0703   1.6803   0.4506
  H  -0.9716   1.0373   1.6552
  C   1.1515   1.1036  -0.5752
  H   0.9722   1.0360  -1.6555
  H   2.0716   1.6787  -0.4512
  C   0.0007   1.9078  -0.0001
  O   0.0012   3.1188  -0.0000
  C  -2.2980  -1.1039   0.8818
  H  -3.2554  -0.5783   0.9267
  H  -2.4606  -2.1035   0.4814
  H  -1.9081  -1.2040   1.8973
  C  -1.8575  -0.2688  -1.4496
  H  -2.8664   0.1522  -1.4668
  H  -1.2282   0.3393  -2.1035
  H  -1.8973  -1.2809  -1.8564
  C   1.8567  -0.2697   1.4499
  H   1.8958  -1.2818   1.8569
  H   2.8658   0.1508   1.4674
  H   1.2273   0.3387   2.1035
  C   2.2977  -1.1056  -0.8812
  H   2.4595  -2.1052  -0.4805
  H   1.9081  -1.2056  -1.8968
  H   3.2554  -0.5805  -0.9257

$NBO FILE=Tempone $END
```

Table 7.3 Calculated natural spin density (NSD) distribution for tempone spin label, showing total atomic spin density at each atomic center (cf. I/O-7.6).

Atom	NSD (e)	Atom	NSD (e)
O1	0.5112	H15^c	−0.0005
N2	0.4652	H16^c	−0.0002
C3	−0.0251	C17	0.0241
C4	−0.0251	H18^d	0.0025
C5	0.0029	H19^d	−0.0008
H6^a	0.0006	H20^d	−0.0010
H7^a	−0.0002	C21	0.0242
C8	0.0029	H22^e	−0.0010
H9^b	−0.0002	H23^e	0.0025
H10^b	0.0006	H24^e	−0.0008
C11	−0.0002	C25	0.0105
O12	0.0002	H26^f	−0.0005
C13	0.0105	H27^f	−0.0002
H14^c	−0.0010	H28^f	−0.0011

^{a}C(5)-bonded.
^{b}C(8)-bonded.
^{c}C(13)-bonded.
^{d}C(17)-bonded.
^{e}C(21)-bonded.
^{f}C(25)-bonded.

As shown in the table, the nitroxide oxygen O(1) is indeed the principal spin density site, but only by a small margin over N(2), which shares nearly equally in carrying unpaired spin. Although these two atoms account for the vast majority (~98%) of the net spin density, interesting smaller contributions are found at other atoms. These include significant *negative* NSD contributions at C(3) and C(4) (totaling about 5%) that are offset by other significant positive contributions at C(13), C(17), C(21), and C(25) (totaling about 7%). At a still finer level of detail, small remnants of radical character are scattered over *all* nuclei (particularly, on methyl protons nearer the nitroxide group), with roughly equal numbers of positive and negative spin polarizations.

How can we understand the chemical origins of these spin density patterns? The starting point is the "different Lewis structures for different spins" NBO description of open-shell systems (cf. Section 4.5), which leads to the two distinct spin NBO representations of the Tempone nitroxide bonding pattern, as shown in (7.6α,β):

$$\overset{\displaystyle\diagdown}{\underset{\displaystyle\diagup}{\bar{N}}}-\underline{\bar{O}}\text{ı} \qquad \overset{\displaystyle\diagdown}{\underset{\displaystyle\diagup}{N}}=\underline{O}\text{ı} \tag{7.6}$$

$$\alpha \qquad\qquad \beta$$

Because all other Lewis structural features coincide in the two spin sets, we can crudely estimate spin density at N and O as the difference between the nonbonding spin NBOs n_N, n_O of the α structure (7.6α) versus the prorated percentages (47.91% O, 52.09% N, according to the respective NBO coefficients) of the π_{NO} spin NBO in the β structure (7.6β), namely,

$$NSD(O) \cong 1 - 0.4791(1) \cong 0.52e \qquad (7.7)$$

$$NSD(N) \cong 1 - 0.5209(1) \cong 0.48e \qquad (7.8)$$

These estimates are in good agreement with NSD entries of Table 7.3, if the latter are suitably "renormalized" as percentages of the actual N, O total $(0.5112 + 0.4652 = 0.9764)$, namely,

$$\%\text{-NSD(O)} \cong (100)0.5112/0.9764 \cong 52\% \qquad (7.9)$$

$$\%\text{-NSD(N)} \cong (100)0.4652/0.9764 \cong 48\% \qquad (7.10)$$

Thus, dominant NSD contributions within the NO moiety can be rather simply understood from the NBO spin Lewis structures (7.6α,β) and the β–π_{NO} bond ionicity.

The weaker secondary NSD contributions of the nitroxide environment can also be qualitatively understood from leading (vicinal) donor–acceptor delocalizations of unique spin NBOs of structures (7.6α,β). Because the local nitroxide frame is essentially planar, the relevant spin NBOs (n_N, n_O in α; π_{NO} in β) are all of local p-π type, perpendicular to the C(3)C(4)N(2)O(1) plane. The leading hyperconjugating candidates are therefore the vicinal C–C bonds of methyl substituents, distinguishable as being oriented strongly [C(3)–C(17) or C(4)–C(21): 82°] versus weakly [C(3)–C(13) or C(4)–C(21): 38°] out-of-plane with respect to the nitroxide moiety.

To a first approximation, we may ignore the α-n_O donor (because it lacks out-of-plane acceptors at the vicinal position) and all donor–acceptor interactions that involve common features of the two spin sets (because they are largely cancelling). The most *unique* donor–acceptor interactions then involve the β–π^*_{NO} acceptor, particularly $\sigma_{C(3)-C(17)}$–π^*_{NO}, $\sigma_{C(4)-C(21)}$–π^*_{NO} ("strong," 1.5 kcal/mol each) and $\sigma_{C(3)-C(13)}$–π^*_{NO}, $\sigma_{C(4)-C(25)}$–π^*_{NO} ("weak," 0.5 kcal/mol each). Because these delocalizations all remove β-spin from the surroundings, they contribute *negative* spin density at surrounding sites, particularly the adjacent C(3), C(4) atoms that have 2:1 statistical preponderance in the four delocalizations. Additional negative spin density is contributed by delocalizations π_{NO}–$\sigma^*_{C(3)-C(17)}$, π_{NO}–$\sigma^*_{C(4)-C(21)}$ (1.3 kcal/mol) and π_{NO}–$\sigma^*_{C(3)-C(13)}$, π_{NO}–$\sigma^*_{C(4)-C(25)}$ (0.3 kcal/mol) of the β-spin set, but these are countered by corresponding positive NSD from delocalizations n_N–$\sigma^*_{C(3)-C(17)}$, n_N–$\sigma^*_{C(4)-C(21)}$ (3.2 kcal/mol) and n_N–$\sigma^*_{C(3)-C(13)}$, n_N–$\sigma^*_{C(4)-C(25)}$ (1.2 kcal/mol) of the α-spin set. The net result of this confusing give and take is that adjacent carbon atoms C(3), C(4) are left with slight negative NSD (ca. $-0.025e$), whereas next-nearest C(17), C(21) ("out-of-plane," ca. 0.024e) and C(13), C(25) ("in-plane," ca. 0.011e) gain slight positive NSD, creating a spin-polarization "wave" radiating outward from the nitroxide radical center. Still weaker nitroxide interactions with the nonvicinal environment (particularly, nearby methyl CH bonds) contribute to the still smaller NSD values found elsewhere in this species. Thus, the subtle spin polarizations detectable by ESR seem to be largely accountable in terms of familiar NBO donor–acceptor patterns that were also found to be prominent in nonradical species.

PROBLEMS AND EXERCISES

7.1.   Consider the hydrocarbon species methane (CH_4, sp^3), ethylene (H_2CCH_2, sp^2), and acetylene (HCCH, sp^1) as prototype examples of sp^n bonding, using idealized Pople–Gordon comparison geometry as shown below:

```
#B3LYP/6-311++G**

methane

0 1
 C
 H  1  1.09
 H  1  1.09  2  109.47
 H  1  1.09  2  109.47  3  120.
 H  1  1.09  2  109.47  3  240.

--Link1--
ethylene

0 1
 C
 C  1  1.34
 H  1  1.08  2  120.
 H  1  1.08  2  120.  3  180.
 H  2  1.08  1  120.  3    0.
 H  2  1.08  1  120.  3  180.

--Link1--
acetylene

0 1
 C  0.  0.  -0.60
 C  0.  0.  +0.60
 H  0.  0.  -1.66
 H  0.  0.  +1.66
```

(a) Which species is expected to exhibit the most downfield-shifted ^{1}H-NMR resonance? Which the most upfield-shifted? Explain your reasoning briefly.

(b) Calculate the ^{1}H-NMR shifts (relative to TMS, 31.98 ppm) for idealized methane, ethylene, and acetylene. Do these shifts vary with hybridization and/or CH bond ionicity (cf. Fig. 7.1) in the expected way? Discuss briefly.

(c) Do the NCS analysis results show evidence for any NL (resonance-type) effects in these species, beyond the inductive effect of bond hybridization and polarization? Comment on any features of the comparative NCS analyses that indicate interesting differences between these species, as signaled by their ^{1}H-NMR shifts.

7.2.   For an idealized ethanol model with variable torsional angle PHI for hydroxyl proton H(4), namely,

```
#B3LYP/6-311++G**

ethanol torsional averaging model

0 1
C
C  1  1.54
O  2  1.43   1  109.47
H  3  0.96   2  109.47   1   PHI
H  2  1.09   1  109.47   3   60.
H  2  1.09   1  109.47   3  -60.
H  1  1.09   2  109.47   3  180.
H  1  1.09   2  109.47   3   60.
H  1  1.09   2  109.47   3  -60.

PHI  180.
```

evaluate the H(4) chemical shift and H(4)−H(8) spin–spin coupling for PHI = 180°, 60°, and −60°. From these results, estimate the torsionally averaged value for each quantity. Compare your calculated $^3J_{H(4),H(8)}$ estimate with the observed value (ca. 5 Hz) in sufficiently pure liquid ethanol. (Why is this J-coupling normally unobservable in aqueous ethanol solutions, particularly if traces of acid are present?)

7.3.   Consider the effect of substituting S for O in a model thioethanol species [for example, by substituting S for O (with $R = 1.81$ Å, $R_{SH} = 1.34$ Å) in the previous problem]. From the electronegativity difference of S versus H (Table 4.2), can you predict, at least qualitatively, the magnitude of SH/OH proton shift in thioethanol versus ethanol? (Hint: see Fig. 7.1.) Similarly, can you predict the direction in which methylene protons might be shifted in CH_3CH_2SH versus CH_3CH_2OH? Calculate the NMR shieldings for thioethanol to test your predictions, and comment on observed NCS differences compared to ethanol.

7.4.   Consider the various fluoroethylene isomeric species $C_2H_nF_{4-n}$ derived by replacing one or more of the H atoms at positions H(4), H(5), or H(6) by F atoms [but leaving at least one proton H(3) fixed in all isomers], as illustrated below for the *cis* $C_2H_2F_2$ isomer:

```
#B3LYP/6-311++G** SCF=TIGHT PO:

C2HnFm fluoroethylene isomers

0 1
C
C  1  1.34
H  1  1.08   2  120.
F  1  RA     2  120.   3  180.
H  2  RB     1  120.   3    0.
F  2  RC     1  120.   3  180.

RA  1.33
RB  1.08
RC  1.33

$NBO FILE=C2H2F2_c NCS $END
```

By making other H/F replacements at positions 4–6, and altering the values of bondlengths RA, RB, RC for CF (1.33 Å) or CH (1.08 Å) accordingly, you can easily generate idealized geometries for all eight $C_2H_nF_{4-n}$ isomers with $n = 1$–4.

(a) How many distinct ^{1}H-NMR chemical shifts do you expect to find in each of the eight $C_2H_nF_{4-n}$ species?

(b) Suppose your NMR sample consisted of a mixture of all eight species, with unknown concentrations of each. Which species (and proton) would you expect to give rise to the most downfield-shifted resonance in the mixture spectrum? Which to the most upfield-shifted resonance? Explain briefly.

(c) To check your answers in Problems (a) and (b), evaluate the proton chemical shifts (relative to TMS, 31.98 ppm) for each isomer, and use the results of NCS analysis to rationalize the proton shifts in each species, including their relationship to shifts in other species. Can you find evidence for dependence of these shifts on electronegativity and ionicity differences, hybridization changes, or stereoelectronic hyperconjugative effects?

(d) From your results in Problem (c) (together with integrated values of spectral peaks), describe a procedure by which you might estimate the unknown concentrations of each isomeric species in the mixture (b), assuming sufficient experimental resolution and no J-coupling complications.

(e) Which of these species (if any) are expected to exhibit measurable $J_{HH'}$ couplings? (Recall that only *in*equivalent protons can lead to experimental splittings.) Calculate the expected $J_{HH'}$ couplings for all such H–H′ pairs, as an aid to identifying NMR resonances of the mixture. Comment on the J-couplings that are most diagnostic of specific bonding relationships between the protons.

7.5. For the three possible difluoroethylene isomers of the previous problem, consider the corresponding $C_2H_2F_2^{+\bullet}$ radical cations produced by vertical (fixed geometry) photo-ionization (i.e., just change the charge/multiplicity entries to "+1 2" in the input file). Where is the radical character located in each species? Describe the details of the localized spin density distribution of each radical, and determine which (if either) of the terms "σ radical" or "π radical" is appropriate in each case.

Chapter 8

Coordination and Hyperbonding

Students sometimes assume (mistakenly) that "chemical bonding" is completed once the electrons are maximally paired up in a closed-shell species of valid Lewis structural form. The error of this assumption was recognized nearly a century ago with discovery of numerous "complexes" that defied Lewis structural formulation, unless written as two (or more) distinct species. Such complexes therefore appear to *violate* the valence rules that usually govern chemical structure and reactivity, apparently involving some type of "extra-valence" (*Nebenvalenz*, in the phrase of German inorganic chemist Alfred Werner) that demands significant extension of Lewis structural concepts. Nowadays, the term "hypervalency" is commonly used to describe species that have "too many bonds" for conventional Lewis structural depiction, or seem to require chemical "association" mechanisms beyond those of closed-shell Lewis structure formation.

A simple and provocative example of such strange association complexes is provided by the bifluoride ion (FHF^-). This species can be formulated perfectly well as the Lewis-compliant HF molecule and F^- fluoride anion,

$$H - \ddot{\underset{\cdot\cdot}{F}}: \quad + \quad :\ddot{\underset{\cdot\cdot}{F}}:^- \qquad (8.1)$$

but not in the bonding diagram that best represents its structural and chemical properties, namely,

$$:\ddot{\underset{\cdot\cdot}{F}} - H - \ddot{\underset{\cdot\cdot}{F}}:^- \qquad (8.2)$$

which has "too many bonds" to hydrogen. G. N. Lewis himself recognized this species as the most challenging exception to his Lewis structural theory of chemical bonding, and speculated on the nature of the hydrogenic "bivalency" responsible for such exceptional "H-bonding" propensity.

Discovering Chemistry With Natural Bond Orbitals, First Edition. Frank Weinhold and Clark R. Landis.
© 2012 John Wiley & Sons, Inc. Published 2012 by John Wiley & Sons, Inc.

Of course, it is initially tempting to characterize FHF^- as some type of "ion–dipole complex" of classical electrostatic origin. However, numerous lines of chemical evidence indicate the superficiality and inaccuracy of such description. These include the following:

- The chemically robust binding energy (>40 kcal/mol, actually *stronger* than that of F_2 itself)

- The *symmetrical* structure (defying any possible distinction between the supposed "ion" and "molecule" ends of the species)

- The distinctive vibrational, NMR, and other spectroscopic signatures (quite unlike those of HF in apparently analogous ion–dipole complexes such as $H-F \cdots Na^+$)

Although a virtual *continuum* of H-bonding strengths and structural parameters can now be identified—ranging from strong, symmetric species such as $F \cdots H \cdots F^-$ or $H_2O \cdots H \cdots OH_2^+$ down to weak, asymmetric species such as $H_2O \cdots HCH_3$—only the weakest and most uninteresting forms of H-bonding seem to manifest appreciable classical electrostatic character.

Still other forms of Lewis-defiant association (not involving H-bonding) were recognized in the ubiquitous "Werner complexes" of transition metal chemistry. Related valency puzzles arose in "adducts" of boron trifluoride and other diamagnetic Group 13 halides with ammonia and other Lewis-compliant species. The latter were clarified by Lewis himself in his famous generalization of acid–base theory, which recognizes BF_3 as a prototype "Lewis acid" and $:NH_3$ as a prototype "Lewis base" that combine in coordinative ("dative") B:N bond formation, leading to a Lewis acid–base adduct that is formally compliant with Lewis structural concepts, namely,

$$BF_3 + :NH_3 \rightarrow \begin{array}{c} \quad F \quad H \\ \quad | \quad | \\ F-B-N-H \\ \quad | \quad | \\ \quad F \quad H \end{array} \tag{8.3}$$

Even though the B:N shared electron pair is formally "donated" from the closed-shell Lewis base $:NH_3$ (rather than, e.g., from two open-shell doublet radical precursors, as in conventional covalent bonding), the chemical stabilization conferred by Lewis-compliant electron pair sharing warrants recognition as a distinctive type of "coordinate covalent" bond. Such coordinative bonding readily rationalizes the distinctive chemical properties of Werner complexes and many other inorganic species. Today, every beginning chemistry student is taught about covalent versus coordinate covalent bond "types" and the additional opportunities provided by suitable Lewis acid species (whether diamagnetic or paramagnetic) for coordinative bonding beyond the standard Lewis structural level.

In this chapter, we wish to explore how the NBO program detects and characterizes such distinctive bond types. This includes how the fundamental Lewis acid–base interactions are manifested in F_3BNH_3 and other main-group species (Section 8.1) as well as in open-shell transition metal species (Section 8.2). We then examine the more profound challenges to Lewis theory presented by three-center, four-electron "hyperbonding" (Section 8.3), including such puzzling species as SF_4 or $SO_4{}^{2-}$. For many years, students were taught to explain main-group hypervalency in terms of "d-orbital participation" and the "electroneutrality principle," but there is now ample *ab initio* evidence that such rationalizations are invalid.

In a sense, the various apparent Lewis structural exceptions considered in this chapter all serve to blur the boundaries between molecular and supramolecular domains. The topics of this chapter therefore infringe on the "intermolecular interactions" of Chapter 9, where H-bonded species will be considered. These exceptions also force further recognition of intra- and intermolecular *fractional* bonding that is the essential feature of the resonance extension of primitive Lewis structure concepts (Chapter 5).

8.1 LEWIS ACID–BASE COMPLEXES

As freshman chemistry students learn, a *Lewis acid* is an electron pair acceptor and a *Lewis base* an electron pair donor. From an orbital-based perspective, this definition focuses direct attention on one-center donor (LP) and acceptor (LP*) NBOs of the valence shell. Most important are the valence-shell vacancies ("LP*" NBOs) that characterize strong Lewis acids. Whereas filled valence LP-type NBOs are ubiquitous features of neutral pnictogen, chalcogen, and halogen compounds, as well as practically all anions, the corresponding unfilled LP*-type NBOs typically occur only in neutral ("hypovalent") or ionic compounds of Groups 11–13, or in open-shell transition metal species. Thus, the exploration of Lewis acid–base interactions leads us into the domain of inorganic and organometallic species, particularly metals with LP*-type "valence holes" that signal unusually strong electron pair acceptor properties.

Important general questions are raised by Lewis acid–base concepts (see W.B. Jensen, *The Lewis Acid–Base Concepts: An Overview* Wiley-Interscience, New York, 1980), such as:

- What are the characteristic electronic signatures of Lewis acid–base interactions?

- How can we distinguish the anomalous "coordinative" bonds from ordinary "covalent" bonds in a general Lewis acid–base adduct?

- How are covalent versus coordinative bonding propensities manifested in characteristic differences between "organic" and "inorganic" chemical species?

In this section, we first address such questions in the framework of main-group chemistry, focusing on simple prototype species (Al_2Cl_6, BH_2NH_2) that exhibit interesting *intra*molecular aspects of the covalent-coordinate dichotomy (Sections 8.1.1 and 8.1.2). We then briefly describe the analogous *inter*molecular aspects of coordinative bonding in the classic $BF_3:NH_3$ adduct (Section 8.1.3).

8.1.1   Coordinative σ-Bonding in Dimers of AlCl$_3$

Let us first consider the simple example of aluminum chloride (AlCl$_3$), a prototype Lewis acid. The leading symmetry-unique valence NBOs of this trigonal species ($R_{AlCl} = 2.0835$ Å) are shown in abridged form in I/O-8.1. These include the σ_{AlCl} (BD-type) polar covalent bond between Al(1) and Cl(2) (NBO 1), the three n_{Cl} (LP-type) lone pairs on Cl(2) (NBOs 24–26), and the n^*_{Al} (LP*-type) vacancy on Al(1) (NBO 33), as illustrated in Fig. 8.1.

As shown in Fig. 8.1a, the "normal" σ_{AlCl} NBO exhibits the expected high polarity, with calculated ionicity ($i_{AlCl} = 0.6238$) close to that expected from Ξ_{Al}, Ξ_{Cl} electronegativity values (Table 4.2)]. The component NHOs of σ_{AlCl}, $h_{Al} \cong sp^{2.0}$ and $h_{Cl} \cong sp^{3.2}$, are also rather unexceptional. The three chlorine lone pairs have the expected distinct forms (Sidebar 4.2) that allow recognition in other complexes, namely, $n_{Cl}^{(\sigma)}$ (Fig. 8.1b), $n_{Cl}^{(y)}$ (Fig. 8.1c), and $n_{Cl}^{(\pi)}$ (Fig. 8.1d). The striking new feature of AlCl$_3$ is the formal non-Lewis LP* NBO (Fig. 8.1e), which is essentially the left-over $(3p_z)_{Al}$ NAO that is unused in skeletal hybridization. Although formally a "vacant" non-Lewis orbital, the LP* NBO 33 is significantly populated (ca. 0.24e) by delocalizations from the three adjacent π-type Cl lone pairs (e.g., NBO 26), each noticeably depleted in occupancy (by ca. 0.08e).

Although AlCl$_3$ is sometimes described as an ionic "salt," its physical and chemical properties are quite unlike those of ordinary ionic solids. Instead of the usual strong interionic forces and high melting point of ionic salts, solid AlCl$_3$ *sublimates* at rather low temperatures ($\sim$180 C) to a gaseous phase composed of dimeric Al$_2$Cl$_6$ species. This behavior can be understood in terms of the expected donor–acceptor interactions between molecular AlCl$_3$ units. Because AlCl$_3$ exhibits *both* Lewis acid (NBO 33) and Lewis base (NBOs 24–26) orbital characteristics, it can pair up with its twin to form cyclic dimeric (AlCl$_3$)$_2$ complexes with *complementary* $n_{Cl} \rightarrow n^*_{Al}$ donor–acceptor interactions, as represented schematically in (8.4):

$$\begin{array}{c} Cl \\ \diagdown Al{-}Cl \\ Cl^{\diagup} \;\uparrow \quad \downarrow \diagup Cl \\ Cl{-}Al \diagdown \\ \qquad Cl \end{array} \qquad (8.4)$$

```
──────────────────  I/O-8.1  ──────────────────
   (Occupancy)   Bond orbital/ Coefficients/ Hybrids
------------------------------------------------------------

   1. (1.98447) BD ( 1)Al   1 -Cl   2
      ( 18.81%)   0.4337*Al   1 s( 33.15%)p 1.95( 64.62%)d 0.07(  2.23%)

      ( 81.19%)   0.9011*Cl   2 s( 23.80%)p 3.19( 76.01%)d 0.01(  0.19%)
   :
  24. (1.98415) LP ( 1)Cl   2          s( 76.18%)p 0.31( 23.81%)d 0.00(  0.01%)

  25. (1.94437) LP ( 2)Cl   2          s(  0.00%)p 1.00( 99.94%)d 0.00(  0.06%)

  26. (1.91454) LP ( 3)Cl   2          s(  0.00%)p 1.00( 99.93%)d 0.00(  0.07%)
   :
  33. (0.23965) LP*( 1)Al   1          s(  0.00%)p 1.00(100.00%)
```

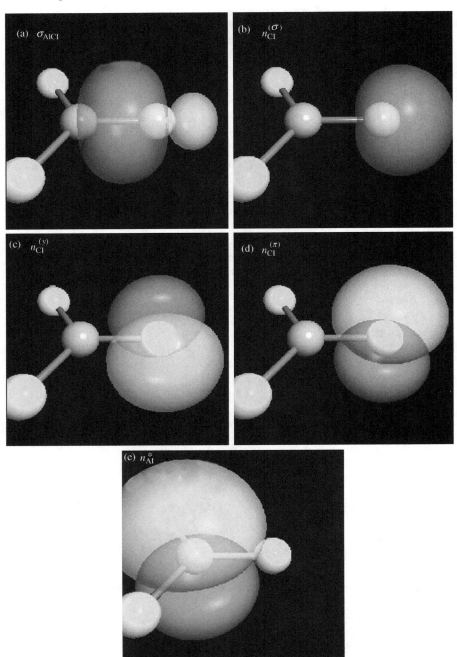

Figure 8.1   Leading NBOs of $AlCl_3$, showing Lewis-type (a) σ_{AlCl} (NBO 1), (b) $n_{Cl}^{(\sigma)}$ (NBO 24), (c) $n_{Cl}^{(y)}$ (NBO 25), (d) $n_{Cl}^{(\pi)}$ (NBO 26), and non-Lewis-type (e) n_{Al}^* (NBO 33), the characteristic "LP*" acceptor of a strong Lewis acid (cf. I/O-8.1). (See the color version of this figure in Color Plates section.)

Each arrow represents a directed two-electron $n_{Cl}^{(\pi)} \rightarrow n^*_{Al}$ donor–acceptor interaction from the filled $n_{Cl}^{(\pi)}$ lone pair of Cl into the unfilled n^*_{Al} "hole" of Al. Such a Lewis acid–base interaction is called a "coordinate covalent" or "dative" bond, and is often symbolized by a double-dot (Al:Cl) or directed arrow (Al←Cl) in the Lewis structure diagram, to distinguish it from an ordinary covalent (Al−Cl) bond-stoke. The distinctive coordinative Al←Cl bond will be written as "$\sigma_{Al:Cl}$" to distinguish it from the covalent "σ_{AlCl}" bond of Fig. 8.1a.

More generally, we may envision an ordinary covalent σ_{AB} bond as forming from two *singly* occupied bonding hybrids h_A, h_B:

$$A^{\uparrow} + {}^{\downarrow}B \rightarrow A - B \ (\sigma_{AB}) \tag{8.5}$$

whereas the coordinative $\sigma_{A:B}$ bond forms from a *doubly* occupied hybrid h_B with *un*occupied h_A:

$$A + {}^{\uparrow\downarrow}B \rightarrow A:B \ (\sigma_{A:B}) \tag{8.6}$$

As a result of its dative-coordinate character, formation of a $\sigma_{A:B}$ bond is associated with *formal charge* separation (A^--B^+ "ylidic character") that is expected to weaken the bonding interaction compared to ordinary covalent interaction. The unusual ylide formal charge pattern is manifested in anomalous ionicity of a coordinate $\sigma_{A:B}$ bond compared to the normal covalent-bond ionicity expected from Ξ_A, Ξ_B electronegativity values.

The student may object that the covalent/dative distinction is merely "in the eye of the beholder," because the final shared electron pair cannot "know" whether it "originated from" (8.5) or (8.6). However, this objection is invalid, because the intrinsic diffuseness and energy of atomic bonding hybrids is known to depend strongly on *occupancy* (Chapter 4). Asymmetries of initial NHO occupancy are therefore expected to be preserved in the shapes, energies, and other details of the final NBOs.

The intrinsic difference between covalent and coordinate bonds can also be recognized from the fact that a given bond will generally exhibit an inherent preference for either homolytic (covalent) or heterolytic (coordinative, "zwitterionic") dissociation. Such disparate modes could be distinguished, for example, by the disparate responses to bond-dissociative distortions in the presence of an external electric field. Consistent with the principle of microscopic reversibility, we naturally choose to envision bond formation as occurring by the reverse heterolytic or homolytic pathway that leads to dissociation. The intrinsic coordinative or covalent character of the bond could therefore be defined in terms of its preferred dissociation (rather than formation) pathway, an experimentally measurable property.

Let us see how the covalent versus coordinate bonding differences are manifested in Al_2Cl_6. Although the idealized bonding picture in (8.4) suggests C_{2h} symmetry and three distinct Al−Cl bond types, the actual optimized D_{2h} structure makes the four endocyclic bonds equivalent, though still recognizably distinct from the four exocyclic bonds, as shown in I/O-8.2.

I/O-8.2

```
#B3LYP/6-311++G** SCF=TIGHT POP=NBORead

Al2Cl6, E=-3246.7192484

   0  1
Al
Cl   1   2.2918
Cl   2   3.2457   1   44.9215
Al   2   2.2919   1   89.8397   3    0.0000
Cl   1   2.0888   2  110.0919   4  248.5181
Cl   4   2.0888   2  110.0595   1  248.4831
Cl   1   2.0888   3  110.0595   4  248.4831
Cl   4   2.0888   2  110.0595   1  111.5119

$NBO file=Al2Cl6 $END
```

Some details of the NBOs of Al_2Cl_6 are presented in I/O-8.3 and Fig. 8.2, allowing comparisons with NBOs of the parent monomer (I/O-8.1, Fig. 8.1), as well as direct comparison of coordinate $\sigma_{Al:Cl}$ (endo) versus covalent σ_{AlCl} (exo) bonds of the dimer. Compared to the exocyclic Al(1)−Cl(5) bond, the endocyclic Al(1):Cl(2) bond exhibits enhanced ionicity (0.7154 versus 0.6094) and significantly higher hybrid p-character at both Al (sp$^{3.97}$ versus sp$^{2.21}$) and Cl (sp$^{3.96}$ versus sp$^{2.63}$), consistent with Bent's rule (Chapter 4). Each exocyclic Cl exhibits the recognizable three lone pairs of the monomer form (e.g., NBOs 53–55), whereas each endocyclic Cl exhibits only the remaining two lone pairs ($n_{Cl}^{(y)}, n_{Cl}^{(\sigma)}$; NBOs 49, 50) after donation of $n_{Cl}^{(\pi)}$ to the coordinative $\sigma_{Al:Cl}$ interaction.

I/O-8.3

```
      (Occupancy)  Bond orbital/ Coefficients/ Hybrids
-----------------------------------------------------------------

   1. (1.96882) BD ( 1)Al   1 -Cl   2
              ( 14.23%)   0.3772*Al   1 s( 19.52%)p 3.97( 77.46%)d 0.15(  3.02%)
              ( 85.77%)   0.9261*Cl   2 s( 20.13%)p 3.96( 79.72%)d 0.01(  0.16%)
   :
   3. (1.98174) BD ( 1)Al   1 -Cl   5
              ( 19.53%)   0.4419*Al   1 s( 30.46%)p 2.21( 67.40%)d 0.07(  2.14%)
              ( 80.47%)   0.8971*Cl   5 s( 27.46%)p 2.63( 72.36%)d 0.01(  0.18%)
   :
  49. (1.98205) LP ( 1)Cl   2              s( 59.75%)p 0.67( 40.23%)d 0.00(  0.02%)
  50. (1.95014) LP ( 2)Cl   2              s(  0.00%)p 1.00( 99.94%)d 0.00(  0.06%)
   :
  53. (1.98175) LP ( 1)Cl   5              s( 72.52%)p 0.38( 27.47%)d 0.00(  0.01%)
  54. (1.93131) LP ( 2)Cl   5              s(  0.00%)p 1.00( 99.93%)d 0.00(  0.06%)
  55. (1.92541) LP ( 3)Cl   5              s(  0.00%)p 1.00( 99.94%)d 0.00(  0.06%)
   :
 233. (0.12290) BD*( 1)Al   1 -Cl   2
              ( 85.77%)   0.9261*Al   1 s( 19.52%)p 3.97( 77.46%)d 0.15(  3.02%)
              ( 14.23%)  -0.3772*Cl   2 s( 20.13%)p 3.96( 79.72%)d 0.01(  0.16%)
   :
 235. (0.06957) BD*( 1)Al   1 -Cl   5
              ( 80.47%)   0.8971*Al   1 s( 30.46%)p 2.21( 67.40%)d 0.07(  2.14%)
              ( 19.53%)  -0.4419*Cl   5 s( 27.46%)p 2.63( 72.36%)d 0.01(  0.18%)
```

Figure 8.2 Endocyclic coordinative $\sigma_{Al:Cl}$ bond of Al_2Cl_6 (NBO 1; cf. I/O-8.1). Except for the "missing" $n_{Cl}^{(\pi)}$ that is the coordinative "parent" of $\sigma_{Al:Cl}$, other NBOs of Al_2Cl_6 closely resemble those shown in Fig. 8.1 for $AlCl_3$. (See the color version of this figure in Color Plates section.)

Note that despite the superficial resemblance to the H-bridging geometry of diborane (B_2H_6, Section 4.4), the optimal NBO description of Al_2Cl_6 reflects a quite different bonding pattern, with *no* appreciable "three-center hypovalent bond" character in the latter case. Thus, Al_2Cl_6 provides a fairly direct intramolecular comparison between two-center σ bonds of distinct covalent versus coordinate character, with the latter bonds responsible for the unusual four-membered ring motif.

Differences can also be seen in I/O-8.3 between covalent versus coordinative antibonds of the dimer. The coordinative antibonds exhibit significantly higher occupancy (0.123 versus 0.070), roughly indicative of the statistical 2:1 advantage in number of vicinal $n_{Cl} \rightarrow \sigma^*_{Al:Cl}$ interactions available to the endocyclic antibonds (each of ca. 6–7 kcal/mol stabilization energy).

It is also interesting that the high polarity of coordinative bonds and antibonds allows modest $\sigma_{Al:Cl} \rightarrow \sigma^*_{Al:Cl'}$ resonance delocalization *within* the coordinative four-membered ring, as depicted in Fig. 8.3.

In contrast to cyclobutadiene, where endocyclic $\sigma–\sigma^*$ and $\pi–\pi^*$ delocalizations vanish by symmetry (see *V&B*, p. 200ff), the polarity of coordinative bonds allows weak $\sigma–\sigma^*$ resonance delocalizations with facing antibonds of the four-membered ring, contributing slight (ca. 2 kcal/mol) stabilization to this unusual structural motif. Thus, the unusual shapes and sizes of coordinative bonds and antibonds provide interesting opportunities for hyperconjugative delocalizations that are weak or absent in the apparently analogous interactions of apolar covalent bonds in organic species.

8.1.2 Coordinative π-Bonding in BH_2NH_2

We can gain a clearer picture of the aptness of the formal distinction in (8.5) versus (8.6) by considering analogous intramolecular π-bonds of covalent (π_{AB}) versus

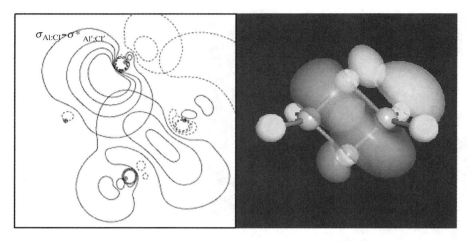

Figure 8.3 Coordinative $\sigma_{Al:Cl}$-$\sigma^*_{Al':Cl'}$ interaction of Al_2Cl_6 in contour and surface plots, showing nonvanishing hyperconjugative overlap ($\Delta E^{(2)} = 2.32$ kcal/mol) despite the unfavorable (cyclobutadiene-like) vicinal bond–antibond alignment. (See the color version of this figure in Color Plates section.)

coordinative ($\pi_{A:B}$) type, which can be "formed" or "dissociated" merely by intramolecular twisting.

To see this distinction, let us compare the ordinary covalent π_{CC} bond of ethylene (CH_2CH_2) with the coordinative $\pi_{B:N}$ bond of its ylidic "cousin" aminoborane (BH_2NH_2). The isolated -BH_2 moiety has the formal LP* valence-vacancy (empty 2p orbital, n^*_B) characteristic of a Lewis acid, while the isolated -NH_2 amine group has the formal LP lone-pair (n_N) characteristic of a Lewis base. Hence, a strong π-type $n_N \rightarrow n^*_B$ interaction is expected to result in formation of a formal $\pi_{B:N}$ bond when n_N and n^*_B orbitals are suitably coaligned in BH_2NH_2. However, perpendicular twisting of CH_2CH_2 versus BH_2NH_2 leads to homolytic versus heterolytic π-bond dissociation, clearly revealing the intrinsic difference between covalent (π_{CC}) versus coordinate ($\pi_{B:N}$) π-bonds in (8.5) versus (8.6).

The strong difference between torsional potential energy surfaces of CH_2CH_2 versus BH_2NH_2 is already revealed by the diradical versus closed-shell character (UHF- versus RHF-type solution; cf. Sidebar 3.2) of the 90°-twisted transition-state geometry of each species. Whereas simple closed-shell RHF-type description is adequate for both species in near-planar geometry ($\varphi \cong 0°$), ethylene undergoes RHF → UHF symmetry-breaking near $\varphi \cong 60°$, as expected for homolytic diradical dissociation. In contrast, BH_2NH_2 twisting proceeds smoothly on the heterolytic RHF-type dissociative pathway, as shown in Fig. 8.4.

The energy difference ΔE between planar ($\varphi = 0$) and perpendicular ($\varphi = 90$) rotamers gives a useful measure of intrinsic π-bond strength for covalent ($\Delta E_{CC} \cong 63.7$ kcal/mol) versus coordinate ($\Delta E_{BN} \cong 32.5$ kcal/mol), confirming the expected weakness of the latter type.

The covalent versus coordinative π-bonds also differ *qualitatively* in a variety of other NBO descriptors. Most characteristic is the induced *formal-charge* separation

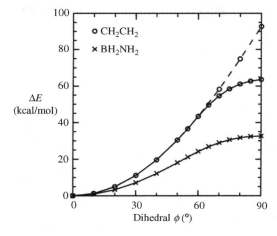

Figure 8.4 Calculated rotation barrier $\Delta E(\varphi)$ for CH_2CH_2 (circles) and BH_2NH_2 (crosses), showing the unphysical RHF-type (dashed line) versus physical UHF-type (solid line) diradical π_{CC}-breaking in the ethylene case. (The $\pi_{B:N}$-breaking in BH_2NH_2 is an RHF-stable heterolytic dissociation at all angles.)

accompanying formation of the coordinative $\pi_{B:N}$ bond, as shown in the Q_B, Q_N atomic charge variations in Fig. 8.5.

As a result of its formal 2e-donor character, the ionicity of the coordinative $\pi_{B:N}$ bond is sharply higher than expected from electronegativity differences, leading to conspicuously different ionicities (0.5094 versus 0.7140) of covalent σ_{BN} versus coordinative $\pi_{B:N}$ bonds. As a consequence, the electrooptical responses associated with BH_2NH_2 torsions are expected to differ markedly from those of CH_2CH_2.

The formation or dissociation of a bond is sometimes pictured (erroneously) as a discontinuous "on–off" process. This misconception is encouraged by excessive reliance on a single Lewis structural model (perforce "bonded" *or* "dissociated"), without regard for resonance-type corrections to the model. For example, if one simply carries out default NBO analysis at each dihedral angle, the NBO Lewis structure for either CH_2CH_2 or BH_2NH_2 will appear to switch discontinuously from double- to single-bonded form at some cross-over angle φ_x (near $\varphi \cong 75°$ for CH_2CH_2 or 66° for BH_2NH_2). As described in Sidebar 8.1, the precise location of this NLS

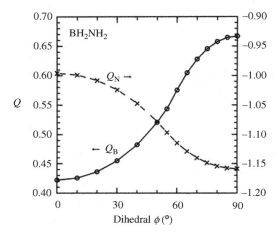

Figure 8.5 Natural atomic charge (Q) variations with torsional angle φ in aminoborane, showing zwitterion-like pattern of charges on B (Q_B, circles, left scale) and N (Q_N, crosses, right scale) as the $\pi_{B:N}$ coordinate bond is broken by 90° twisting.

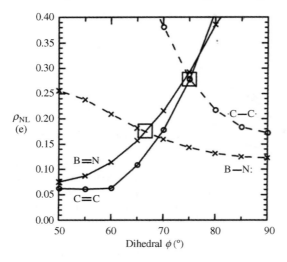

Figure 8.6   Torsional variations of non-Lewis "error" (ρ_{NL}) for single-bonded (dashed line) versus double-bonded (solid line) natural Lewis structures of CH_2CH_2 (circles) and BH_2NH_2 (crosses), showing the NLS crossing point for each species (marked by a box) with crossing angles $\varphi_x \cong 66°$ and 75° for BH_2NH_2 and CH_2CH_2, respectively.

switch depends on the chosen NBO criterion for distinguishing a highly ionic 2-c "bond" from a strongly delocalized 1-c "lone pair" (e.g., at least 5% of the NBO on each center to be called "two-center"), which inevitably involves a somewhat arbitrary cut-off criterion. But the large non-Lewis error ρ_{NL} of *either* description warns against taking this jump as a meaningful physical discontinuity. Figure 8.6 displays the variations of $\rho_{NL}(\varphi)$ for single- versus double-bonded NLS representations of CH_2CH_2, BH_2NH_2, showing that the only special feature of the cross-over angle φ_x is that the two distinct NLS depictions become equally *poor* at this point.

The more general resonance-type description of covalent or coordinate bond formation reveals the essential physical *continuity* (and *fractional* bond orders) of all bonding processes. Figure 8.7 exhibits the calculated NRT bond orders for dihedral twisting of CH_2CH_2 and BH_2NH_2, showing the smooth behavior through the "half-π-bonded" intermediate (dotted line) between idealized single- and double-bonded limits. [Note that the slight "break" in b_{CC} near 60° is due to wavefunction bifurcation to open-shell diradical character (RHF → UHF-type), not to numerical artifacts of the NRT method.]

The figure shows that the covalent π_{CC} and coordinate $\pi_{B:N}$ bonds differ somewhat in the "abruptness" of the bond order transition, but the essential *continuity* of π-bond formation is clearly demonstrated in each case. To estimate a single angle φ_h at which the π-bond is "half-formed," we choose the value for which $b = 1.50$, which gives the estimates

$$\varphi_h(\text{BN}) \cong 67° \tag{8.7}$$

$$\varphi_h(\text{CC}) \cong 76° \tag{8.8}$$

very close to the corresponding φ_x estimates of Fig. 8.6. The b_{NRT} transition profiles of Fig. 8.7 confirm the previous conclusion that coordinate $\pi_{B:N}$-bond formation (starting from the twisted rotamer) of BH_2NH_2 is somewhat "softer," "later," and "weaker" than the corresponding covalent π_{CC}-bond formation of CH_2CH_2.

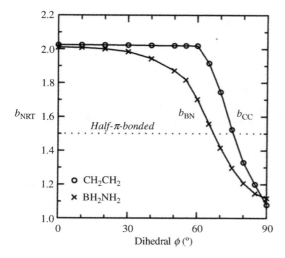

Figure 8.7 Torsional variations of NRT bond order (b_{NRT}) for b_{BN} of BH_2NH_2 (crosses) and b_{CC} of CH_2CH_2 (circles), showing the *continuous* changes of fractional bond order around the "half-π-bonded" value (dotted line), which is reached near $\varphi_h \cong 67°$, 76° for BH_2NH_2, CH_2CH_2, respectively (cf. Fig. 8.6).

The pronounced variations of NRT bond orders are of course reflected in many other structural properties of these species, particularly the R_{CC} or R_{BN} bond length variations. Figure 8.8a exhibits the $R(\varphi)$ dependence for both species, showing the expected strong contractions of R_{CC}, R_{BN} with π-bond formation. Figure 8.8b similarly displays the curve of bond length R with bond order b_{NRT}, showing the expected smooth bond-order–bond-length correlations for both covalently and coordinatively π-bonded species.

Although the b_{NRT} values derive purely from information in the first-order density matrix (*V&B*, p. 21ff), with *no* molecular geometry input, it is evident from Fig. 8.8b that bond-order–bond-length curves can be used to "read" bond order from given R_{CC} or R_{BN} distances, particularly in the broad intermediate region of near-linear correlation. Thus, b_{NRT} serves as the single most useful theoretical descriptor of

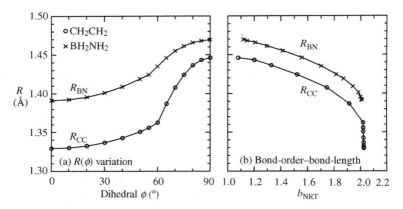

Figure 8.8 Covalent (R_{CC}) and coordinate (R_{BN}) bond-length variations in CH_2CH_2 (circles) and BH_2NH_2 (crosses), shown (a) versus dihedral twisting angle φ, and (b) versus NRT bond order b_{NRT}.

π-bonding in these species, closely correlated with a host of structural, reactivity, and electrooptical properties of interest.

SIDEBAR 8.1   *POLAR BOND OR DELOCALIZED LONE PAIR?*

In principle, the general expression for a normalized two-center σ_{AB} bond NBO

$$\sigma_{AB} = c_A h_A + c_B h_B \tag{8.9}$$

allows description of bonds of any desired ionicity i_{AB} in the interval

$$-1 \leq i_{AB} \leq +1 \tag{8.10}$$

with polarization coefficients c_A, c_B chosen to satisfy

$$i_{AB} = |c_A|^2 - |c_B|^2 \tag{8.11}$$

However, the extreme ionicity limits $i_{AB} = \pm 1$ cannot be achieved by any "two-center" NBO, because the corresponding coefficients c_A, c_B required by (8.11)

$$i_{AB} = +1 : |c_A| = 1, \ |c_B| = 0 \tag{8.12}$$

$$i_{AB} = -1 : |c_A| = 0, \ |c_B| = 1 \tag{8.13}$$

would reduce (8.9) to one-center form.

However, merely excluding the extreme ionic limits $i_{AB} = \pm 1$ cannot solve the problem of distinguishing "bonds" and "lone pairs" in a chemically meaningful way. As discussed in Section 5.4, the NLMO for a chemical lone pair generally includes weak delocalization tails on other centers. For example, if n_A is an idealized lone-pair NBO on center A, weak delocalization into acceptor n^*_B on adjacent center B will lead to a normalized NLMO $^{sl}n_A$ of the form

$$^{sl}n_A = (1 - \lambda^2)^{1/2} n_A + \lambda n^*_B \tag{8.14}$$

which is merely a special case of (8.9) with $c_A = (1-\lambda^2)^{1/2}$, $c_B = \lambda$.

Maintenance of a chemically meaningful distinction between "polar bond" and "delocalized lone pair" therefore requires further restriction of (8.10)

$$-i_{max} \leq i_{AB} \leq +i_{max} \tag{8.15}$$

expressed in terms of a numerical "ionicity threshold" i_{max} ($0 < i_{max} < 1$) that is chosen to correspond to general chemical usage. The NBO program chooses this threshold to require at least 5% of the "two-center" NBO density to be on each center, that is,

$$|c_A|^2 \leq 0.95, \ |c_B|^2 \geq 0.05 \tag{8.16}$$

or

$$i_{max} = 0.90 \tag{8.17}$$

The threshold in (8.16) and (8.17) generally leads to assigned "BD" (two-center) versus "LP" (one-center) labels that are consistent with common chemical usage. The 5%-limit on delocalization tail density is also consistent with the default 1.90e threshold for satisfactory NBO "pair occupancy"(i.e., missing no more than 5% of 2e). Although somewhat arbitrary, threshold (8.17) seems to adequately represent the fuzzy boundary between polar bonds and lone pairs.

As a result of this threshold, a delocalized LP-type NBO may be abruptly relabeled as a BD-type NBO if variation of a physical parameter brings the i_{AB} into the allowed two-center range. To preserve a consistent comparison for all such parameter variations, use the $CHOOSE keylist to override the default BD/LP labeling.

A related consistency problem occurs in default NRT analysis of coordinative bonding. A proposed NRT reference structure that merely represents the ionic limit of an existing reference structure (such as the long-range lone-pair limit of a coordinate covalent bond) is not included as an independent contribution to the NRT expansion, unless specifically requested by inclusion of a $NRTSTR keylist. [The default procedure implicitly treats all such "covalent-ionic resonance" (Sidebar 5.4) in terms of covalent versus ionic contributions to total NRT bond order, thereby avoiding the exponential proliferation of resonance structures and associated numerical instabilities.] For the examples of Section 8.1.1, we therefore included the $NRTSTR keylist as illustrated below to insure balanced weighting of both $H_2B=NH_2$ (STR3) and H_2B-NH_2 (STR2) structures at all angles.

```
#N b3lyp/6-311++g** nosymm  POP=NBORead scf=tight

BH2NH2 (aminoborane), E(0)=-82.074182

  0  1
B
N  1  1.3914
x  1  1.0     2  90.
x  2  1.0     1  90.  3  phi
H  1  1.1922  3  90.  2   118.99
H  1  1.1922  3  90.  2  -118.99
H  2  1.0087  4  90.  1   123.26
H  2  1.0087  4  90.  1  -123.26

phi    0.

$NBO file=bh2nh2_0 nrt $END
$NRTSTR
   STR1         ! :B-N
     LONE 1 1 END
     BOND S 1 2 S 1 3 S 1 4 S 2 5 S 2 6 END
   END
   STR2         ! B-N:
     LONE 2 1 END
     BOND S 1 2 S 1 3 S 1 4 S 2 5 S 2 6 END
   END
   STR3         ! B=N
     BOND D 1 2  S 1 3  S 1 4  S 2 5  S 2 6 END
   END
$END
```

For completeness, we included H_2B-NH_2 (STR1) in the keylist, but this structure received *no* weighting at any φ, consistent with chemical expectations. Analogous $NRTSTR keylists were included in CH_2CH_2 jobs to insure consistent inclusion of $H_2C=CH_2$ and H_2C-CH_2 reference structures. (See *NBO Manual*, p. B77ff for further details of $NRTSTR keylist construction.)

8.1.3 Coordinative σ-Bonding in BF₃:NH₃

For completeness, let us briefly describe coordinative σ-bonding in the classic BF_3:NH_3 Lewis acid–base adduct (8.3). A simple computational model of the R-dependence of coordinative $\sigma_{B:N}$ in BF_3:NH_3 is illustrated in I/O-8.4, for comparison with the corresponding φ-dependence of $\pi_{B:N}$ bonding in aminoborane (Section 8.1.2).

Figure 8.9 presents the plot of binding energy, $\Delta E(R)$, for dissociation of the Lewis acid–base adduct. As shown in the figure, the intermolecular $\sigma_{B:N}$ bond of BF_3NH_2 is rather weak (ca. 22 kcal/mol), weaker even than the intramolecular $\pi_{B:N}$ bond of BH_2NH_2 (cf. Fig. 8.9). The equilibrium bond length ($R_{BN} \cong 1.68$ Å) is also unusually long, considerably beyond that, for example, in perpendicularly twisted BH_2NH_2 (Fig. 8.8a). The relative weakness, softness, and elongation of coordination

--------- *I/O-8.4* ---------

```
#B3LYP/6-311++G** SCF=TIGHT POP=NBORead

BF3...NH3, Ropt=1.6799, Eopt=-381.2812084
(+0.00, -21.55 kcal/mol)

0 1
B
N   1   bn
F   1   bf    2   nbf
F   1   bf    2   nbf    3   120.
F   1   bf    2   nbf    3   240.
H   2   nh    1   bnh    3   180.
H   2   nh    1   bnh    6   120.
H   2   nh    1   bnh    6   240.

bf   1.3786
nh   1.0184
nbf  104.19
bnh  110.25
bn   1.6799

$nbo file=bf3nh3 nrt $end
$nrtstr
 str1  ! F3B-NH3
   lone 3 3  4 3  5 3 end
   bond s 1 3  s 1 4  s 1 5
        s 2 6  s 2 7  s 2 8  s 1 2 end
 end
 str2  ! F3B :NH3
   lone 3 3  4 3  5 3  2 1 end
   bond s 1 3  s 1 4  s 1 5
        s 2 6  s 2 7  s 2 8 end
 end
$end
```

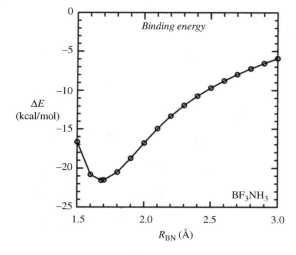

Figure 8.9 Calculated binding energy ΔE for dissociation of the coordinative $\sigma_{B:N}$ bond of BF_3NH_3, showing the energy minimum ($\Delta E = -21.55$ kcal/mol) at $R_{BN} = 1.6799$ Å.

bonds are expected to generally distinguish dative ("push–push") and covalent ("push–pull") bonding mechanisms (cf. *V&B*, p. 177ff).

Many of the NBO/NRT characteristics of intermolecular $\sigma_{B:N}$-bonding could also be anticipated from analogies to the intramolecular $\pi_{B:N}$ results of Section 8.1.2. Figure 8.10 displays the natural atomic charges $Q_B(R)$, $Q_N(R)$ of boron and nitrogen atoms in the bonding region, exhibiting the familiar "mirror image" variations that are characteristic of dative charge transfer (cf. Fig. 8.5).

Figure 8.11 displays the NRT bond order variations $b_{B:N}(R)$ for coordinative bond dissociation in BF_3NH_3. The bond-order–bond-length correlation exhibits the expected smooth decay from short- to long-range separation, with fractional

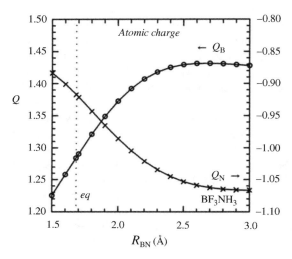

Figure 8.10 Natural atomic charge (Q) variations with distance R_{BN} in BF_3NH_3, showing zwitterionic-like pattern of charges on B (Q_B, circles, left scale) and N (Q_N, crosses, right scale) as the $\sigma_{B:N}$ coordinate bond is broken by dissociation (cf. Fig. 8.5). The equilibrium bond length is shown by the vertical dotted line.

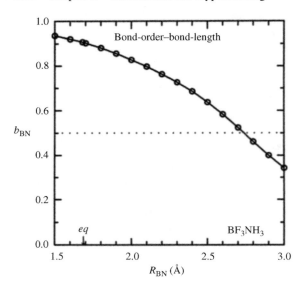

Figure 8.11   NRT bond-order–bond-length correlation, $b_{BN}(R)$, for coordinative $\sigma_{B:N}$ of BF_3NH_3, showing the continuous fractional decreases ($b_{BN} \rightarrow 0$) as $R_{BN} \rightarrow \infty$. The "half-bonded" distance ($R_h \cong 2.74$ Å) is marked by the dotted line intersection, and the equilibrium bond length (~1.68 Å) by a small arrow.

bond order that varies near-linearly around the "half-bonded" transition value $b_{B:N} = 1/2$ (dotted line). The fractional bond orders reduce smoothly toward $b_{B:N} \rightarrow 0$ as $R_{BN} \rightarrow \infty$, requiring the bond-order–bond-length curve to "level out" at large R (beyond the values shown in Fig. 8.11) as well as near the small-R equilibrium limit.

Although the physical bond order variations of Fig. 8.11 are continuous over the entire dissociation range, it may be useful to identify a single characteristic distance that can be associated with the "bond-breaking" transition. For this purpose, the half-bonded distance ($R_h \cong 2.74$ Å) is recommended, because it takes account of factors other than the approximate equal weighting ($w_{B-N} \cong w_{B...N}$) of idealized bonded and nonbonded resonance forms. Figure 8.12 displays the NRT weightings for these leading resonance structures, which lead to a crossing point near $R \cong 2.86$ Å. This equal-w value is slightly beyond the b_{NRT}-based value of Fig. 8.11 (dotted line), and significantly beyond the NLS ρ_{NL}-crossing (determined in analogy with Fig. 8.6). The latter is rather arbitrary, and occurs in this case about 0.4 Å inside the more reliable NRT-based values. (For separations within this range, the default NLS structure will not be the resonance structure of highest weighting, but this should not provoke undue concern.)

It is physical fiction to suppose that bond order has only *integer* values that undergo discontinuous transitions at some envisioned bond-breaking distance. The bond-order–bond-length relationship for coordinative Lewis acid–base adducts shows that bond order should be considered a *continuously variable* measure of chemical bonding interactions, with *fractional* values in the range of *inter*molecular interactions ($0 \leq b \leq 1$) as well as the familiar range of intramolecular resonance phenomena ($b \geq 1$). The observed continuity of NRT-based description in these cases is in accord with the wave mechanical electronic continuity that must be expected on physical grounds.

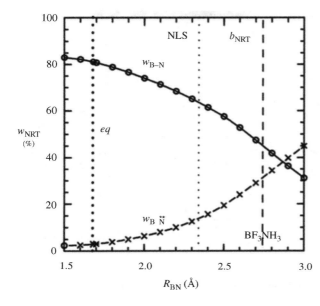

Figure 8.12 Variations of NRT weightings w_{NRT} for bonded (w_{B-N}, circles) and nonbonded ($w_{B..N}$, crosses) resonance structures of BF_3NH_3 (cf. $NRTSTR key-list of I/O-8.4). Alternative b_{NRT} based (dashed line) or NLS-based (light dotted line) criteria for "bond-breaking" are compared with the equilibrium bond distance (heavy dotted line) and the equal-w crossing point near 2.86 Å.

8.2 TRANSITION METAL COORDINATE BONDING

Transition metal (TM) species offer spectacular opportunities for coordinative bonding, due to the presence of both donor (LP) and acceptor (LP*) functionality in the metal valence shell. The unique shapes and symmetries of sd^m-based TM hybrids also offer highly unusual covalent geometries and delocalization patterns (cf. *V&B*, Chapter 4), quite unlike those of common organic species. The present section only hints at the richness of TM covalent and coordinate bonding phenomena that offer one of the most exciting frontiers of modern chemical research.

As a simple example, let us consider nickel (Ni, $Z = 28$), a common constituent of metallic alloys. Although the nominal configuration of an isolated Ni atom is $(3d)^8(4s)^2$, nickel easily achieves the "promoted" $(3d)^{10}(4s)^0$ configuration, which is its primary identity in the molecular coordination complexes to be described below.

[Note that multiconfigurational coordination bonding raises a number of difficult technical and computational issues, as discussed in Sidebar 8.2. In this section, we seek to bypass technical issues as far as possible, focusing instead on qualitative aspects of coordination bonding that seem to be adequately described by the DFT-based methods employed throughout this book. However, the student is forewarned that explorations beyond the relatively simple examples described below will typically require dealing with RHF/UHF instability, spin contamination, and other issues alluded to in Sidebar 8.2.]

Nickel is found to make "sticky" complexes with virtually *any* small lone-pair-bearing molecular species ("ligand," Lig) one might choose, such as

$$Lig = H_2O, HF, CO, NH_3, PH_3 \tag{8.18}$$

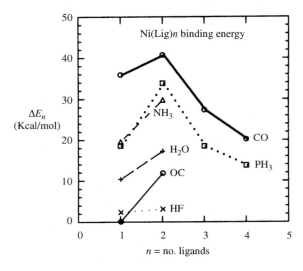

Figure 8.13 Calculated binding energy ΔE_n (kcal/mol) for successive additions of Ni(Lig)$_n$ coordinative ligands: CO (circles, heavy solid line); PH$_3$ (squares, heavy dotted line); NH$_3$ (triangles, dashed line); H$_2$O (plusses, long-dashed line); OC (circles, light solid line); HF (crosses, dotted line). Note that Ni(OC) is unbound, but Ni(OC)$_2$ is bound by ca. 12 kcal/mol.

Indeed, nickel generally forms robust Ni(Lig)$_n$ complexes with all these ligands in *multiple* coordination stoichiometries,

$$\text{Ni(Lig)}, \ \text{Ni(Lig)}_2, \ldots, \text{Ni(Lig)}_n \tag{8.19}$$

up to $n = 2$ or $n = 4$, with typical binding energies per ligand in the 10–40 kcal/mol range. Figure 8.13 shows the calculated binding energy $(\Delta E)_n$ trends for stable Ni(Lig)$_n$ complexes of ligands (8.18), illustrating the interesting growth and saturation patterns for each aggregation sequence. (Of course, if excess ligand is present, only the final member of the sequence will be found in the reaction pot.) Whereas H$_2$O, HF, and NH$_3$ coordinate with Ni only up to $n = 2$, both PH$_3$ and CO continue aggregation up to $n = 4$. Figure 8.14 displays corresponding plots of

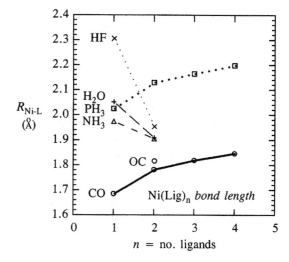

Figure 8.14 Similar to Fig. 8.13, for Ni-ligand bond lengths $R_{\text{Ni-L}}$ (Å).

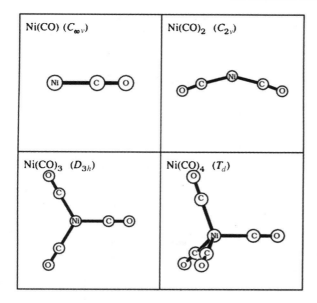

Ni(CO) ($C_{\infty v}$)	Ni(CO)$_2$ (C_{2v})
Ni(CO)$_3$ (D_{3h})	Ni(CO)$_4$ (T_d)

Figure 8.15 Ni(CO)$_n$ coordination complexes, $n = 1$–4, showing high symmetry geometry (and point-group symbols) for each species.

metal–ligand bond length (R_{Ni-L}), showing that weaker bond energies tend to display the expected correlation with longer bond length as aggregation proceeds, but the first members ($n = 1$) of the CO, PH$_3$ sequences are anomalous. What's going on here?

The geometries of successive coordination complexes tend to adopt "VSEPR-like" structures of high symmetry. This is illustrated for Ni(CO)$_n$ complexes in Fig. 8.15, where the successive aggregates form linear ($C_{\infty v}$; $n = 1$), slightly bent digonal (C_{2v}; $n = 2$), trigonal (D_{3h}; $n = 3$), and tetrahedral (T_d; $n = 4$) symmetry species as complexation proceeds.

The VSEPR-like geometries may superficially suggest "electrostatic" or "steric" influences, but a little reflection shows that this line of reasoning is unproductive. The essential *un*importance of the dipole moment (or dipole-associated electrostatic properties) can already be inferred from the observation (Fig. 8.13) that CO forms isomeric complexes of *both* Ni(CO)$_2$ and Ni(OC)$_2$ type, *in*dependent of which end of the CO dipole is oriented toward the metal. Figure 8.16 displays the optimized structures of monomer and dimer complexes for HF, H$_2$O, and NH$_3$ ligands, which further confound electrostatic or steric expectations. The essential *in*difference to dipolar or steric factors is particularly apparent in the curiously canted structures of Ni(HF), Ni(HF)$_2$ (Fig. 8.16a,b), which seem to be the *least-*expected geometry from either dipole-induced dipole or steric-repulsions viewpoint. The equilibrium bond lengths of Fig. 8.14 usually lie far inside the sum of empirical van der Waals radii further suggesting the dominant role of quantum mechanical valency forces rather than the classical-like forces of the exchange-free long-range limit.

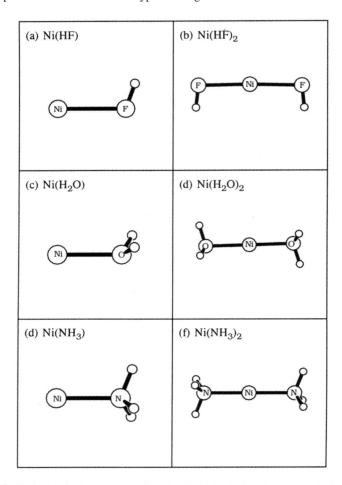

Figure 8.16   Ni(Lig)$_n$ complexes, $n = 1$–2, for Lig $=$ HF (a,b), H$_2$O (c,d), and NH$_3$ (e,f), showing the bent geometry of Ni(HF), Ni(H$_2$O) monomers and puzzling pattern of eclipsed and staggered dimer conformations that challenge superficial VSEPR-type (electrostatic or steric) rationalizations.

As described in Sidebar 8.2, the key to each ligand's success in coordinate bonding is its ability to participate in complementary donor–acceptor interactions with both the unfilled LP* (4s) and filled LP (3d) orbitals of the metal atom. The primary metal–ligand interaction is that between the ligand lone pair (n_{Lig}) and the metal acceptor (n^*_{Ni}) orbital, the ligand-to-metal (L $\rightarrow$ M*) $n_{Lig} \rightarrow n^*_{Ni}$ *σ-bonding* interaction. However, ligands with suitable acceptor orbitals (a^*_{Lig}) are also able to participate in secondary metal-to-ligand (M $\rightarrow$ L*) $n_{Ni} \rightarrow a^*_{Lig}$ *back-bonding* interactions. Such back-bonding serves to counteract capacitive charge buildup induced by the primary dative σ-bonding, thus reinforcing *both* interactions in cooperative fashion.

Which ligands offer the best combination of donor (n_{Lig}) and acceptor (a^*_{Lig}) orbitals? Sigma-bonding strength of donor lone pairs is expected to be (inversely) correlated with electronegativity, with,

$$n_C > n_N > n_O > n_F \qquad (8.20)$$

among first-row ligands. Back-bonding strength of ligand acceptors (valence anti-bonds) is expected to vary with symmetry type (favoring π^*-type over σ^*-type) and polarity (favoring high a^*_{Lig} amplitude nearest the metal). Thus, among the ligands (8.18), acceptor strength is expected to vary in the order

$$\pi^*_{CO} \gg \sigma^*_{PH} > \sigma^*_{NH} > \sigma^*_{OH} > \sigma^*_{FH} \qquad (8.21)$$

"Backward" Ni(OC) is clearly inferior to Ni(CO), because n_O is inferior to n_C as a donor, and the π^*_{CO} acceptor is polarized *away* from the contact point with Ni, thus diminishing its advantage as a π^*-acceptor relative, to H_2O or other n_O donors. Thus, from (8.20) and (8.21) we can anticipate the qualitative order of ligand coordinative strength, consistent with the well-known empirical "spectrochemical series" of ligands.

To obtain more tangible NBO evidence for these qualitative chemical concepts, we consider the combined second-order perturbative estimate (Section 5.2) for the sum of $L \rightarrow M^*$ and $M \rightarrow L^*$ interactions,

$$\Delta E^{(2)}_{D \rightarrow A} = \Delta E^{(2)}_{L \rightarrow M^*} + \Delta E^{(2)}_{M \rightarrow L^*} \qquad (8.22)$$

Figure 8.17 compares the R-dependence of $\Delta E^{(2)}_{D \rightarrow A}$ for different ligands in the form of a bar graph for three specific R_{Ni-L} distances (2.0, 2.5, 3.0 Å) along the Ni $\cdots$ Lig reaction coordinate, with each bar partitioned into $L \rightarrow M^*$ σ-bonding (shaded) and $M \rightarrow L^*$ back-bonding (unshaded) contributions. (PH$_3$ is omitted from these R_{Ni-L} comparisons, because the atomic radius of P differs markedly from the first-row ligands compared in Fig. 8.17.)

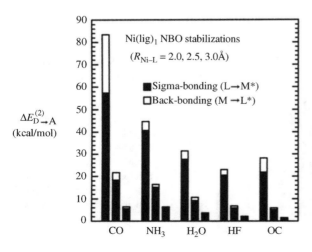

Figure 8.17 Bar graph for NBO donor–acceptor stabilizations ($\Delta E^{(2)}_{D \rightarrow A}$) in Ni(Lig)$_1$ complexes showing principal $L \rightarrow M^*$ σ-bonding (filled-bar) and $M \rightarrow L^*$ back-bonding (open-bar) contributions at $R =$ 2.0 Å (left), 2.5 Å (middle), 3.0 Å (right) for first-row ligands CO, NH$_3$, H$_2$O, HF, and OC (cf. Fig. 8.22). Note the increasing percentage of back-bonding for the strong CO ligand.

As shown by the steeply rising bar heights in Fig. 8.17, the metal–ligand donor–acceptor interactions increase exponentially with R_{Ni-L} approach for all ligands, consistent with the expected exchange-type nature of the interactions. However, CO is seen to be the most powerful coordinator at each R_{Ni-L}, both in $L \rightarrow M^*$ bonding and $M \rightarrow L^*$ back-bonding. The advantage of CO's π^* back-bonding is clearly exhibited at each R_{Ni-L}, and moreover, the *proportion* of $M \rightarrow L^*$ back-bonding is also steeply increasing at smaller R_{Ni-L}, giving CO an ever-increasing advantage over the hydride ligands. Among the hydride σ^* back-bonders, the three σ^*_{NH} orbitals of NH_3 give a slight advantage over the two σ^*_{OH} orbitals of H_2O or the single σ^*_{FH} orbital of HF, but $M \rightarrow L^*$ back-bonding is a relatively minor contribution to overall coordination strength in these ligands. All aspects of the $\Delta E_{D \rightarrow A}^{(2)}$ comparisons are seen to be in good qualitative agreement with the anticipated trends in (8.20–8.22).

Figure 8.18 presents an alternative bar graph comparison of relative coordination strength $\Delta E_{D \rightarrow A}^{(2)}$ for the entire set of $Ni(Lig)_n$ complexes in their equilibrium geometry, using the same convention for σ-bonding (shaded) and back-bonding (unshaded) contributions. CO and PH_3 are seen to tower over NH_3, H_2O, or HF in coordinative strength, even in terminal ($n = 4$) members of the coordination sequence. As usual, the attractive donor–acceptor interactions estimated by $\Delta E_{D \rightarrow A}^{(2)}$ must be offset against repulsive donor–donor interactions (Section 6.1) and configurational promotion (Sidebar 8.2) to obtain corresponding estimates of net binding energy (Fig. 8.13). However, the essential "driving force" provided by $L \rightarrow M^*$ σ-bonding and $M \rightarrow L^*$ back-bonding is clearly indicated in Figs. 8.17 and 8.18, consistent with the well-known empirical concepts of coordination bonding.

Further details of bonding and back-bonding interactions in equilibrium Ni $(CO)_2$, $Ni(PH_3)_2$, and $Ni(NH_3)_2$ complexes are shown in the NBO contour plots of Fig. 8.19a–f. In this figure the leading $L \rightarrow M^*$ bonding (left panel) and $M \rightarrow L^*$ back-bonding (right panel) NBO interactions are shown with corresponding

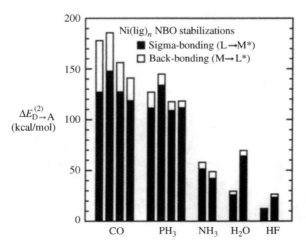

Figure 8.18 Bar graph for NBO donor–acceptor stabilizations ($\Delta E_{D \rightarrow A}^{(2)}$) in equilibrium $Ni(Lig)_n$ complexes ($n = 1$–4 or $n = 1$–2), showing principal $L \rightarrow M^*$ σ-bonding (filled-bar) and $M \rightarrow L^*$ back-bonding (open-bar) contributions for CO, PH_3, NH_3, H_2O, and HF ligands. (For consistency, the \$CHOOSE option was employed to describe Ni and ligands as separate "molecular units" in all species.)

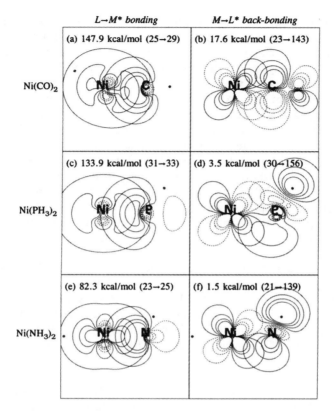

Figure 8.19 NBO orbital contour diagrams for leading $n_{Lig} \rightarrow n^*_{Ni}$ σ-bonding (L → M*; left panel) and $n_{Ni} \rightarrow a^*_{Lig}$ back-bonding (M → L*; right panel) interactions in Ni(CO)$_2$ (upper), Ni(PH$_3$)$_2$ (middle), and Ni(NH$_3$)$_2$ (lower) coordination complexes, with corresponding perturbative stabilization estimates (and associated NBO numbers) shown in each panel (cf. Table 8.1). (In each case, the $CHOOSE option was employed to force recognition of Ni as a nonbonded atom, whereas default NBO analysis describes the $n_{Lig} \rightarrow n^*_{Ni}$ interaction as a highly polarized σ_{Ni-L} coordination bond.)

$\Delta E^{(2)}_{L \rightarrow M^*}$, $\Delta E^{(2)}_{M \rightarrow L^*}$ perturbative stabilization estimates (and associated NBO numbers) for each species, allowing direct visual comparisons of bonding and back-bonding orbital overlap in each case. As shown in the contour diagrams, coordinative bonding involves primary L → M* donation from the n_{Lig} lone pair of the ligand into the 4s-dominated LP* orbital on Ni [panels (a), (c), (e)], with secondary M → L* donation from 3d-dominated n_{Ni} orbitals of Ni into a^*_{Lig} acceptor orbitals of the adjacent ligand [π^*_{CO} for CO, panel (b); σ^*_{PH} for PH$_3$, panel (d); σ^*_{NH} for NH$_3$, panel (f)].

One can see from visual comparisons of panels (d) and (f) in Fig. 8.19 that the $3d_{Ni}-\sigma^*_{PH}$ NBOs of Ni(PH$_3$)$_2$ are better "matched" in size for favorable overlap than are the corresponding $3d_{Ni}-\sigma^*_{NH}$ NBOs of Ni(NH$_3$)$_2$. However, neither can compete with the much more favorable $3d_{Ni}-\pi^*_{CO}$ overlap of Ni(CO)$_2$ shown in panel (b), which accordingly leads to far stronger back-bonding stabilization. One can also see that $L \rightarrow M^*$ coordination with the "$4s_{Ni}$"-type LP* on Ni is complicated by d-type admixtures to the metal acceptor orbital, particularly prominent for $Ni(NH_3)_2$ [panel (e)]. Apparently, the slight admixture of d-character is also responsible for the bending of $Ni(CO)_2$, where slight $n_C–n^*_{Ni}$ twisting serves to weaken the unfavorable inner-lobe overlap near the Ni center, as depicted in Fig. 8.19a. Differences of energy and shape in both n^*_{Ni} acceptor orbitals and n_P versus n_N donor orbitals evidently contribute to the greatly reduced $\Delta E^{(2)}_{L \rightarrow M^*}$ stabilizations (134 versus 82 kcal/mol) estimated for $Ni(PH_3)_2$ versus $Ni(NH_3)_2$.

Table 8.1 shows some further details of the leading $n_{Lig}–n^*_{Ni}$ NBOs and associated $\Delta E^{(2)}_{L \rightarrow M^*}$ stabilization estimates depicted in Fig. 8.19, to indicate how the pronounced differences in $L \rightarrow M^*$ stabilizations arise. The table entries include hybrid composition (%-p for donor n_{Lig}, %-s for acceptor n^*_{Ni}) and occupancy (e) of donor and acceptor NBOs, and the associated F_{ij} Fock-matrix elements and $\Delta \varepsilon_{ij}$ orbital energy differences (a.u.) that lead to the $\Delta E^{(2)}_{L \rightarrow M^*}$ perturbative stabilization estimate [cf. Eq. (5.18)].

As shown in Table 8.1, the $L \rightarrow M^*$ interaction in $Ni(NH_3)_2$ exhibits both weaker $F_{L \rightarrow M^*}$ Fock-matrix element and larger $\Delta \varepsilon_{L \rightarrow M^*}$ energy gap, leading to significantly weaker stabilization compared to $Ni(PH_3)_2$ or $Ni(CO)_2$. Although NH_3 has a slight advantage over PH_3 in %-p character of its n_{Lig} orbital (as expected from Bent's rule; Section 4.3), the latter ligand has considerable advantage in other chemical factors (including $3d_{Ni}–\sigma^*_{PH}$ back-bonding; Fig. 8.19) that dictate overall coordinative success.

The examples given above illustrate how we can "make sense" of complex $L \rightarrow M^*$, $M \rightarrow L^*$ patterns by peering into details of NBO bonding and back-bonding interactions. These donor–acceptor stabilizations, when combined with opposing steric repulsions (Section 6.1), result in the complex patterns of coordination binding energies and geometries exhibited in Figs. 8.13–8.18. Although details of coordinative TM bonding are admittedly far more complex than those of main-group

Table 8.1 Hybrid composition (%-p or %-s) and occupancy (occ, e) of donor (n_{Lig}) and acceptor (n^*_{Ni}) NBOs for given species, showing associated Fock-matrix element ($F_{L \rightarrow M^*}$, a.u.) and orbital energy difference ($\Delta \varepsilon_{L \rightarrow M^*}$, a.u.) that lead to the perturbative estimate [$\Delta E^{(2)}_{L \rightarrow M^*}$, kcal/mol; cf. Eq. (5.18)] of $L \rightarrow M^*$ stabilization (cf. Fig. 8.19). Note that the \$CHOOSE-forced perturbative stabilization estimate is only a general guide to chemical trends and should not be taken literally, because the coordinative interaction is too strong to be considered a mathematically small "perturbation."

Species	n_{Lig}		n^*_{Ni}		$F_{L \rightarrow M^*}$	$\Delta \varepsilon_{L \rightarrow M^*}$	$\Delta E^{(2)}_{L \rightarrow M^*}$
	%-p	Occ	%-s	Occ			
$Ni(CO)_2$	35.5	1.74	91.4	0.49	0.25	0.45	148
$Ni(PH_3)_2$	58.4	1.75	92.2	0.47	0.22	0.40	134
$Ni(NH_3)_2$	61.2	1.80	61.6	0.35	0.18	0.47	82

counterparts, TM chemistry offers a rich treasure trove of structural and catalytic phenomena to reward the chemical explorer. In the present era, *ab initio* computational methods and analysis are expected to play an increasingly important role in pushing back the frontiers of organometallic chemistry and guiding successful discovery of catalytic principles that govern many commercial industrial applications as well as key aspects of photosynthesis and other life processes.

SIDEBAR 8.2   *MULTICONFIGURATIONAL ISSUES IN MONOLIGATED Ni COMPLEXES*

Unlike its periodic neighbor Cu, the Ni atom obeys the Madelung *Aufbau* configuration rule, leading to the "expected" $(3d)^8(4s)^2$ (a^3F) triplet ground state. However, near-degenerate alternative $(3d)^9(4s)^1$ and $(3d)^{10}(4s)^0$ configurations occur as low-lying excited states that are easily involved in chemical interactions, leading to complex multiconfigurational character in the potential energy surfaces for Ni and other TM species. This in turn leads to aberrant numerical behavior of DFT and related theoretical methods (Sidebars 2.2 and 3.2), which will be described briefly in this sidebar. The $Ni(Lig)_n$ complexes present edifying examples of how TM calculations and orbital interpretation are challenged by multiconfigurational issues.

The $Ni(Lig)_n$ species of Figs. 8.13–8.16 are all spin singlets, as are the ligands themselves. Ground-state ^3F Ni is therefore spin-forbidden from forming these complexes, but the onset of reactive singlet Ni states lies only slightly ($\sim$10 kcal/mol) above the unreactive triplet ground state. According to Moore's spectroscopic tables (C.E. Moore, *Atomic Energy Levels*, Vol. II, NBS Circular 467, U.S. Government Printing Office, Washington D.C., 1952, p. 98), three important configurations of the singlet manifold are bunched within ca. 30 kcal/mol of singlet onset, as described in Table 8.2.

Table 8.2   Spectroscopic labels, configurational assignments, energy level ΔE (kcal/mol, relative to a^1D ground singlet), and shell-character description for low-lying singlet states of Ni atom.

Label	Configuration	ΔE	Description
a^1D	$(3d)^9(4s)^1$	0.0	Anisotropic open-shell diradical
b^1D	$(3d)^8(4s)^2$	28.9	Anisotropic open-shell diradical
a^1S	$(3d)^{10}(4s)^0$	32.3	Spherical closed-shell

Although the lowest-energy singlet UB3LYP solution for an isolated Ni atom is indeed the diradical $(3d)^9(4s)^1 a^1D$ state, the effective Nickel NEC (Section 3.2) of the more robust $Ni(Lig)_n$ coordination species corresponds to the promoted a^1S-like $(3d)^{10}(4s)^0$ configuration, with closed-shell RHF-type solution. Accordingly, initial coordination ($n=1$) typically requires $\sim$30 kcal/mol investment to surmount the promotion barrier, thereby leading to an initial ΔE_1 binding energy that appears "too low" compared to the trend for higher n values (cf. Fig. 8.13). However, once promoted to the coordinatively active ^1S configuration, the Ni atom can coordinate additional ligands in an exoergic diminishing-returns pattern, as shown in Fig. 8.13.

What are the key bonding characteristics of the coordinatively active ^{1}S configuration? As shown in the entries of Table 8.2, both the unfilled 4s orbital (LP*) and the filled shell of 3d (LP) orbitals of ^{1}S Ni are spherically symmetric. The isotropic nickel LP* suggests the high-symmetry (VSEPR-like) patterns of successive $n_{Lig} \to n^*_{Ni}$ coordinative attacks on the metal n^*_{Ni} orbital, the primary ligand $\to$ metal σ-bonding interaction that drives complexation. However, the filled shell of $3d_{Ni}$ LP orbitals also suggests the isotropic flexibility to exploit reciprocal $n_{Ni} \to a^*_{lig}$ back-bonding interactions with ligand acceptor orbitals (a^*_{lig}) of practically any symmetry. Thus, while ligand donor strength (as gauged, e.g., by electronegativity of the lone-pair-bearing atom) is the primary factor controlling the primary $n_{lig} \to n^*_{Ni}$ σ-bonding interaction, ligand *acceptor* strength (as gauged, e.g., by the number of σ- or π-type a^*_{lig} orbitals) will also play an important role in controlling the secondary $n_{Ni} \to a^*_{lig}$ back-bonding interactions that dictate overall coordinative success. Because secondary metal $\to$ ligand back-bonding intrinsically relieves the capacitive charge polarization "pressure" induced by primary ligand $\to$ metal bonding, it reinforces overall coordinative strength in a highly cooperative manner. The details of optimally matching cooperative $n_{lig} \to 4s_{Ni}$ bonding and $3d_{Ni} \to a^*_{lig}$ back-bonding interactions with the LP, LP* orbitals of the isotropic ^{1}S Ni configuration thus become the key to maximizing overall metal–ligand coordination.

Although the concept of a $(3d)^9(4s)^1 \to (3d)^{10}(4s)^0$ promotion barrier to access the Ni ^{1}S bonding configuration appears rather simple, its computational description by DFT and related single-configuration methods is far from straightforward. The open-shell diradical $(3d)^9(4s)^1$ character of the long-range Ni$\cdots$Lig limit is expected to mix at intermediate R_{Ni-L} with Ni $(3d)^8(4s)^2$ configurational character before emerging on the Ni $(3d)^{10}(4s)^0$ bonding surface in the short-range limit. Such configurational mixing is often manifested in strong "spin contamination" (i.e., total squared spin angular momentum expectation values far from the value $\langle S^2 \rangle = 0$ expected for a spin singlet state) and/or pathological SCF convergence problems. Figure 8.20 displays calculated UB3LYP/6-311++G** potential curves $\Delta E_1(R)$ for binding of the initial ligand in the approach region, $3.5 \leq R \leq 1.5$ Å. In the

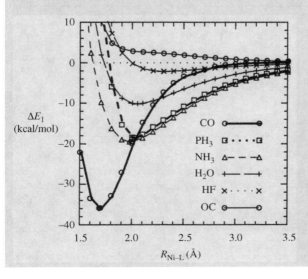

Figure 8.20 Binding energy curves $\Delta E_1(R)$ for monoligated Ni(Lig)$_1$ species (cf. Fig. 8.13), showing shallow binding minimum (or feeble barrier) for weaker ligands with equilibrium bond lengths beyond $R \cong 2$ Å, associated with significant spin contamination ($\langle S^2 \rangle \cong 1$) and other symptoms of multiconfigurational mixing.

shallow wells or rises approaching $R \cong 2$ Å, DFT spin-contamination values are typically around $\langle S^2 \rangle \cong 1$, reflecting strong multiconfigurational character.

How can we follow the configurational changes that accompany coordinative bond formation in Fig. 8.20? The multiconfigurational character of the longer-range region is most directly manifested in the total squared spin angular momentum $\langle S^2 \rangle$ "spin contamination" values plotted in Fig. 8.21. As shown in the figure, the $\langle S^2 \rangle$ values increase to ca. 1.0 beyond ca. 2.5 Å, reflecting strong spin contamination and mixed singlet–triplet diradical character of the long-range open-shell limit. However, $\langle S^2 \rangle$ plummets toward 0.0 at shorter range, corresponding to replacement of the long-range diradical UHF-type solution by the closed-shell RHF-type (Ni ^{1}S) solution around 2 Å. The UHF-RHF cross-over occurs just *in*side R_{eq} for weaker ligands such as NH_3, H_2O or HF, which therefore lead to equilibrium monoligated $Ni(Lig)_1$ species exhibiting weak open-shell character. For the stronger ligands PH_3 and (especially) CO, the UHF-bifurcation occurs *out*side R_{eq}, and the equilibrium Ni $(Lig)_1$ species exhibit closed-shell character (as do all higher-n complexes).

What drives the configurational promotion depicted in Fig. 8.21? The net binding-energy curves of Fig. 8.20 give the superficial visual impression that metal–ligand interactions are relatively weak ($<10–15$ kcal/mol) beyond ca. 2 Å. However, the NBO donor-acceptor interactions of bonding ($L \rightarrow M^*$) and back-bonding ($M \rightarrow L^*$) type paint an entirely different picture. Figure 8.22 plots the total "weak" ($E^{(2)}_{D \rightarrow A}$ donor–acceptor stabilization (sum of $\Delta E^{(2)}_{L \rightarrow M^*}$ bonding and $\Delta E^{(2)}_{M \rightarrow L^*}$ back-bonding interactions) for long-range ligand approach, showing the onset of strong attractive (stabilizing) forces in the 2–3 Å region, roughly coinciding with the region in which diradical spin contamination (Fig. 8.21) is being quenched.

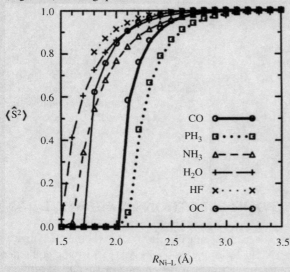

Figure 8.21 Total squared spin angular momentum expectation values $\langle S^2 \rangle$ for monoligated $Ni(Lig)_1$ complexes (cf. Fig. 8.13), showing reduction of diradical spin contamination ("promotion" to closed-shell ^{1}S-like Ni configuration) as incoming ligand approaches $R_{Ni-L} \cong 2$ Å.

As shown in the figure, each ligand obtains sufficient $\Delta E^{(2)}_{D \rightarrow A}$ stabilization to eventually overcome most or all of the $\sim$30 kcal/mol promotion barrier and yield net binding (except for OC) before final R_{eq} is achieved. Most successful in this respect are PH_3 and CO, which surmount the promotion barrier near $R_{Ni-L} \cong 2.6$ Å, leading to the "earliest" extinction of diradical $\langle S^2 \rangle$ spin contamination in Fig. 8.21. Taken together, Figs. 8.20–8.22 exhibit a fairly coherent picture of coordinative promotion in the monoligated $Ni(Lig)_1$ species, showing how differences in long-range donor–acceptor stabilizations ($\Delta E^{(2)}_{D \to A}$; Fig. 8.22) overcome the promotion barrier to singlet closed-shell formation (reduction to $\langle S^2 \rangle \to 0$; Fig. 8.21) and lead to net binding energy curves (ΔE_1; Fig. 8.20) that reflect the overall competition between promotion and closed-shell coordination. (Chemical origins of the $\Delta E^{(2)}_{D \to A}$ differences and further breakdown into $\Delta E^{(2)}_{L \to M^*}$, $\Delta E^{(2)}_{M \to L^*}$ contributions are discussed in the main text.)

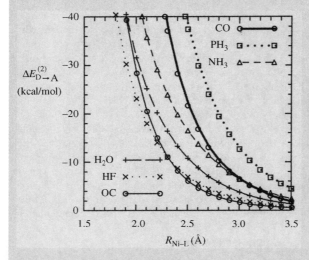

Figure 8.22 Distance-dependence of perturbative stabilizations $\Delta E^{(2)}_{D \to A}$ due to leading NBO donor–acceptor interactions in monoligated Ni(Lig)$_1$ complexes (cf. Fig. 8.13). Total $\Delta E^{(2)}_{D \to A} = \Delta E^{(2)}_{L \to M^*} + \Delta E^{(2)}_{M \to L^*}$ is evaluated as sum of leading $n_{Lig} \to n^*_{Ni}$ (σ-bonding) and $n_{Ni} \to a^*_{Lig}$ (back-bonding) interactions, neglecting many smaller contributions.

Fortunately, beyond the troublesome $n = 1$ cases, the final Ni(Lig)$_n$ equilibrium species tend to be of robust single-configurational character, well described by RHF-type DFT solutions that are stable both with respect to wavefunction form (passing the STABLE=OPT test) and geometric distortions (passing the all-positive FREQ test for vibrational frequencies). Accordingly, the analysis of bonding in these species can be carried out without undue concern for the multiconfigurational technicalities that tend to plague DFT calculations for monoligated Ni complexes.

8.3 THREE-CENTER, FOUR-ELECTRON HYPERBONDING

Still stranger violations of Lewis structural sensibilities occur when a coordinating ligand (L:) competes with an existing coordinative bond (A:L′) involving equivalent (or near-equivalent) ligand L′:, giving rise to near-degenerate dissociative pathways, namely,

$$
L:\cdots A\cdots:L' \begin{cases} L: + A:L' & \quad \textbf{I} \\ \\ L:A + :L' & \quad \textbf{II} \end{cases} \qquad (8.23)
$$

The intact molecular species on the left appears to have "too many electrons" to be viewed as a conventional Lewis structure, whereas neither dissociated form **I**, **II** on the right is objectionable. As first recognized by Coulson, the alternative formulations on the right of (8.23) can be sensibly regarded as the intermolecular *resonance structures* that contribute (near-)equal weightings w_I, w_{II} and approximate "half-bonds" to the "resonance hybrid" on the left, namely,

$$w_I \cong w_{II} \tag{8.24}$$

$$b_{AL} \cong b_{AL'} \cong 1/2, \ b_{LL'} \cong 0 \tag{8.25}$$

As expressed in the prescient phrase of Nobelist Gerhard Herzberg, the resonance *is* the binding.

Because the resonance mixing in (8.23) involves three atomic centers competing for two electron pairs, it is also described as "three-center, four-electron (3c/4e) hypervalency." The phenomenon can also be identified as "hyperbonding" and denoted by a distinctive stroke-symbol ($L \vdash A \rightarrow L'$) and ω-bond notation that suggests its unique electronic character, namely,

$$\omega_{LA:L'} \quad [\text{2c/2e } L \vdash A \text{ bond, with proximal } L'] \tag{8.26}$$

$$\omega_{L:AL'} \quad [\text{2c/2e } A \rightarrow L' \text{ bond, with proximal } \ddot{L}] \tag{8.27}$$

$$\omega_{LAL'} \quad [\text{3c/4e } L \vdash A \rightarrow L' \text{ } hyperbond; \text{ resonance hybrid of (8.26 and 8.27)}] \tag{8.28}$$

As shown by Pimentel, Rundle, and Coulson (cf. *V&B*, p. 278ff), hyperbonded 3c/4e species can also be rationalized in molecular orbital terms, but the resonance-hybrid formulation (8.23–8.28) seems to capture the essence of ω-bonding most elegantly and succinctly.

In exceptional cases, a still weirder resonance form of 3c/4e "long-bonding" (denoted with a caret or connecting "brad" symbol) may become dominant, as depicted schematically in (8.29):

$$L:\cdots A\cdots:L' \longrightarrow \ \overset{\frown}{L} \ \ddot{A} \ L' \quad (\ddot{A} + \overline{L \ L'}) \tag{8.29}$$
$$\mathbf{III}$$

The two-center electron pair of (8.29) is found to be shared between the *terminal* L, L' atoms, constituting a valid L–L' "chemical bond" in the Lewis sense, even though these atoms are spatially *separated* by the central A atom. In such a limiting case, the NRT weightings and bond orders contrast sharply with those in (8.24) and (8.25)

$$w_{III} \cong 1 \tag{8.30}$$
$$b_{LL'} \cong 1, \ b_{AL} \cong b_{AL'} \cong 0 \tag{8.31}$$

and the electronic properties of the 3c/4e-triad (8.29) are expected to differ sharply from those of the ω-bonded triad (8.10). However, further exploration of such long-bonding phenomena would carry us into the metallic domain, beyond the scope of this book.

PROBLEMS AND EXERCISES

8.1. As illustrated in (8.2), a strong Lewis acid with 1c "LP*" (n^*_A) acceptor orbital will attack a 1c "LP" (n_B) donor orbital of a Lewis base to yield a 2c dative "BD" ($\sigma_{A:B}$), expressed as the "orbital reaction"

$$n^*_A + n_B \rightarrow \sigma_{A:B} \text{ (coordinate 2c-bond formation)}$$

However, in the absence of a lone pair, the n^*_A acceptor orbital might alternatively attack a 2c bond (σ_{BC} or π_{BC}) donor orbital to yield a three-center bond (Section 4.4),

$$n^*_A + \sigma_{BC} \rightarrow \tau_{BAC} \text{ (hypovalent 3c-bond formation)}$$

a so-called "agostic interaction." As a simple example, consider the agostic interaction of BH_3 with H_2, as modeled in the following Gaussian input file:

```
#B3LYP/6-311++G** POP=NBOREAD OPT=Z-MATRIX

Model BH3 + H2 reaction for chosen DIST

0 1
x
B  1  DIST
H  2  bh1  1  ang1
H  2  bh2  1  ang2  3  dihed
H  2  bh2  1  ang2  3  -dihed
H  1  sep  2  90.0  3  90.0
H  1  sep  2  90.0  3  -90.0

bh1   1.0
bh2   1.0
ang1  90.0
ang2  90.0
dihed 120.
sep   0.37

DIST  5.0

$NBO file=BH3_H2 3CBOND $END
```

At large DIST (such as the starting distance 5.0Å shown in the file), NBO analysis is expected to lead to reactant $BH_3 + H_2$ molecular units, with LP* orbital on the BH_3 unit. However, at smaller DIST the default analysis leads to a single BH_5 product unit, with 3c τ_{HBH} bond.

(a) Starting from DIST values 5.0, 4.5, 4.0, . . . , 1.0, evaluate an approximate potential curve for the $BH_3 + H_2$ agostic interaction.

(b) Locate the approximate DIST at which the NBO description "switches" from reactant-like (with LP* on BH_3) to product-like (with 3c cyclic bridge bond). (Note that the 3CBOND search should be carried out at each DIST, as shown in the $NBO keylist.) Obtain NBOView plots of the interacting $\sigma_{HH}-n^*_B$ PNBOs (at larger DIST) and 3c τ_{HBH} NBO (at smaller DIST) on either side of this "transition state" DIST to visualize how the agostic interaction evolves to hypovalent three-center bonding in continuous fashion.

(c) From the NPA atomic charges, evaluate the total charge Q_{BH3}, Q_{H2} on each monomeric fragment, and check whether these agostic charge variations vary with DIST in the "mirror image" pattern expected for coordinative bonding (cf. Fig. 8.5).

(d) Briefly summarize your conclusions concerning the underlying orbital relationships between (i) coordinate bonding, (ii) agostic interactions, and (iii) hypovalent three-center bond formation.

(To check your results and see corresponding results for agostic π-bond interactions, see *V&B*, p. 317ff.)

8.2. Following the pattern of I/O-8.5, one can easily obtain the X_3^- series for first few trihalides ($X = F$, Cl, Br) or trialkides ($X = Li$, Na, K) that are covered by the 6-311++G** basis set. For each member of either series (or both, if you are a zealot):

(a) Evaluate the 3c binding energy ΔE_{3c} with respect to $X^- + X\text{-}X$ dissociation products.

(b) Evaluate the ω-type ($w_\omega = w_I = w_{II}$) and long-bond-type (w_{III}) NRT weightings.

(c) Plot ΔE_{3c} and w_{III} values as functions of electronegativity Ξ_X for series members. Describe your general conclusions about how 3c bonding strength and propensity for long-bonding seem to vary with "metallic" character of X.

(d) The hypothetical trihydride anion H_3^- might be visualized as a member of either sequence. Do you expect H_3^- to be a stable species? Why or why not? [If in doubt, repeat part (a) for $X = H$.]

8.3. The resonance forms **I**, **II**, **III** provide the basis for describing charge-transfer "conduction" of an electron from the left end (**I**: $X^- + X_2$) to the right end (**II**: $X_2 + X^-$) of the X_3^- triad, under the influence of an external electric field F. As a benchmark, we might say that the electron has transferred to the right (i.e., to atom X_R) whenever $w_{II} \geq 2\,w_I$, or (expressed as a more direct charge-based criterion) whenever the terminal atomic charges Q_L, Q_R satisfy $|Q_R| \geq 2\,|Q_L|$ for critical transfer field $F = F_c$. Suppose $Q_R^{(0)}$ denotes the initial (equal) charge on each terminal atom in the field-free limit. The critical field F_c to achieve the $Q_R(F_c) \geq 2\,Q_L(F_c)$ criterion must therefore correspond to net transfer of charge

$$\delta Q_c \equiv Q_R - Q_R^{(0)} \geq Q_R^{(0)}/2 \quad \text{for } F \geq F_c$$

The critical field F_c to transfer δQ_c can serve as a qualitative measure of the "barrier" to elementary charge transfer in X_3^-. The goal of this problem is to determine F_c and associated energetic and geometrical barriers for "conduction" in the triad.

A Gaussian input file to evaluate the reoptimized X_3^- triad in the presence of an electric FIELD of strength "10" (multiples of 0.0001 a.u.) along the interatomic z-direction is shown below for $X = Li$:

```
#B3LYP/6-311++G** SCF=(QC,TIGHT) NOSYMM FIELD=X+10 OPT=Z-MATRIX


Model Li3(-) "conduction" in external FIELD of strength F(X)=0.001 a.u.


-1 1
x
Li  1  1.0
Li  2  r12  1  90.0
Li  2  r23  1  ang   3  180.

r12  1.5
r23  1.5
ang  90.
```

Choose your favorite X_3^- species from the trialkide series. Reevaluate the X_3^- energy, geometry, and NBO/NRT properties for applied fields of strength $F = 0.001$, $0.002, \ldots, 0.005$ a.u. (or other field-strengths you believe to be informative). From your computational results:

(a) Plot the charges Q_1, Q_2, Q_3 of each atom as functions of field strength, and verify that negative charge ΔQ gradually "transfers" toward X_3 (X_R) as F increases. (You can also replace "+10" by "−10" in the FIELD keyword to reverse field direction and drive charge-transfer toward the opposite end.) Determine the critical field strength F_c for which the criterion of charge transfer (as given above) is satisfied.

(b) Plot the triad energy $E(F)$ as a function of field strength and evaluate the energy barrier $\Delta E_c = \Delta E(F_c)$ for the critical charge-transfer field strength F_c.

(c) Similarly, plot the geometrical variations $r_{12}(F)$, $r_{23}(F)$, and $r_{13}(F)$ as functions of field strength. Evaluate the maximum geometry change Δr_c in overall r_{13} triad length (anywhere in the interval $0 \leq F \leq F_c$) needed to accommodate the critical transfer of charge δQ_c.

(d) Carry out Problem 8.3a–c for one or more alternative X_3^- species. Can you see any relationship between the relative "nanoconduction" barriers (Δr_c, ΔE_c) for the triatomic anions and the measured conductivity differences for the bulk metals? Discuss briefly.

Chapter 9

Intermolecular Interactions

It is a surprising fact that current quantum chemical understanding of molecule formation and aggregation rests largely on work of a little-known physicist—Fritz London (1900–1954) of Breslau, Berlin, and Durham, North Carolina—who pioneered both the theory of chemical bonding (W. Heitler and F. London, *Z. Phys.* **44**, 455, 1927) and the theory of long-range intermolecular interactions (F. London, *Z. Physik. Chem.* **B11**, 222, 1930). The complexity of the former theory prevented significant chemical applications until advent of the computer era in the 1960–1970s. However, the simplicity of London's theory of long-range intermolecular forces permitted a broad range of early applications, culminating in the Hirschfelder–Curtiss–Bird "Green Bible" (J. O. Hirschfelder, C. F. Curtiss, and R. B. Bird, *Molecular Theory of Gases and Liquids*, John Wiley, New York, 1954) that virtually defined the research agenda of theoretical chemical physics throughout the precomputer era and continues to shape chemical pedagogy to this day.

The simplicity of London's long-range theory rests on a drastic approximation—neglect of Pauli exchange ("overlap-type") interactions between molecules—that permitted leading intermolecular interactions to be reduced to familiar classical electrostatic form. The widespread perception that intermolecular forces are "only electrostatics" in turn stimulated electrostatics-based empirical potentials for molecular dynamics (MD) simulations of liquids, such as the SPC ("simple point-charge") model for liquid water. Numerous scientists (and textbooks) adopted the presumption that the forces of attraction between molecules were essentially classical electrostatic in nature, clearly distinguishable from the "chemical" (exchange-type) forces of attraction between atoms. However, recent studies have increasingly called this presumption into question.

Chapter 8 has already introduced numerous examples of association between closed-shell molecules that originate in resonance-type donor–acceptor interactions. This chapter provides additional examples of intermolecular resonance bonding as well as additional NBO-based analysis tools for decomposing intermolecular interactions into components of classical and nonclassical origin. The examples suggest that proper treatment of exchange-type donor–acceptor interactions is a prerequisite for realistic description of practically *all* condensation and solvation phenomena of chemical relevance, because the benign power-law behavior of the classical

Discovering Chemistry With Natural Bond Orbitals, First Edition. Frank Weinhold and Clark R. Landis.
© 2012 John Wiley & Sons, Inc. Published 2012 by John Wiley & Sons, Inc.

long-range limit is trumped by the *exponential* onset of exchange-type intermolecular interactions in the density range of practical interest.

9.1 HYDROGEN-BONDED COMPLEXES

As numerous NBO studies have shown (see *V&B*, pp. 593–661 and references therein), hydrogen-bonded complexes of generic formula L:$\cdots$H-L$'$ may be generally characterized as donor–acceptor complexes of "n_L–$\sigma^*_{HL'}$" type, driven by intermolecular resonance delocalization from donor lone pair n_L of one monomer into the acceptor $\sigma^*_{HL'}$ orbital of the other.

Alternatively (and equivalently), L:$\cdots$H-L$'$ may be regarded as a special case of resonance hybrid (8.23) for A=H, namely,

$$\text{L:}\cdots\text{H}\cdots\text{:L}'\left\langle\begin{array}{l}\text{L: + H:L}' \\ \quad\quad\textbf{I} \\ \text{L:H + :L}' \\ \quad\quad\textbf{II}\end{array}\right. \tag{9.1}$$

with $w_I \geq w_{II}$. [If $w_{II} > w_I$, we merely reverse the identification of "covalent bond" (solid line) and "H-bond" (dotted line) in the L:$\cdots$H-L$'$ complex.] Figure 9.1 depicts the intermolecular generalization of Fig. 5.6, showing the association between a 2e-delocalization (Fig. 5.1) and the corresponding resonance diagram that underlies (9.1).

In terms of resonance hybrid (9.1) or the equivalent NBO donor–acceptor characterization, we can alternatively describe the H-bonded L:$\cdots$H-L$'$ complex as:

- Partial proton-*sharing* between lone-pair-bearing Lewis bases :L, :L$'$ (or other 2e donors)
- Partial 3c/4e *hyperbonding* between a hydride Lewis acid H-L$'$ and Lewis base: L (or other 2e donor)
- Binary association driven by partial intermolecular *charge-transfer* delocalization from the lone pair (n_L) of the :L donor into the hydride antibond ($\sigma^*_{HL'}$) of the H-L$'$ acceptor

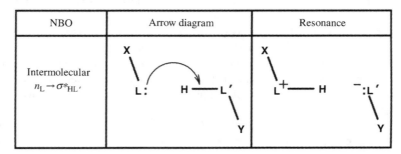

NBO	Arrow diagram	Resonance
Intermolecular $n_L \rightarrow \sigma^*_{HL'}$		

Figure 9.1 Generic "arrow pushing" diagram (left) and secondary resonance structure (right) for intermolecular $n_L \rightarrow \sigma^*_{HL'}$ NBO donor–acceptor interaction (cf. Fig. 5.6).

Implicit in all these descriptions is the dominance of the quantum mechanical "2e-stabilizing" interaction (cf. Fig. 5.1) and the specific intermolecular n_L–$\sigma^*_{HL'}$ interaction matrix element ($F_{n\sigma^*}$) in dictating H-bonding energetics and geometry. The degree of "partial" sharing, hyperbonding, and charge transfer can be quantified in terms of the relative NRT weightings (w_I versus w_{II}) or bond orders (b_{HL} versus $b_{HL'}$) in the resonance hybrid in (9.1).

It hardly needs to be stated that H-bonding interactions fall into the realm of coordination and hyperbonding as discussed in Chapter 8. Indeed, strong H-bonded species such as bifluoride anion (8.2) could be included seamlessly in Table 8.2 as examples of strong 3c/4e hypervalency. Moreover, the many formal orbital analogies between bonding in bifluoride HF_2^- and trifluoride F_3^- (Problem 8.2) serve as a representative example of the much-discussed relationship between "hydrogen bonding" and "halogen bonding" (see, for example, S. V. Rosokha et al., *Heteroatom Chem.* **17**, 449, 2006; P. Metrangolo and G. Resnati, *Science* **321**, 918, 2008). We leave it as a student exercise to show how the most powerful H-bonded species such as FHF^- or $H_5O_2^+$ (with binding energies in the 40 kcal/mol range; see *V&B*, p. 618ff) fit into the general framework of strong, symmetric 3c/4e ω-bonding as outlined in Chapter 8.

As a more representative example of weaker neutral H-bonded species, let us consider the $(HF)_2$ dimer, which offers a particularly clear contrast to the "dipole–dipole" expectations of classical electrostatics. The $(HF)_2$ species is bound by about 5 kcal/mol (in the same range as water dimer and many other common H-bonded species) and exhibits a curiously bent equilibrium geometry, as shown in I/O-9.1. Although HF has a robust dipole moment (calculated as $\mu = 1.92$ Debye) and $F^-\cdots HF$ has the linear geometry expected for an electrostatic ion–dipole complex, the nonlinear geometry of $(HF)_2$ clearly differs from the expected linear geometry of a dipole–dipole model. What's going on here?

Second-order perturbative analysis of $(HF)_2$ shows the leading intermolecular interaction to be the expected $n_F \rightarrow \sigma^*_{HF}$ delocalization from donor monomer

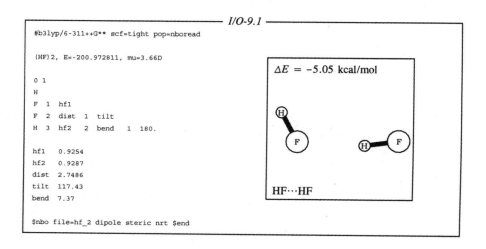

I/O-9.1

```
#b3lyp/6-311++G** scf=tight pop=nboread

(HF)2, E=-200.972811, mu=3.66D

0 1
H
F  1  hf1
F  2  dist  1  tilt
H  3  hf2   2  bend  1  180.

hf1   0.9254
hf2   0.9287
dist  2.7486
tilt  117.43
bend  7.37


$nbo file=hf_2 dipole steric nrt $end
```

$\Delta E = -5.05$ kcal/mol

HF$\cdots$HF

─────── *I/O-9.2* ───────

SECOND ORDER PERTURBATION THEORY ANALYSIS OF FOCK MATRIX IN NBO BASIS

Donor NBO (i)	Acceptor NBO (j)	E(2) kcal/mol	E(j)-E(i) a.u.	F(i,j) a.u.
from unit 1 to unit 2				
1. BD (1) H 1- F 2	53. RY*(3) H 4	0.07	2.86	0.013
1. BD (1) H 1- F 2	58. BD*(1) F 3- H 4	0.07	1.35	0.009
5. LP (1) F 2	58. BD*(1) F 3- H 4	0.21	1.44	0.015
7. LP (3) F 2	40. RY*(7) F 3	0.06	1.46	0.009
7. LP (3) F 2	58. BD*(1) F 3- H 4	6.64	1.06	0.075

H(1)-F(2) ("unit 1") to acceptor monomer F(3)-H(4) ("unit 2"), with estimated stabilization energy:

$$n_{F(2)} \to \sigma^*_{H(4)F(3)}: \quad \Delta E^{(2)}_{D \to A} = 6.64 \, \text{kcal/mol} \qquad (9.2)$$

as shown in I/O-9.2. The preferred donor orbital [NBO 7, LP(3) on F(2)] is found to be the expected off-axis p-rich $n_F^{(\pi)}$ lone pair (Sidebar 4.2), because the alternative on-axis $n_F^{(\sigma)}$ [NBO 5, LP(1) on F(2)] is too contracted and low in energy to serve as an effective intermolecular donor.

Figure 9.2 displays the overlap contour diagram for the strongly interacting $n_F^{(\pi)}$–σ^*_{HF} NBOs of $(HF)_2$. The NBO diagram immediately suggests the strong propensity for linear F$\cdots$H−F H-bonding and L-shaped dimer geometry, both of which serve to maximize $n_F^{(\pi)}$–σ^*_{HF} overlap. (The slight H-bond bending might be attributed to residual dipole–dipole forces that are expected to oppose the powerful $n_F^{(\pi)}$–σ^*_{HF} geometrical preference for L-shaped geometry.) The NRT weighting ($w_{II} = 0.62\%$) and bond order ($b_{F\cdots H} = 0.006$) associated with this n_F–σ^*_{HF} delocalization may appear miniscule, as does the intermolecular charge transfer

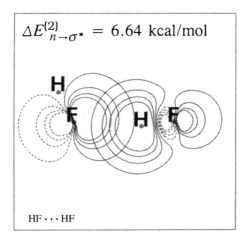

$$\Delta E^{(2)}_{n \to \sigma^*} = 6.64 \, \text{kcal/mol}$$

HF $\cdots$ HF

Figure 9.2   NBO contour diagram for leading n_F–σ^*_{HF} donor–acceptor interaction in $(HF)_2$ (with estimated second-order stabilization energy).

($Q_{CT} = 0.012\,e$) between donor and acceptor monomers. Nevertheless, as shown in (9.2), these small resonance delocalizations are fully adequate to account for the net H-bond attraction in $(HF)_2$, and the numerical relationships between w_{II}, Q_{CT}, and $\Delta E^{(2)}{}_{D\rightarrow A}$ are fully consistent with those found for a wide variety of H-bond complexes (see *V&B*, pp. 622–624).

The stabilizing donor–acceptor contribution (9.2) must of course be offset against destabilizing donor–donor (steric) interactions and combined with possible long-range electrostatic attractions to obtain the net H-bonding energy at equilibrium geometry ($\Delta E_{HB} = -5.05$ kcal/mol; cf. *I/O-9.1*). Sidebar 9.1 describes further details of the competition between Lewis (steric, electrostatic) and non-Lewis (n_F–σ^*_{HF} resonance) contributions to H-bond energy and its variation with intermolecular separation. More detailed analysis of energy components is described in Section 9.3.

Although the strangely bent geometry of $(HF)_2$ is strongly suggestive of non-electrostatic influences, how can we be assured that the nonclassical n_F–σ^*_{HF} "charge transfer" interaction, rather than classical-like "dipole–dipole" interaction, is really the essential origin of H-bond geometry and energetics?

One approach is to eliminate dipole–dipole influences by *zeroing* the dipole moment of one or both monomers of the H-bonded complex, for example, in species such as $F_2\cdots HF$ ($\mu_D = 0$), $HF\cdots H_2$ ($\mu_A = 0$), or $F_2\cdots F_2$ (both $\mu_D = \mu_A = 0$). As an example, Fig. 9.3 summarizes relevant comparisons for $F_2\cdots HF$, showing the optimized structure and binding energy (left panel) and the leading NBO donor–acceptor interaction and $\Delta E^{(2)}{}_{D\rightarrow A}$ stabilization (right panel). Net binding is significantly weakened (to $\Delta E \cong -1$ kcal/mol), as expected from the significant reduction of anionic n_F character (and Lewis base strength) in F–F versus H–F. Nevertheless, the evident similarities in geometry (cf. I/O-9.1) and NBO n_F–σ^*_{HF} overlap (cf. Fig. 9.2) show that characteristic H-bonding properties persist in $F_2\cdots HF$ (including correlations with experimental H-bond signatures), relatively *in*sensitive to erasure of the leading classical electrostatic contribution. Qualitatively similar results are found for the other dipole-challenged complexes mentioned above.

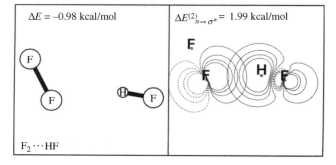

Figure 9.3 Optimized geometry (left) and leading NBO donor–acceptor interaction (right) of $F_2\cdots HF$, showing evident similarities to $HF\cdots HF$ (cf. I/O-9.1, Fig. 9.2) despite removal of leading dipole–dipole contribution (equilibrium $r_{F\cdots F} = 3.02$ Å in $F_2\cdots HF$, 2.75 Å in $HF\cdots HF$).

An alternative approach is to eliminate n_F–σ^*_{HF} or other non-Lewis influences by zeroing either the single $F_{n\sigma^*}$ matrix element or *all* non-Lewis contributions with the $DEL keylist options of the NBO program (Section 5.3; *NBO Manual*, p. B-17ff). Figure 9.4 shows the result of full NOSTAR reoptimization of $(HF)_2$, with all possible non-Lewis interactions removed. The high accuracy ($>99.9\%$) of the NLS description (Section 5.1) insures that all steric and electrostatic properties of the monomers are accurately preserved in the $E($DEL$)$ reoptimization, and indeed, the reoptimized geometry has the linear form expected for a "dipole–dipole complex." However, deletion of the non-classical n_F–σ^*_{HF} contribution has *qualitatively* altered the geometry and energetics of the complex, with $r_{F\cdots F}$ increased by ~0.6 Å (essentially, to van der Waals contact distance) and binding energy reduced to less than half its value in the actual H-bonded species. Although classical Coulomb-type forces (Section 6.2) may contribute incrementally to the H-bonding phenomenon, they are evidently unable by themselves to yield the characteristically short approach distances (typically 0.5 Å or more *in*side van der Waals contact), valence-like bending angles (leading to characteristically open three-dimensional structures), and strong binding energies (ranging up to ~40 kcal/mol) that are representative of observed H-bonded systems.

Other NBO analysis options are available to explore the unique causal relationship between the $n_F \rightarrow \sigma^*_{HF}$ charge-transfer interaction and distinctive H-bonding properties. These include the following:

- Characteristic downfield ^{1}H-NMR chemical shifts of H-bonded nuclei, investigated by NCS analysis (Section 7.1);

- Characteristic elongation (and red-shifting of IR vibrational frequencies) for H-bonded hydride bonds, investigated by $DEL-deletion techniques (Section 5.3);

- Characteristic dipole shifts and vibrational intensity enhancements associated with intermolecular charge transfer, investigated by DIPOLE analysis (Section 6.2).

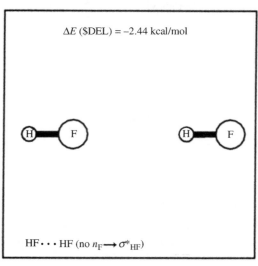

ΔE ($DEL) = -2.44 kcal/mol

HF $\cdots$ HF (no $n_F \longrightarrow \sigma^*_{HF}$)

Figure 9.4   Reoptimized NOSTAR structure of $(HF)_2$, with $DEL keylist options used to delete all non-Lewis interactions (esp., the intermolecular n_F–σ^*_{HF} charge transfer).

It is also easy to explore details of monomer *polarizability* in the induced changes of NBO hybridization and polarization coefficients (Section 4.3) as H-bond formation proceeds. With the help of NAONBO = R/W keywords (see *NBO Manual*, p. B-8), one can even "freeze" the NBO hybridizations and polarizations in their isolated monomer forms, thereby *suppressing* classical-type "polarizability effects" during H-bond formation. Of course, strong changes of overall $(HF)_2$ polarizability are found to accompany H-bond formation, but these are due primarily to the nonclassical inter-molecular charge transfer (occupancy shifts between n_F and σ^*_{HF} NBOs) rather than the mildly distortive orbital shape-changes *within* monomers that could be associated with the classical polarizability concept. One can also investigate the important *cooperativity* (nonpairwise additivity) effects of H-bonding (see *V&B*, p. 635ff), which are strictly absent in classical Coulomb electrostatics. Some further aspects of H-bond charge shifts and cooperativity are pursued in the Problems and Exercises.

SIDEBAR 9.1 | *LEWIS AND NON-LEWIS CONTRIBUTIONS TO H-BONDING*

The $DEL-keylist NOSTAR option (Section 5.3) allows direct separation of total energy E into Lewis (L) and non-Lewis (NL) components as follows:

$$E(R) = E_L(R) + E_{NL}(R) \qquad (9.3)$$

for any chosen monomer separation R. This allows the potential energy curve for H-bond formation to be visualized in terms of the *competition* between E_L (generally repulsive) and E_{NL} (*always* attractive) components. As described in Section 5.1, E_L typically includes >99.9% of total energy, including all significant steric and electrostatic features of the monomer electron density. However, the miniscule $E_{NL}(R)$ contribution (incorporating the nonclassical intermolecular charge transfer) is generally found to provide the decisive "driving force" for H-bond formation.

We can illustrate this decomposition with a simple model HF···HF potential. If we constrain the F···HF H-bond to be linear (a good simplifying approximation), only the *r12* (donor HF) and *r34* (acceptor HF) bond length and F···HF *tilt* angle variables require optimization at each R_{FF} *dist* value. Each point of the relaxed potential scan can then be decomposed into E_L, E_{NL} components as described in Section 5.3.

[*Practical Note*: The following Gaussian input file will calculate the relaxed (geometry-optimized) potential curve for values of *dist* R_{FF} in the range 2.0–4.0 Å (in increments of 0.01 Å):

```
#b3lyp/6-311++g** scf=tight opt=z-matrix


model HF...HF potential curve (constrained linear F...HF) for L/NL decomposition

0 1
H
F   1   r12
F   2   dist    1   tilt
x   3   1.0     2   90.0    1   180.0
H   3   r34     4   90.0    2   0.0
```

```
r12    0.92
r34    0.92
tilt   110.
dist   4.0  S 20 -0.1
```

For each chosen *dist* value, include the Gaussian "POP=NBODEL" keyword and attach the $DEL keylist (cf. I/O-5.2) to the end of the file containing the optimized geometry variables to obtain the E_L, E_{NL} components, as described in Section 5.3.]

The table below summarizes values of the optimized geometrical variables and E_L, E_{NL} components at selected distances in the range of interest.

Distance	Relaxed geometry (Å,)				L/NL energy components (a.u.)	
	r12	r34	Tilt	E_{tot}	E_L	E_{NL}
∞	0.9224	0.9224	180.0	−200.964755	−200.929495	0.035372
4.0	0.9229	0.9241	150.2	−200.967597	−200.931701	0.035896
3.5	0.9236	0.9261	135.2	−200.969270	−200.932367	0.036903
3.0	0.9244	0.9280	128.9	−200.971872	−200.925991	0.045881
2.5	0.9265	0.9269	123.9	−200.970505	−200.859170	0.111335

As shown in the table, the nonclassical E_{NL} component is vastly smaller than E_L. The latter contains all significant classical-type steric and electrostatic effects, as well as the energy of Lewis-type covalent bonding, and is typically >99.9% of E_{tot}. Yet as shown in the more complete potential curves of Fig. 9.5, E_L exhibits only feeble net attraction at large distances (ca. 2 kcal/mol near 3.5 Å), then turns steeply *repulsive* well beyond the distance of equilibrium H-bonding.

The steep rise of E_L coincides with the expected onset of severe steric repulsions at the van der Waals contact distance, where the monomers begin to encounter the steric

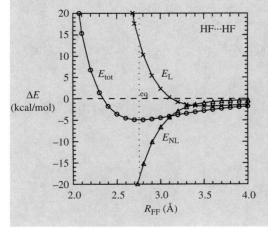

Figure 9.5 Potential energy curves $\Delta E(R)$ showing variations of Lewis (E_L, x's) and non-Lewis (E_{NL}, triangles) components of total H-bond energy (E_{tot}, circles) with intermolecular R_{FF} distance in HF···HF (constrained linear F···H−F model). The vertical dotted line marks the equilibrium distance, and the horizontal dashed line marks the limit of infinitely separated monomers.

exchange-type (exponential) donor–donor interactions of filled valence-shell orbitals. However, the "miniscule" E_{NL} component also begins to gain exponential *attractive* strength in this region, leading to an extended attractive well in overall E_{tot}. At still smaller R_{FF} the E_L steric repulsions overwhelm the competing E_{NL} donor–acceptor attractions, giving rise to the final steep repulsive behavior of E_{tot} that parallels the E_L curve, but lies about 0.6 Å *inside* the latter. Hence, one sees in the nonclassical E_{NL} attractions the essential driving force for deep penetration into the repulsive barrier presented by E_L, leading to the unusually short equilibrium distance associated with H-bonding.

The classical-type long-range Coulombic interactions also persist into the near-equilibrium H-bonding region, but they seem little more than passive bystanders to the primary competition between exponential charge transfer attractions and steric repulsions. Classical electrostatic-type contributions evaluated at the equilibrium H-bond distance are therefore somewhat misleading, because these contributions alone could never bring the monomers to the characteristic short distances of H-bonding (cf. Fig. 9.4).

It is even more misleading to use the unlimited flexibility of multipole (or distributed multipole) series to *fit* the final E_{tot} potential. With sufficient effort, such numerical fitting is assured of success for any chosen set of data values (whether of chemical origin or not). Successful numerical fitting should not be mistaken for conceptual validity or predictive reliability.

9.2 OTHER DONOR–ACCEPTOR COMPLEXES

Hydrogen-bonded clusters are an important subset of the large number of neutral and ionic binary complexes (often called "van der Waals molecules") that can now be characterized by modern gas-phase spectroscopic and molecular beam techniques. These complexes often exhibit puzzling structural properties that seem to defy simple rationalization or prediction. As an illustrative example, we consider here the simple "n–π^*" binary complexes formed from isoelectronic closed-shell diatomic species CO (carbon monoxide) and NO^+ (nitrosyl cation),

$$: C \equiv O: + :N \equiv O:^+ \rightarrow [CO \cdots NO]^+ \tag{9.4}$$

that are found to be important participants in numerous atmospheric and biological phenomena.

Compared to HF (Section 9.1), carbon monoxide has only a weak dipole moment ($\mu_{CO} \cong 0.1$ D), with its *negative* end oriented toward C (as might be expected from the formal charges on C and O in the Lewis structure diagram). The cationic charge on NO^+ (ca. 77% concentrated on N) leads to strong classical-type "ion–dipole" interactions with CO that might be expected to dominate $CO \cdots NO^+$ structure. However, the Lewis-acid strength of the NO^+ acceptor is also enhanced by its cationic character, leading to stronger nonclassical donor–acceptor interactions than those of $(HF)_2$. What are the expectations of each line of reasoning, and which type of interaction—classical electrostatic or nonclassical donor–acceptor—proves most important in dictating the equilibrium $CO \cdots NO^+$ structure?

Let us first try to predict $CO \cdots NO^+$ structure from the classical electrostatic perspective. A simple geometrical model of $CO \cdots NO^+$ angular shape is shown in (9.5), specified in terms of the angle θ between fixed monomers, with the negative end of the CO dipole oriented toward the midpoint of the NO^+ cation at distance R (optimized for each θ):

$$ \text{(9.5)} $$

From classical electrostatics we expect binding energy $\Delta E(\theta)$ to be maximized (most negative) for linear dipole–dipole alignment at $\theta = 0°$ or $180°$ (particularly the former, which points the CO dipole toward the more cationic N end). However, if dipole reasoning somehow fails, electrostatic attention turns instead to the quadrupole moment ("last refuge of the scoundrel"), which predicts maximum binding in T-shaped geometry near $\theta \cong 0°$. Thus, our electrostatics-based reasoning leads us to expect deepest $\Delta E(\theta)$ binding wells at $\theta = 0°$ and/or $180°$ (or failing that, at $\theta \cong 90°$).

The actual angular potential $\Delta E(\theta)$ is shown in Fig. 9.6, demonstrating the spectacular *failure* of electrostatics-based structural reasoning. As seen in Fig. 9.6, the predicted electrostatic "wells" at $\theta = 0°$, $180°$ (or $180°$) all turn out to be *maxima*, and the actual potential minima (near $\theta \cong 55°$, $135°$) are oriented nearly as far as possible from predicted electrostatic angles. Of course, the angular dependence of Fig. 9.6

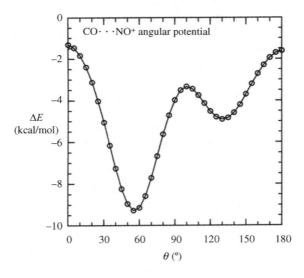

Figure 9.6 Angular potential for $CO \cdots NO^+$ interaction model (Eq. 9.5), showing binding energy $\Delta E(\theta)$ for rotation of CO dipole about the midpoint of the NO^+ cation. In this simplified model, monomer bond lengths and CO orientation toward the NO^+ midpoint are held fixed, but intermolecular distance is optimized for each point of the angular scan.

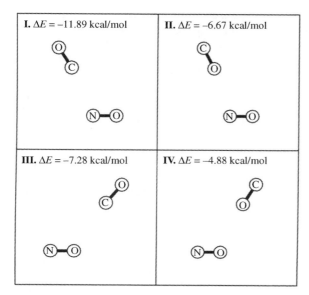

Figure 9.7   Stable isomers of CO$\cdots$NO$^+$, with calculated net binding energy ΔE (relative to CO $+$ NO$^+$).

(or an arbitrarily drawn angular potential) could be fitted with an infinite series of multipole or distributed-multipole terms, but such numerical fitting would not add appreciably to our conceptual understanding of the actual forces dictating CO$\cdots$NO$^+$ angular shape.

The failures of dipole–dipole-type reasoning become even more alarming when we explore the full CO$\cdots$NO$^+$ potential energy surface. As shown in Fig. 9.7, we find *four* distinct energetic isomers **I–IV**, all of which exhibit the strangely bent "anti-electrostatic" angles of Fig. 9.6, but with two of the isomers (**II, IV**) having the "wrong" end of the CO dipole oriented toward the cation. Table 9.1 compares calculated geometrical, energetic, and vibrational properties of the fully optimized isomers, showing that all four species are locally stable, with binding energies in the 5–12 kcal/mol range. [Although all four isomers coexist in low-temperature beam conditions, only isomer **I** (with $\Delta G^\circ = -4.05$ kcal/mol) is expected to be thermally stable at ambient T, P.] These results all suggest the general *un*reliability of classical electrostatic reasoning for predicting or rationalizing the structures of such complexes.

Let us instead consider the CO$\cdots$NO$^+$ structure(s) from the donor–acceptor perspective. It is easy to guess that NO$^+$ cation is the Lewis acid "acceptor" and CO the Lewis base "donor" for complexation. We can also guess that the n_C and n_O lone pairs of CO and the π^*_{NO} antibonds of NO$^+$ are the leading candidate NBOs for n–π^* donor–acceptor interactions. From the expected "four-leaf clover" shape of the π^*_{NO} acceptor orbital, we can easily anticipate that n_C–π^*_{NO} or n_O–π^*_{NO} overlap is maximized in structures such as **I–IV**, as shown by the NBO contour diagrams of Fig. 9.8. From the relative amplitudes of the π^*_{NO} lobes (which suggest larger acceptor strength at the N end of the antibond) and the relative electronegativities of

Table 9.1 Geometrical, energetic, charge-transfer, and vibrational properties of $CO\cdots NO^+$ isomers, showing geometrical variables, binding energy and free energy (ΔE, ΔG° relative to $CO + NO^+$ monomers; kcal/mol), net intermolecular charge transfer (Q_{CT}; e), and vibrational frequencies (ν_i; cm^{-1}), for each isomer **I–IV** (cf. Fig. 9.7).

	I	**II**	**III**	**IV**
r_{NO}	1.0710	1.0642	1.0642	1.0614
r_{NC}	2.4846	3.5890	3.3978	4.2801
r_{CO}	1.1181	1.1378	1.1202	1.1356
θ_{ONC}	113.67	110.81	42.53	50.82
θ_{NCO}	171.87	5.78	172.70	7.20
dihed	180.00	0.00	180.00	0.00
ΔE	−11.89	−6.67	−7.28	−4.88
ΔG°	−4.05	+0.20	+10.04	+1.58
Q_{CT}	0.1628	0.0567	0.0749	0.0302
ν_1	140	104	124	93
ν_2	161	112	139	108
ν_3	173	133	141	110
ν_4	350	221	220	149
ν_5	2292	2128	2275	2146
ν_6	2362	2423	2414	2442

C, O (which suggest greater donor strength of the n_C lone pair), we can also guess the qualitative ordering of n–π^* interaction strength in **I–IV**, as shown in the $\Delta E^{(2)}_{n \to \pi^*}$ values in each figure panel. Although numerous other intermolecular interactions are contributing to net binding and charge transfer (cf. Table 9.1), those shown in Fig. 9.8 exert decisive control over the structure and energetics of each isomer (as might be tested by $DEL reoptimizations analogous to those carried out in Section 9.1).

Several unique features distinguish these n–π^* complexes from H-bonded complexes of n–σ^* type. As shown by comparison of stabilization energy $\Delta E^{(2)}_{n \to \pi^*}$ (Fig. 9.8) and net binding energy ΔE (Fig. 9.7), the leading $n \to \pi^*$ charge-transfer (CT) interactions cannot account for the full attraction between monomers. Also, the n–π^* complexes exhibit conspicuously larger intermolecular CT (cf. Table 9.1) for the given $\Delta E^{(2)}_{n \to \pi^*}$. These two features are evidently connected. The enhanced $Q_{n \to \pi^*}$ is due essentially to the unusually small energy separation between the cationic π^* acceptor and neutral lone pair donor orbitals in these species, which alters the usual perturbative proportionality factor between $Q_{n \to \pi^*}$ and $\Delta E^{(2)}_{n \to \pi^*}$ [cf. *V&B*, Eq. (2.18)]. As a result of such CT-induced charge delocalization, Coulomb-type repulsions *within* the cation are significantly reduced, conferring significant electrostatic stabilization on the complex. This additional "CT-induced electrostatic stabilization" provides an instructive example of the symbiotic interplay between classical and nonclassical contributions to binding energy, which intrinsically makes the separation into independent "energy components" somewhat problematic.

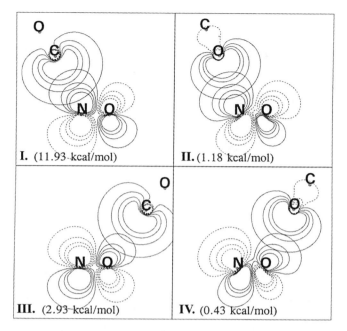

Figure 9.8 NBO contour diagrams (and estimated $\Delta E^{(2)}_{n \to \pi^*}$ stabilizations) for leading $n_{C/O}-\pi^*_{NO}$ donor–acceptor interactions in CO$\cdots$NO$^+$ isomers **I–IV** (cf. Fig. 9.6, Table 9.1).

Finally, let us consider complexation by a second CO molecule to form a NO$^+$(CO)$_2$ trimer. Two reasonable donor–acceptor possibilities may be considered:

(1) The n_C lone pair of the second CO may attack the other large-amplitude (N-based) lobe of the π^*_{NO} antibond (cf. **I** in Fig. 9.8), giving a Y-shaped planar structure that minimizes the steric repulsion between CO molecules. However, such competitive double-donor attack on the same acceptor orbital is anti-cooperative (see *V&B*, p. 635ff), so the incremental binding energy for the second CO will be considerably less than that (11.89 kcal/mol; cf. Fig. 9.7) for the first, making this structure suboptimal.

(2) The n_C lone pair of the second CO may attack a large-amplitude N-based lobe of the *other* π^*_{NO} antibond, in a plane perpendicular to that of the first. This leads to a pyramidal "folded" geometry of C_{2v} symmetry, as shown in the optimized structure of I/O-9.3. Although this structure appears quite strange from both electrostatic and steric viewpoints, it takes maximal advantage of the *two* powerful π^*_{NO} acceptor orbitals of nitrosyl cation and clearly makes sense from the donor–acceptor perspective. The calculated binding energy, $\Delta E = -20.07$ kcal/mol, corresponds to incremental $\Delta\Delta E = -8.18$ kcal/mol, showing the expected anti-cooperativity of a "busy" acceptor monomer (even if using distinct acceptor NBOs). Note that a planar-constrained Y-shaped geometry [as described in (1)] is found to lie about 0.9 kcal/mol

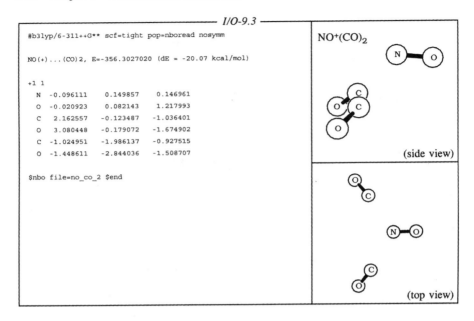

─────────── *I/O-9.3* ───────────

```
#b3lyp/6-311++G** scf=tight pop=nboread nosymm

NO(+)...(CO)2, E=-356.3027020 (dE = -20.07 kcal/mol)

+1 1
  N   -0.096111    0.149857    0.146961
  O   -0.020923    0.082143    1.217993
  C    2.162557   -0.123487   -1.036401
  O    3.080448   -0.179072   -1.674902
  C   -1.024951   -1.986137   -0.927515
  O   -1.448611   -2.844036   -1.508707

$nbo file=no_co_2 $end
```

NO⁺(CO)₂

(side view)

(top view)

higher than the folded geometry shown in I/O-9.3, representing a transition state between inverted pyramidal structures.

Figure 9.9 illustrates one of the two near-equivalent $n_C-\pi^*{}_{NO}$ interactions of NO$^+$(CO)$_2$ in two-dimensional contour and three-dimensional surface plots for comparison with the analogous interaction **I** of Fig. 9.8, showing the anti-cooperative weakening of $\Delta E^{(2)}{}_{n \to \pi^*}$ in the trimeric complex. Note that the dihedral fold-angle

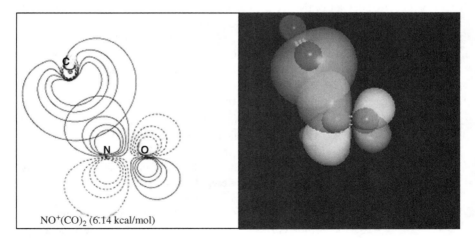

NO⁺(CO)₂ (6.14 kcal/mol)

Figure 9.9 NBO contour diagram (left) and surface plot (right) of $n_C-\pi^*{}_{NO}$ donor–acceptor interaction in NO$^+$(CO)$_2$ (with estimated $\Delta E^{(2)}{}_{n \to \pi^*}$ stabilization). The O atom of CO lies slightly out of the contour plane in the left panel. (See the color version of this figure in the Color Plates section.)

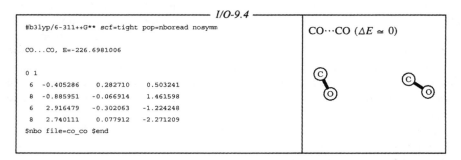

I/O-9.4

```
#b3lyp/6-311++G** scf=tight pop=nboread nosymm

CO...CO, E=-226.6981006

0 1
  6   -0.405286    0.282710    0.503241
  8   -0.885951   -0.066914    1.461598
  6    2.916479   -0.302063   -1.224248
  8    2.740111    0.077912   -2.271209
$nbo file=co_co $end
```

$CO \cdots CO$ ($\Delta E \simeq 0$)

($\sim 102°$) of the optimum structure places the two CO groups slightly out of alignment with the orthogonal planes of the π^*_{NO} NBOs, so that each n_C has weak secondary delocalization (namely, 2.52 kcal/mol for the n_C shown in Fig. 9.9) with the "other" π^*_{NO} orbital.

The interesting cooperative and anti-cooperative aspects of donor–acceptor complexation in trimeric and higher clusters provide rich opportunities for further chemical exploration, beyond the scope of present discussion. Note that Coulomb's law of classical electrostatics epitomizes the pairwise-additive limit in which cooperativity effects (of either sign) are strictly *absent*. Classical-type electrostatic reasoning therefore serves insidiously to divert attention from *concerted*-CT effects in donor–acceptor networks (including H-bond circuits) that appear highly significant for overall structure and reactivity. Even the simplest trimeric species demonstrate that CT-type cooperativity effects require accurate recognition (and numerical modeling) if realistic chemical understanding of complex higher-order cluster networks is to be achieved.

Although it sounds paradoxical, increasing the Coulombic charge of a monomer apparently enhances donor–acceptor interactions *faster* than classical electrostatic interactions (although, of course, *both* are significantly enhanced by net ionic charge). The difference is illustrated, for example, by comparing the charged $NO^+ \cdots CO$ complex with the isoelectronic neutral $CO \cdots CO$ complex. As shown in I/O-9.4, the latter optimizes to a feebly bound dimer ($R_{CO \cdots CO} > 3.8$ Å) that vaguely resembles an electrostatic "dipole–quadrupole complex" but lacks any vestige of donor–acceptor bonding. The relative ineffectiveness of $n–\pi^*$ complexation compared to H-bonding $n–\sigma^*$ interactions can be traced to the much stronger steric barriers presented by the inner cores of π-bonding atoms (whereas the H atom is uniquely *free* of such core repulsions; see *V&B*, pp. 660–661). In the absence of ionic enhancement, the $n–\pi^*$ interactions of $CO \cdots CO$ cannot overcome the usual steric barriers associated with short-range (sub-R_{vdW}) approach, and the species reverts to the type of long-range "electrostatic complex" that is accurately described by London's theory.

9.3 NATURAL ENERGY DECOMPOSITION ANALYSIS

A much more sophisticated and thorough analysis of intermolecular interactions is provided by the *natural energy decomposition analysis* (NEDA) module of the NBO program. The NEDA keyword is implemented as an optional feature of the $DEL keylist (Section 5.3) and, like other $DEL options, requires a complex series of interactive tasks with the host ESS program. At present, NEDA is fully implemented only for GAMESS and NWCHEM programs as the host ESS.

Full description of NEDA method and usage involves advanced concepts beyond the scope of the present work. The illustrative applications of NEDA provided in the *NBO Manual* (p. B-104ff) are quite extensive, and should be consulted before attempting research-level use of this keyword. This discussion provides only a qualitative physical description of NEDA energy components and their evaluation for a simple case requiring only the most primitive form of NEDA keyword input (for default NBO molecular units), namely,

```
$DEL NEDA END $END
```

However, the enthusiastic chemical explorer is encouraged to investigate the many options for alternative dissections of the target supramolecular species and the deeper quantum mechanical subtleties of the underlying NEDA mathematical formalism (E. D. Glendening and A. Streitwieser, *J. Chem. Phys.* **100**, 2900, 1994; E. D. Glendening, *J. Am. Chem. Soc.* **118**, 2473, 1996; G. K. Schenter and E. D. Glendening, *J. Phys. Chem.* **100**, 17152, 1996).

For a given supramolecular A···B complex in given geometry, described by wavefunction Ψ, the goal of NEDA is to calculate the binding energy ΔE and its decomposition into well-defined electrical (*EL*), charge-transfer (*CT*), and residual core-repulsion (*CORE*) contributions of clear physical origin, namely,

$$\Delta E = \Delta E_{EL} + \Delta E_{CT} + \Delta E_{CORE} \tag{9.6}$$

NEDA evaluates ΔE by first performing separate wavefunction calculations Ψ_A, Ψ_B on each monomer A, B (in its geometry in the complex) with the full dimer basis set, corresponding to the "counterpoise-corrected" binding energy (as defined by S. F. Boys and F. Bernardi, *Mol. Phys.* **19**, 553, 1970), namely,

$$\Delta E = E(\Psi) - [E(\Psi_A) + E(\Psi_B)] \tag{9.7}$$

The key step of NEDA decomposition is to evaluate for each monomer a "deformed" wavefunction (Ψ_A^{def}, Ψ_B^{def}, constructed from block eigenvectors of the NBO Fock matrix with intermolecular-CT elements deleted) that includes all the Lewis-type influences due to electric fields and steric pressure of the surrounding monomer(s) (as well as all intra-monomer non-Lewis effects), but *deletes* intermolecular CT. The antisymmetrized product of Ψ_A^{def}, Ψ_B^{def} then provides the "localized" (CT-suppressed) dimer wavefunction Ψ_{loc} ($= \det|\Psi_A^{def}\Psi_B^{def}|$) that allows identification of the charge-transfer component ΔE_{CT} as the energy difference

$$\Delta E_{CT} \equiv E(\Psi) - E(\Psi_{loc}) \tag{9.8}$$

The difference between dimer $E(\Psi_{loc})$ and the sum of monomer $E(\Psi_A^{def})$, $E(\Psi_B^{def})$ may now be attributed to all remaining (non-CT) interactions between monomers, including the exchange-type (*EX*) effects of overall antisymmetry, the electrostatic (*ES*) interaction between monomer charge distributions, and the polarizations (*POL*) induced by the fields from each monomer on the charge distribution of the other, as expressed by the following equation:

$$E(\Psi_{\text{loc}}) = [E(\Psi_{\text{A}}^{\text{def}}) + E(\Psi_{\text{B}}^{\text{def}})] \equiv \Delta E_{\text{EX}} + \Delta E_{\text{ES}} + \Delta E_{\text{POL}} \tag{9.9}$$

In addition, the overall "deformation energy" ΔE_{DEF} of $A \cdots B$ formation is obtained by summing the energetic difference between $\Psi_{\text{A}}^{\text{def}}$, Ψ_{A} for each monomer,

$$\Delta E_{\text{DEF}} \equiv [E(\Psi_{\text{A}}^{\text{def}}) - E(\Psi_{\text{A}})] + [E(\Psi_{\text{B}}^{\text{def}}) - E(\Psi_{\text{B}})] \tag{9.10}$$

representing all monomer distortions induced by presence of the other monomer, due to steric pressure, electric fields, or any other envisioned intermolecular influence (except CT).

From the base Equations (9.8)–(9.10), one can now extract the various quantities contributing to each NEDA component in (9.6). The "electrical" term ΔE_{EL} arises from classical electrostatic (ΔE_{ES}) and induction ($\Delta E_{\text{POL}} + \Delta E_{\text{SE}}$) contributions that are well described by electrodynamics,

$$\Delta E_{\text{EL}} \equiv \Delta E_{\text{ES}} + \Delta E_{\text{POL}} + \Delta E_{\text{SE}} \tag{9.11}$$

The "self-energy" (ΔE_{SE}) term in (9.11) is computed as the linear response (energy penalty) of polarization. The "core" contribution arises principally from intermolecular exchange interactions (ΔE_{EX}) and deformations (ΔE_{DEF}) due to distortion of monomer wavefunctions by fields from other monomers (but neglecting the self-energy term that was previously accounted to ΔE_{EL}), namely,

$$\Delta E_{\text{CORE}} \equiv \Delta E_{\text{EX}} + \Delta E_{\text{DEF}} - \Delta E_{\text{SE}} \tag{9.12}$$

Together with the starting definition of ΔE_{CT} and subsidiary definitions extracted from (9.9) and (9.10), one readily verifies that (9.8), (9.11), and (9.12) sum identically to give

$$\Delta E = \Delta E_{\text{CT}} + \Delta E_{\text{EL}} + \Delta E_{\text{CORE}} \tag{9.13}$$

which is the basic NEDA decomposition.

Because the decomposition is performed at equilibrium geometry, the "classical-type" ΔE_{EL} term includes many contributions from resonance-type (CT-induced) interactions of nonclassical origin. As described in Sidebar 9.1, the long-range classical electrostatic-type interactions alone are unable to bring monomers to near-equilibrium geometry, so their extrapolation to this limit is somewhat misleading. However, the form of Equation (9.13) emulates that of alternative "energy decomposition analysis" schemes (Sidebar 9.2) and corrects for the common conclusion that "electrical" or "electrostatic" components are dominant contributions to H-bonding and other intermolecular interactions. For $(H_2O)_2$ dimer, some detailed numerical comparisons of NEDA components with those of the early Kitaura–Morokuma method (*Int. J. Quantum Chem.* **10**, 325, 1976) are presented in the NEDA website tutorial (http://www.chem.wisc.edu/~nbo5/tut_neda.htm).

More general aspects of the comparison of alternative EDA methods with NAO/NBO-based methods are summarized in Sidebar 9.2.

SIDEBAR 9.2	*THE MANY VARIETIES OF "ENERGY DECOMPOSITION ANALYSIS"*

A considerable variety of "energy decomposition analysis" (EDA) schemes are found in the literature, originating in the Kitaura–Morokuma method (K. Morokuma, *Acc. Chem. Res.* **10**, 294, 1977), but including the more recent Bickelhaupt–Baerends (F. M. Bickelhaupt and E. J. Baerends, *Rev. Comput. Phys.* **15**, 1, 2000), "Block-Localized Wavefunction" (BLW-EDA: K. Nakashima, X. Zhang, M. Xiang, Y. Lin, M. Lin, and Y. Mo, *J. Chem. Theory Comput.* **7**, 639, 2008), and related "Absolutely Localized Molecular Orbital" (ALMO-EDA: R. Z. Khaliulin, R. Lochan, E. Cobar, A. T. Bell, and M. Head-Gordon, *J. Phys. Chem. A* **111**, 873, 2007) variants. The related "intermolecular perturbation theory" or "symmetry-adapted perturbation theory" of Stone and coworkers (IPT/SAPT: I. C. Hayes and A. J. Stone, *Mol. Phys.* **53**, 69, 1984); A.D. Buckingham, P. W. Fowler, and A. J. Stone, *Int. Rev. Phys. Chem.* **5**, 107, 1986) are also representative of this group.

Although details of these methods vary slightly, all are characterized by *overlap dependencies* that lead to sharp disagreements with NAO/NBO-based and other overlap-free methods. Such overlap-dependence intrinsically leads to ambiguity in assigning electron overlap density to one atomic center or another, thus rendering identification of "charge transfer" essentially arbitrary and subjective. The unphysical aspects of overlap-dependent charge assignments are widely recognized in the pathologies of Mulliken population analysis (see Sidebar 3.1), but the equivalent pathologies are less obvious when only the purported "charge transfer energy" (rather than quantity of charge itself) is presented in EDA output. (For additional discussion of overlap artifacts in wavefunction analysis, see *V&B*, pp. 229–234; F. Weinhold and J. E. Carpenter, *J. Mol. Struct. (Theochem)* **165**, 189, 1988; F. Weinhold, *Angew. Chem. Int. Ed.* **42**, 4188, 2003; www.chem.wisc.edu/~nbo5/tut_neda.htm, and references therein.)

In the perturbative framework, such overlap-dependence also runs afoul of the mathematical requirement that non-degenerate eigenfunctions of physical Hermitian operators (such as the presumed "unperturbed Hamiltonian" operator underlying IPT/SAPT) are necessarily orthonormal, thus implying non-Hermitian probability nonconservation, Pauli violations, and other aberrations. Indeed, IPT-based formulations fail to recognize "charge transfer" as a valid physical effect, claiming instead that it is "ill-defined" and "part of the induction (polarization)" component that "vanishes in the limit of a complete basis" [A. J. Stone. *Chem. Phys. Lett.* **211**, 101, 1993]. The disagreements between overlap-dependent and overlap-free methods could hardly be stronger.

Still other methodological issues surround Mo's BLW-EDA, which is closely tied to the original Heitler–London formulation of covalent bond formation with its demonstrated numerical failures for systems other than H_2 (J.M. Norbeck and G. A. Gallup, *J. Am. Chem. Soc.* **96**, 3386, 1974; cf. Sidebar 5.3).

The overlap-sensitivity of EDA conclusions can be easily tested by reformulation in terms of NAOs or other overlap-free orbitals. Indeed, the recent ETS-NOCV ("Extended Transition State – Natural Orbital for Chemical Valence") method of Ziegler and coworkers (M. P. Mitoraj, A. M. Michalak, and T. Ziegler, *J. Chem. Theory Comput.* **5**, 962, 2009) corrects for normalization and Pauli-violation errors by a simple AO symmetric orthogonalization procedure, and its conclusions closely resemble those of NAO/NBO-based methods.

Many lines of theoretical and experimental evidence now support the conclusion that covalent-type charge transfer interactions (rather than classical-type electrostatic interactions) are the ubiquitous defining feature of all H-bonding phenomena (see S. J. Grabowski, *Chem. Rev.* **111**, 2597–2625, 2011). Although electrostatic-type rationalizations based on overlap-dependent EDA methods continue to appear in the literature, their divergent conclusions with respect to overlap-free NAO/NBO analysis methods should be considered neither surprising nor informative. *Caveat emptor.*

PROBLEMS AND EXERCISES

9.1. Solid carbon dioxide ("dry ice") has a relatively high melting point ($-78°C$), indicative of appreciable intermolecular interactions despite lack of a dipole moment. One might instead surmise that attractions between CO_2 molecules are related to the nonvanishing quadrupole moment, which favors T-shaped geometry of interaction. However, after considerable theoretical and experimental controversy the actual geometry of $CO_2\cdots CO_2$ is now known to be the surprising "slipped parallel" structure shown below (with corresponding z-matrix input):

```
#b3lyp/6-311++G** scf=tight

(CO2)2 dimer

0 1
8    -2.3213    1.6403    0.0000
6    -1.1616    1.6245    0.0000
8     0.0000    1.6146    0.0000
8    -0.0000   -1.6205    0.0000
6     1.1616   -1.6245    0.0000
8     2.3214   -1.6344    0.0000
```

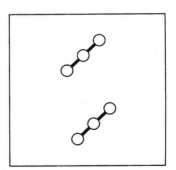

(a) What is the calculated net binding energy of the dimer? What is the effective point-group symmetry of the equilibrium dimer complex? How does the closest intermolecular $O\cdots O$ distance compare with expected van der Waals contact distance?

(b) From NBO analysis of the equilibrium dimer species, find the intermolecular NBO donor–acceptor interactions that seem to be primarily responsible for dimer formation. Plot two-dimensional contour and three-dimensional surface views of these interactions. Does the optimized geometry make sense in terms of maximizing these donor–acceptor attractions? Explain briefly.

(c) From the initial z-matrix shown below, obtain the optimized geometrical, energetic, and vibrational properties of the hypothetical T-shaped dimer that would be suggested by quadrupole–quadrupole interactions. From the vibrational analysis, verify that T-shaped geometry actually corresponds to a *transition* state, not an alternative equilibrium isomer.

```
#b3lyp/6-311++G** opt freq

(CO2)2 T-shaped isomer?

0 1
x
C  1  1.
O  2  co   1  90.
O  2  co   1  90.  3  180.
C  2  cc   3  90.  1  180.
x  5  1.   2  90.  4  180.
O  5  co   6  90.  1  180.
O  5  co   6  90.  7  180.

co   1.2
cc   3.
```

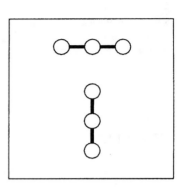

(d) Summarize your overall conclusions concerning the relative importance of quantal donor–acceptor forces versus classical quadrupole–quadrupole forces in the structure and energetics of the CO_2 dimer. Compare your conclusions with a recent research article (J. A. Gomez Castano, A. Fantoni, and R. M. Roman., *J. Mol. Struct.* **881**, 68–75, 2008) that also examines the $CO_2\cdots N_2$ complex.

9.2. Carbon monoxide (CO) and hydrogen fluoride (HF) are elementary diatomic species with nonvanishing dipole moments, calculated as follows:

$$\mu_{CO} = 0.0716 \, D[C^-O^+]$$
$$\mu_{HF} = 1.9864 \, D[H^+F^-]$$

(As mentioned in the text and indicated in brackets, CO has a surprisingly "reversed" polarity compared to that expected from atomic electronegativity differences, but that is a question for another chapter.) The $CO\cdots HF$ interaction might therefore be reasonably expected to epitomize the idealized limit of classical dipole–dipole forces.

(a) Calculate the binding energy (ΔE_{OC-HF}), equilibrium distance (R_{OC-HF}), and dipole moment (μ_{OC-HF}) of the $OC\cdots HF$ species with *co*-alignment of μ_{CO}, μ_{HF} dipoles (as expected from classical electrostatics). Suggested z-matrix input is shown below:

```
#b3lyp/6-311++g** opt

OC...HF (co-aligned dipoles)

0 1
O  0.  0.  0.
C  0.  0.  d1
H  0.  0.  d1d2
F  0.  0.  d1d2d3

d1     1.2
d1d2   4.2
d1d2d3 5.2
```

(b) Similarly, calculate the binding energy (ΔE_{CO-HF}), equilibrium distance (R_{CO-HF}), and dipole moment (μ_{CO-HF}) of the alternative CO$\cdots$HF isomer with *anti*-alignment of monomer dipoles. Does the *repulsive* dipole–dipole force prevent formation of an equilibrium dimer?

(c) Despite their diametrically opposed dipole–dipole forces, the "pro-dipole" (OC$\cdots$HF) and "anti-dipole" (CO$\cdots$HF) isomers seem to exhibit many comparable features. Identify the principal NBO donor–acceptor interaction that seems to best account for the formation of each complex and show comparison two-dimensional and three-dimensional graphical plots to illustrate their analogous features. By considering the various contributions to the second-order perturbative estimates of OC$\cdots$HF versus CO$\cdots$HF interaction strength, describe (as specifically as possible) the most important *difference* that seems to favor this interaction in the pro-dipole versus the anti-dipole isomer.

(d) In classical electrostatics, the monomer dipoles are expected to add vectorially to give the resultant dipole of the complex, namely,

$$\mu_{OC-HF} = \mu_{HF} + \mu_{CO}$$
$$\mu_{CO-HF} = \mu_{HF} - \mu_{CO}$$

Compare this classical expectation with the actual quantal behavior. What is the percentage error of the classical dipole–additivity in each case? From the results of DIPOLE analysis (Chapter 6), try to characterize the reasons for the enormous dipole–additivity errors as clearly as possible. Do you find evidence for large changes of individual monomer dipole moments ("polarization") in the complex? If not, where does the strong dipolar rearrangement originate?

9.3. The elementary HF$\cdots$HF hydrogen bond might also be studied in larger HF clusters, namely,

$$(HF)_n, \quad n = 2, 3, 4, \ldots, 8$$

Rather than the linear arrays suggested by classical dipole theory, the quantum clusters strongly prefer *cyclic* topologies. The input z-matrix shown below suggests how one can obtain the optimized 8-mer:

```
%chk=hf_8
#b3lyp/6-311++g** opt

HF(8), E=, dE=

0 1
H
F    1   .92
H    2   hb    1   ang
F    3   .92   2   178.   1   5.
H    4   hb    3   ang    2   3.
F    5   .92   4   177.   3   1.
H    6   hb    5   ang    4   6.
F    7   .92   6   174.   5  -4.
H    8   hb    7   ang    6   3.
F    9   .92   8   176.   7   2.
H   10   hb    9   ang    8  -5.
F   11   .92  10   178.   9   5.
H   12   hb   11   ang   10  -3.
F   13   .92  12   176.  11  -1.
H   14   hb   13   ang   12   5.
F   15   .92  14   178.  13  -2.

hb    2.
ang  139.
```

By removing the last two atoms and reducing *ang* as appropriate [e.g., to 130 ($n = 7$), 120 ($n = 6$), 108 ($n = 5$), 90 ($n = 4$), or 60 ($n = 3$)], one can easily generate each member of the sequence.

(a) Calculate the overall binding energy $\Delta E(n)$ and average H-bond strength

$$\Delta E_{HB}(n) = \Delta E(n)/n, \quad n = 3\text{-}8$$

for each cyclic cluster. Plot $\Delta E_{HB}(n)$ versus n to show the trend in H-bond strength with increase in cluster size. Describe the principal features of the observed trend. Does it correspond to the *constant* value that would be expected in a pairwise-additive potential such as Coulomb's law?

(b) Similarly, evaluate the average H-bond distance $R_{HB} = R_{H\cdots F}$ in each cluster, and plot $R_{HB}(n)$ versus n. Describe the relationship to the trends in H-bond energy seen in Problem (a).

(c) Similarly, evaluate the average second-order interaction energy $\Delta E^{(2)}{}_{n \to \sigma^*}(n)$ and σ^*_{HF} occupancy in each cluster, and show plots of each quantity versus n. Describe the relationship to the trends seen in Problems (a) and (b).

(d) Beyond $n = 5\text{-}6$, the clusters increasingly *buckle* out of planarity, and the sharp cooperative increases in binding energy appear to saturate. What is so favorable about $(HF)_5$ or $(HF)_6$ clustering, and why do larger cyclic clusters appear to increasingly resist planarity? Provide a clear orbital rationale for these striking features of the nonlinear trend lines for larger n.

Chapter 10

Transition State Species and Chemical Reactions

The theoretical challenge of characterizing the bonding of a molecular or supramolecular species becomes considerably more complex when the species undergoes chemical reaction. A primary goal of chemical theorists is to elucidate the macroscopic reaction thermodynamics and kinetics in terms of "elementary" reactions that compose the "mechanism" of overall chemical transformation. Such elementary reactions are typically of unimolecular

$$A \rightarrow [A*]^{\ddagger} \rightarrow B \qquad (10.1)$$

or bimolecular type

$$A + B \rightarrow [A \cdots B]^{\ddagger} \rightarrow C + D \qquad (10.2)$$

but may be expressed more generically in terms of reactant (R) and product (P) species

$$R \rightarrow [TS]^{\ddagger} \rightarrow P \qquad (10.3)$$

where $[TS]^{\ddagger}$ denotes the *transition state* (TS) species (originally called the "activated complex" in Arrhenius reaction theory). Our broad theoretical objective is to characterize the energetics and dynamics of each reaction type in terms of electronic-level understanding of the underlying potential energy surface(s), particularly in the TS region.

For a reaction profile exhibiting a typical Arrhenius-type energy barrier, the TS species can be formally identified as the topological *saddle-point* (stationary point of order one) that separates reactant and product minima on the potential energy surface. In the more descriptive language of mountain travel, we can equivalently describe the energetic saddle-point as the lowest-energy "pass" that allows crossing from one "valley" (reactant) to another (product). The unique minimum-energy pathway or *intrinsic reaction coordinate* (IRC) that connects the energetic TS to associated reactant and product species can be symbolized by a

Discovering Chemistry With Natural Bond Orbitals, First Edition. Frank Weinhold and Clark R. Landis.
© 2012 John Wiley & Sons, Inc. Published 2012 by John Wiley & Sons, Inc.

collective coordinate s that originates at the TS (where $s = 0$) and ranges through positive ("forward") s-values toward products, or negative ("backward") s-values toward reactants,

$$s_{IRC}(\text{reactant}) < 0, \quad s_{IRC}(\text{TS}) = 0, \quad s_{IRC}(\text{product}) > 0 \qquad (10.4)$$

Numerical methods for determining the energy and geometry of the saddle-point TS species or other intermediate species along the IRC pathway are implemented in popular electronic structure programs, but further discussion of these methods is beyond the scope of this discussion [see, e.g., J. B. Foresman and A. Frisch, *Exploring Chemistry With Electronic Structure Methods*, Gaussian, Pittsburgh PA, 1996, and references therein]. Our primary goal in this chapter is to explore the IRC pathway and provide a descriptive roadmap of the TS and other principal features of the chemical reaction landscape.

Despite their central role in chemical reaction theory, TS species challenge conventional structural characterization by experimental means. Modern *ab initio* methods therefore provide a uniquely valuable source of detailed TS information that can significantly advance understanding of chemical reactivity. Given the fact that accurate TS wavefunctions and IRC profiles are now routinely available for a variety of chemical reactions, our aim is to extend NBO/NRT-based analysis techniques to characterize TS and other IRC species in simple Lewis structural and resonance theoretic terms, analogous to those found useful for equilibrium species.

10.1   AMBIVALENT LEWIS STRUCTURES: THE TRANSITION-STATE LIMIT

From the conventional representation (10.3), it is apparent that the mysterious $[TS]^{\ddagger}$ species is to be regarded as neither R-like nor P-like in its chemical bonding pattern, but as some type of "intermediate," "hybrid," or "crossing point" of R/P characteristics. [The word "intermediate" too strongly suggests a (meta)stable chemical species occupying a shallow local minimum atop the reaction barrier, whereas we wish to focus on a true saddle-point TS species.] Such characterization immediately suggests the aptness of a *resonance* description of the TS species, expressing its *ambivalence* toward either R-type or P-type Lewis structural depiction. Indeed, in NRT terms (Section 5.6), we might identify the TS limit with the expected transition from predominant R-like weighting (with $w_R > w_P$) to P-like weighting (with $w_P > w_R$), leading to definition of the *natural transition state* (NTS) along a chosen reaction coordinate s as follows:

$$\text{"Natural" TS}: \quad w_R = w_P \text{ ("half-reacted") at } s_{IRC}(\text{NTS}) \qquad (10.5)$$

The NTS criterion (10.5) may serve as a useful alternative to the ETS saddle-point criterion for the case of barrierless reactions, and the two criteria will be compared in what follows.

As simple prototype examples, let us first consider two elementary gas-phase reactions that were previously encountered in Problem 4.1, both leading to production of formaldehyde (H_2CO, product species P_I). The first involves bimolecular reaction of dihydrogen and carbon monoxide ($H_2 \cdots CO$, reactant species R_{II}), namely,

$$H_2 + CO \rightarrow H_2CO \tag{10.6}$$

with associated transition state species TS_{I-II},

$$R_{II} \rightarrow [TS_{I-II}]^{\ddagger} \rightarrow P_I \tag{10.7}$$

The second involves unimolecular isomerization (hydride transfer) of hydroxymethylene (HCOH, reactant species R_{III}), namely,

$$HCOH \rightarrow H_2CO \tag{10.8}$$

with associated TS_{I-III},

$$R_{III} \rightarrow [TS_{I-III}]^{\ddagger} \rightarrow P_I \tag{10.9}$$

A sample Gaussian input file is shown in I/O-10.1 illustrating use of the QST3 method to find the optimized TS_{I-III} for hydroxymethylene decomposition (10.8), with input specification of optimized structures for R_{III}, P_I and an initial guess for TS_{I-III}:

─── *I/O-10.1* ───

```
#b3lyp/6-311++g** scf=tight nosymm opt=qst3

reactant isomer III (hydroxymethylene): E=-114.4581361

0 1
C
O  1  1.3113
H  1  1.1150  2  102.24
H  1  1.8689  2   29.33   3  180.

product isomer I (formaldehyde): E=-114.5418487

0 1
C
O  1  1.2019
H  1  1.1080  2  121.97
H  1  1.1080  2  121.97   3  180.

I-III transition state: E(ts13)=?

0 1
C
O  1  1.2
H  1  1.1     2  115.
H  1  1.5     2   75.    3  180.
```

─────────────────────── *I/O-10.2* ───────────────────────

```
#b3lyp/6-311++g** scf=tight nosymm
irc=(maxpoints=50,stepsize=10,calcfc)

singlet-COH2 isomer I-III transition state: E(ts13)=-114.4030993

0 1
C
O   1   1.3003
H   1   1.1127   2   114.70
H   1   1.2579   2    54.44   3   180.
```

The corresponding Gaussian input file to find points along the intrinsic reaction coordinate (IRC) for reaction in (10.8) and (10.9) is shown in I/O-10.2.

Figure 10.1 shows the fully optimized structures (and relative energies, in kcal/mol) for product P_I (formaldehyde), reactants R_{II} (long-range $H_2\cdots CO$ complex) and R_{III} (hydroxymethylene), and transition-state species TS_{I-II}, TS_{I-III}

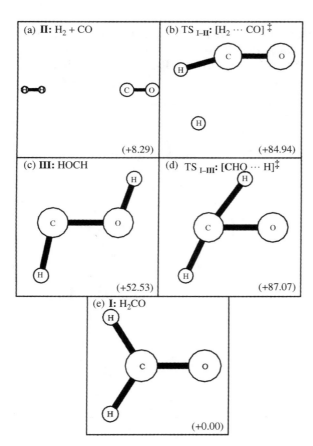

Figure 10.1   Equilibrium geometries for reactants (a), (c), transition states (b), (d), and product (e) species for chemical reactions discussed in the text (cf. Table 10.1), with relative energy (ΔE, kcal/mol) for each species ("bond" sticks of transition species are drawn rather arbitrarily).

Table 10.1 Optimized geometrical parameters (R_{CO}, R_{CH}, $R_{CH'}$, $\angle_{OCH}$, $\angle_{OCH'}$; Å,°) and point-group symmetry of product (**I**), reactant **II**, **III**), and transition state species for bimolecular $H_2 + CO \rightarrow H_2CO$ and unimolecular $HCHO \rightarrow H_2CO$ reactions (cf. Fig. 10.1).

Species	Symmetry	R_{CO}	R_{CH}	$R_{CH'}$	$\angle_{OCH}$	$\angle_{OCH'}$
I. H_2CO	C_{2v}	1.2019	1.1080	1.1080	121.97	121.97
II. $H_2\cdots CO$	$C_{\infty v}$	1.1276	3.7804	4.4916	180.00	180.00
III. HCHO	C_s	1.3113	1.1150	1.8689	102.24	29.33
I–II‡	C_s	1.1604	1.0940	1.6800	163.64	112.62
I–III‡	C_s	1.3003	1.1127	1.2579	114.70	54.44

for reactions in (10.6) and (10.8). Table 10.1 further specifies the symmetry and geometrical parameters of each species.

From the starting transition-state species **I–II**‡ and **I–III**‡, we can proceed to generate the minimum-energy IRC pathway for each reaction. Figure 10.2 shows the calculated IRC reaction energy profile for bimolecular reaction in (10.6), and Fig. 10.3 shows the corresponding profile for the unimolecular reaction in (10.8). In each case, the IRC is found to lead uniquely to the desired reactant (in the backward direction) and product (in the forward direction), confirming the "elementary" character of each reaction. Both bimolecular and unimolecular reactions are seen to be reasonably exothermic (by ca. 8 kcal/mol and 52 kcal/mol, respectively) but rather strongly "forbidden" by high activation barriers (ca. 77 kcal/mol and 35 kcal/mol, respectively).

The Arrhenius-like IRC reaction profiles of Figs. 10.2 and 10.3 exemplify the types of calculations that can now be routinely performed with current ESS program systems. The qualitative features of such diagrams (augmented with vibrational

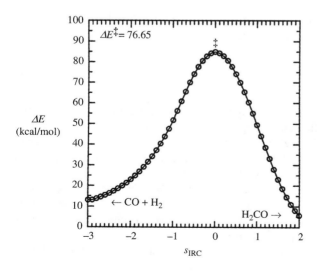

Figure 10.2 Calculated energy profile $\Delta E(s)$ along the intrinsic reaction coordinate (s_{IRC}) for bimolecular $H_2 + CO \rightarrow H_2CO$ chemical reaction ($\Delta E_{rxn} = -8.29$ kcal/mol), showing transition state ($s_{IRC} = 0$) at $\Delta E^\ddagger = 76.65$ kcal/mol above reactants, or 84.94 kcal/mol above product.

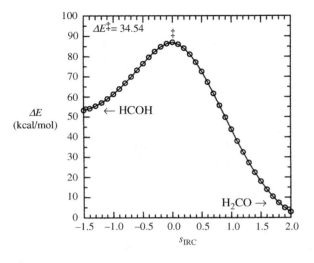

Figure 10.3   Similar to Fig. 10.2, for unimolecular HCOH → H_2CO chemical decomposition of hydroxymethylene ($\Delta E_{rxn} = -52.53$ kcal/mol), showing transition state at $\Delta E^{\ddagger} = 34.54$ kcal/mol above reactant, or 87.07 kcal/mol above product.

corrections at key stationary features) allow direct estimates of thermochemical exothermicity and kinetic reaction rate for many chemical reactions of interest. Such IRC profiles therefore provide valuable information about the energetic landscape along the reaction pathway, but they leave many unanswered questions concerning the details of bond rearrangements, the key electronic factors governing the heights of activation barriers (reaction "allowedness"), and the like. In ensuing Sections 10.2–10.4, we now wish to explore the deeper electronic features of such representative IRC pathways with the tools of NBO/NRT analysis.

10.2   EXAMPLE: BIMOLECULAR FORMATION OF FORMALDEHYDE

Let us first consider the bimolecular formation reaction in (10.6) for formaldehyde, which can be written in greater electronic detail as

$$
\text{H—H} \; + \; \text{:C}{\equiv}\text{O:} \; \rightarrow \quad
\begin{array}{c} \text{H} \\ \diagdown \\ \text{C}{=}\ddot{\text{O}}\text{:} \\ \diagup \\ \text{H} \end{array}
\tag{10.10}
$$

From the Lewis structure diagrams, we recognize that the reaction involves formal transformation of *two* bonding electron pairs, namely, dissociation of $\sigma_{HH'}$ and π_{CO} reactant bonds (formally reducing bond order $b_{HH'}$ from 1 to 0 and b_{CO} from 3 to 2) and formation of two new σ_{CH} product bonds (formally increasing each b_{CH}, $b_{CH'}$ from 0 to 1). How can this chemical magic occur?

To address this question, we first examine how the NRT bond orders vary along the reaction pathway. Figure 10.4 exhibits the computed b_{AB} variations along the IRC pathway in the range $-1.5 \leq s_{IRC} \leq +1.5$, showing the expected smooth changes (despite slight vacillations in b_{CH}, $b_{CH'}$, perhaps due to numerical near-degeneracy problems in the NRT variational algorithm) that are expected from the formal Lewis

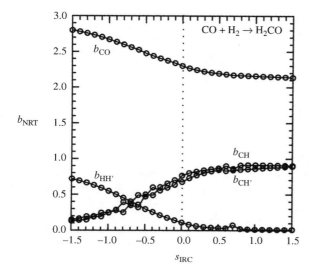

Figure 10.4 NRT bond orders (b_{AB}) along the intrinsic reaction coordinate (IRC) for the bimolecular $CO + H_2 \rightarrow H_2CO$ reaction. A dotted vertical line marks the energetic saddle-point (TS‡) at IRC $= 0$.

structure depictions in (10.10). However, even a cursory glance at Fig. 10.4 indicates that the various bond switches are largely completed *before* the nominal "transition state" (vertical dotted line) at $s_{IRC} = 0$. Although each bond shift appears to be "half-completed" at slightly different IRC values (i.e., $b_{HH'}$ near $s_{IRC} = -0.9$, b_{CO} near -0.6, $b_{CH'}$ near -0.5, and b_{CH} near -0.3), all these shifts appear to be centered in a relatively "early" IRC region (near -0.6, beyond the range of default Gaussian IRC calculation) and to occur more or less simultaneously (concertedly) rather than in pronounced sequence. Thus, primary electronic attention will focus on the region that *precedes* the final saddle-point crossing into the product valley.

Further details of the leading NRT resonance weighting contributions are shown in Fig. 10.5 for the region of principal interest ($-1.5 \leq s_{IRC} \leq 0.5$). As shown in the figure, five leading resonance structure forms contribute significantly to reaction in (10.10) in this IRC region, namely,

(i)   Reactants $H_2 + CO$ (circles)

(ii) Product H_2CO (squares)

(iii) Proton-transfer species $H^- + HCO^+$ (triangles)

(iv) Nucleophilic hydride addition species $H^+ + HCO^-$ (plusses)

(v) H_2 heterolytic dissociation species $H^- + H^+ + CO$ (crosses)

Among the secondary intermediates (iii)–(v), the H_2 dissociation species (v) achieve maximum weighting in the early stages of reaction ($s_{IRC} \cong -0.8$), whereas the hydride addition [$\mathbf{H-\ddot{C}=\ddot{O}}$:$^-$; (iv)] and proton transfer [$\mathbf{H^+ -C \equiv O}$: ; (iii)] species achieve highest weighting near $s_{IRC} \cong 0$. However, the most rapid "transition" of primary resonance weightings is centered around the $w_R = w_P$ crossing near $s_{IRC} \cong -0.6$, well before the energetic saddle-point (as anticipated from the bond-order variations of Fig. 10.4).

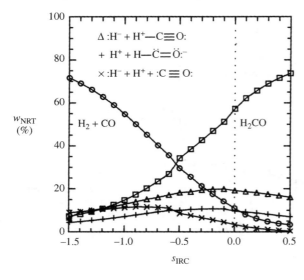

Figure 10.5 NRT weightings (w_{NRT}, %) along the IRC reaction path for $H_2 + CO$ reactants (circles), H_2CO product (squares), and lesser resonance contributions (see inset captions).

Figure 10.6 provides further details of NBO analysis along this "early" portion of the IRC pathway. The top margin of the figure identifies the default Lewis structure (NLS) in each region (separated by vertical heavy-dashed lines), while the vertical scale shows the leading second-order donor–acceptor delocalizations ($\Delta E_{DA}^{(2)}$) in each NLS-region. As expected, the NLS of the left-most region corresponds to the $H_2 + CO$ reactant structure (NLS$_R$), with delocalization corrections that diminish leftward toward the reactant equilibrium limit, but rise

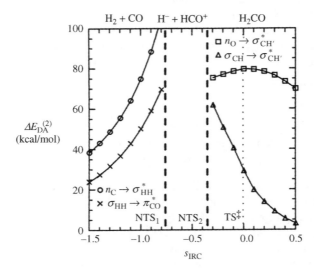

Figure 10.6 Successive NBO Lewis structures (upper border) and leading second-order donor–acceptor stabilizations ($\Delta E_{DA}^{(2)}$) along the IRC pathway (s_{IRC}), showing dominant NBO delocalizations (identified by inset captions) in reactant-like (NLS$_R$) and product-like (NLS$_P$) regions, separated by an intermediate "$H^- + HCO^+$" (NLS$_I$) region. Approximate natural transition boundaries NTS$_1$ (NLS$_R$ $\rightarrow$ NLS$_I$) and NTS$_2$ (NLS$_I$ $\rightarrow$ NLS$_P$) are marked by vertical heavy-dash lines, while the energetic TS saddle-point ($s_{IRC} = 0$) is similarly marked with a vertical dotted line.

spectacularly toward the first NLS-transition around $s_{IRC} \cong -0.75$, which we designate as "NTS$_1$." The two charge-transfer (CT) delocalizations in question

$$n_C \rightarrow \sigma^*_{HH'} \text{ ("push")} \tag{10.11}$$

$$\sigma_{HH'} \rightarrow \pi^*_{CO} \text{ ("pull")} \tag{10.12}$$

are seen to provide a concerted push–pull tandem attack on the H$_2$ bond, in which the carbon lone pair (n_C) donor "pushes" electrons into the $\sigma^*_{HH'}$ antibond while the carbonyl antibond (π^*_{CO}) acceptor "pulls" electrons from the $\sigma_{HH'}$ bond. Both interactions act to dissociate the H–H$'$ bond heterolytically, thereby reducing $b_{HH'}$ and promoting (activating) proton transfer to form the hydride anion (H$^-$) and formyl cation (HCO$^+$) of the intermediate NLS region (NLS$_I$, ca $-0.75 < s_{IRC} < -0.35$).

The intermediate NLS$_I$ region of Fig. 10.5 (separating NTS$_1$ and NTS$_2$) is characterized by a confusing succession of strange NLS structures that can all be loosely classified as "H$^-$ + HCO$^+$" (NLS$_2$) structures, such as

$$\text{H—H} \ + \ \text{:C=Ö:} \ \rightarrow \ \text{H$^+$—C}\equiv\text{O:} \ + \ \text{:H$^-$} \ \rightarrow \ \overset{\overset{\textstyle H}{\diagdown *}}{\underset{\diagup}{\underset{\textstyle H}{}}}\text{C}\equiv\text{O:} \tag{10.13}$$

The "*" in the right-most structure of (10.13) indicates that the "CH bond" is actually of *out*-of-phase (antibonding) character, only slightly more highly occupied than the corresponding in-phase combination, and thus indicative of essential nonbonding character (namely, **H$^+$– C $\equiv$ O:** + **:H$^-$**), as in the middle NLS. The extreme stresses of these tortured "Lewis structures" are indicated by astronomically high second-order "corrections" (ca. 200–400 kcal/mol) that are far beyond the vertical scale of Fig. 10.5 and accurate "perturbative" estimation. This intermediate region culminates near $s_{IRC} \cong -0.35$ in the final transition (designated "NTS$_2$") to recognizable product (NLS$_P$) bonding pattern of H$_2$CO.

Let us first try to obtain a "reactant-like" perspective on the IRC pathway. We can employ the \$CHOOSE keylist (Section 5.5) to specify a reactant-like bond pattern, and thereby continue to follow the progress of the NBO push–pull delocalizations (10.11 and 10.12) whose $\Delta E_{DA}^{(2)}$ values are plotted at the left of Fig. 10.6. As the reaction progresses, we expect to see the reactant geometry rearranging to enhance the two principal push–pull NBO interactions and associated resonance contributions:

(i)   The $n_C \rightarrow \sigma^*_{HH'}$ (push) delocalization corresponds to a contributing *proton transfer* resonance structure, as shown at the right in (10.14),

$$\textbf{H – H} \ \ \textbf{:C} \equiv \textbf{O:} \ \leftrightarrow \ \ \textbf{H:$^-$} \ + \ \textbf{H$^+$–C} \equiv \textbf{O:} \ \text{(proton transfer)} \tag{10.14}$$

(ii) The $\sigma_{HH'} \rightarrow \pi^*_{CO}$ (pull) delocalization corresponds to *nucleophilic hydride addition* to form the final product structure, as shown at the right in (10.15),

$$
\begin{array}{c}
\text{H}^+\!\!-\!\!\text{C}\!\!\equiv\!\!\text{O:} \longleftrightarrow \quad
\begin{array}{c}
\text{H} \\
\diagdown \\
\text{C}\!\!=\!\!\ddot{\text{O}}: \\
\diagup \\
\text{H}
\end{array}
\end{array}
\qquad (10.15)
$$

(nucleophilic hydride addition)

As suggested by the resonance structure depictions on the left, proton transfer delocalization (10.14) is favored in collinear geometry ($\angle_{H\cdots CO} \cong 180°$, which maximizes overlap of the n_C donor NBO with the $\sigma^*_{HH'}$ acceptor NBO), whereas hydride-addition delocalization (10.15) is favored by bent angular geometry ($\angle_{H\cdots CO} \cong 135°$, which maximizes overlap of the $n_{H'}$ donor NBO with one of the four "cloverleaf" lobes of the π^*_{CO} acceptor NBO). These qualitative NBO considerations dictate the qualitative features of transition state geometry (Fig. 10.2), with the "cationic" H atom nearly coaligned with the CO axis and the "anionic" H' atom canted away from this axis to coalign with the adjacent cloverleaf lobe of the π^*_{CO} orbital.

Figure 10.7 displays quantitative features of the geometry and leading NBO interactions for the key NTS_1, NTS_2, and $TS^\ddagger$ species along the IRC pathway. As the reaction progresses (from left to right), the H_2 moiety elongates to bring the proximal H atom into better alignment with the n_C donor orbital (enhancing the $n_C \rightarrow \sigma^*_{HH'}$ "push" interaction; middle panels), while the distal H' remains aligned with a cloverleaf lobe of the π^*_{CO} acceptor orbital (enhancing the $\sigma_{HH'} \rightarrow \pi^*_{CO}$ "pull" interaction; lower panels). A particularly conspicuous feature of the orbital overlap diagrams is the progressive *polarization* of the H_2 moiety, which simultaneously increases orbital amplitude at the H end of the $\sigma^*_{HH'}$ antibond (thereby enhancing n_C–$\sigma^*_{HH'}$ overlap; middle panels) *and* the H' end of the $\sigma_{HH'}$ bond (thereby enhancing $\sigma_{HH'}$–π^*_{CO} overlap; lower panels). These features well illustrate the synergistic (cooperative) aspect of NBO push–pull interactions, which is the key to overcoming the unfavorable steric clashes and H_2 bond weakening that oppose passage through the transition state geometry.

The progress of H_2 polarization toward heterolytic dissociation and addition to CO can also be followed through the natural charges (q_H) of the two H atoms, as shown in Fig. 10.8. As the initial NTS_1 region of the IRC is approached (vertical dashed line near $s_{IRC} = -0.75$), the H_2 bond polarity is seen to increase sharply, with net ionicity $\Delta q = q_H - q_{H'}$ finally exceeding $0.5e$ near the reaction $TS^\ddagger$ (vertical dotted line).

Alternately, we can examine the same features of the IRC from the *product* point of view, using a $CHOOSE keylist to specify product-like NBOs throughout the reactive region (to continue the trends seen at the right edge of Fig. 10.6). As shown in Fig. 10.6, the two principal product-type NBO delocalizations

$$n_O \rightarrow \sigma^*_{CH} \quad \text{(vicinal "push")} \qquad (10.16)$$

$$\sigma_{CH} \rightarrow \sigma^*_{CH'} \quad \text{(geminal "pull")} \qquad (10.17)$$

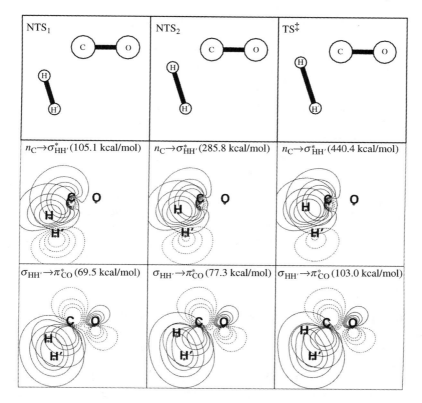

Figure 10.7 Optimized geometry (upper panels) and orbital contour diagrams for leading "push"-type ($n_C \rightarrow \sigma^*_{HH'}$, middle panels) and "pull"-type ($\sigma_{HH'} \rightarrow \pi^*_{CO}$, lower panels) donor–acceptor delocalizations (with $\Delta E_{DA}^{(2)}$ estimates in parentheses) for NTS$_1$ (left), NTS$_2$ (middle), and TS‡ (right) species along the IRC for $H_2 + CO \rightarrow H_2CO$ formation reaction (cf. Fig. 10.6).

correspond to tandem push–pull attack on the σ_{CH} bond. Seen from the product side, the problem is to *maximize* the two delocalizations (10.16 and 10.17) to attain saddle-point TS‡ geometry.

Of course, in the region of the formaldehyde equilibrium minimum (far to the right of Fig. 10.6), the vicinal $n_O \rightarrow \sigma^*_{CH}$ delocalization (hyperconjugation) is significant, but *equivalent* to the $n_O \rightarrow \sigma^*_{CH'}$ delocalization that competes with (10.17). Furthermore, the desired delocalization (10.17) exemplifies *geminal*-type interaction (i.e., "neighbor bond" delocalization $\sigma_{AB} \rightarrow \sigma^*_{BC}$ in A–B–C bond connectivity), which is usually quite negligible compared to vicinal-type interactions (i.e., "next-neighbor bond" delocalizations of $\sigma_{AB} \rightarrow \sigma^*_{CD}$ or $n_B \rightarrow \sigma^*_{CD}$ type in A–B–C–D bond connectivity). How can the desired delocalization pattern be achieved?

As discussed in *V&B*, p. 263ff (cf. Fig. 5.6), geminal $\sigma_{AB} \rightarrow \sigma^*_{CD}$ delocalization typically requires significant $A^{\delta-}$–B–$C^{\delta+}$ polarity and asymmetrical geometry distortions to achieve significant magnitude. Accordingly, the reactive pathway

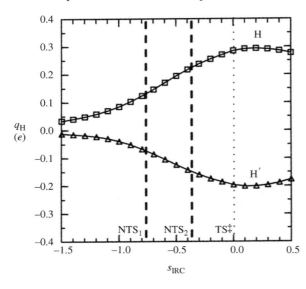

Figure 10.8   Natural atomic charges (q_H) of proximal H (squares) and distal H′ (triangles) along the IRC of reaction in (10.8) and (10.9), showing the increasingly cationic (H) versus anionic (H′) character resulting from heterolytic "push–pull" dissociation interactions (cf. Fig. 10.7).

(10.16 and 10.17) toward TS‡ formation requires *breaking* the symmetry of the CHH′ group to achieve *in*equivalent CH, CH′ bonds of maximally distinct polarity and geometry. This is most simply achieved by *rehybridizing* the carbon atom to put maximal p-character toward H′ and s-character toward H (because high s-character maximizes effective electronegativity and minimizes bonding radius of the hybrid, whereas high p-character has opposite effect). Thus, seen from the viewpoint of the product formaldehyde species, the TS‡ barrier is essentially a *rehybridization* barrier at the carbon atom.

To see this rehybridization aspect of TS‡ formation most directly, we can examine the percentage p-character (%-p) of the H-directed ($h_{C(H)}$) versus H′-directed ($h_{C(H')}$) carbon NHOs of formaldehyde along the IRC, as plotted in Fig. 10.9. As shown toward the right-edge of the figure, the $h_{C(H)}$ and $h_{C(H')}$ hybrids converge toward the expected sp^2 equivalency (67% p-character) at the equilibrium H$_2$CO limit. However, the $h_{C(H')}$ NHO is seen to increase steeply in p-character toward the pure-p (π-type) limit, while the $h_{C(H)}$ NHO simultaneously drops toward s-rich sp (σ-type) character as the TS‡ barrier is approached. The difficult transition to σ/π-type hybrids (with accompanying near-linear $\angle_{H\cdots CO}$ and near-perpendicular $\angle_{H'\cdots CO}$ geometry) is apparently completed near $s_{IRC} \cong 0.3$, allowing H to cross to the same half-plane as H′. Thereafter, the hybrids of the resulting HCO^{+} fragment (which is effectively "detached" from H′) relax toward the bent geometry characteristic of the NLS$_I$ region (cf. upper panels of Fig. 10.7). [Note that farther to the reactant side, near $s_{IRC} \cong -0.7$, the hybridization shifts abruptly as the "CH′ bond" shifts to *out*-of-phase $\sigma^*_{CH'}$ character; such equal $\sigma_{CH'}$, $\sigma^*_{CH'}$ occupancy merely signifies the essential *non*bonding character ($b_{CH'} \cong 0$) of the \$CHOOSE-forced NBOs, and should cause no concern.]

Thus, from both ends of the IRC, we are led to a picture of strong H, H′ asymmetry, with the tightly held "proton-like" H in near-linear (σ-type) geometry and the "hydride-like" H′ in near-perpendicular (π-type) geometry. Both

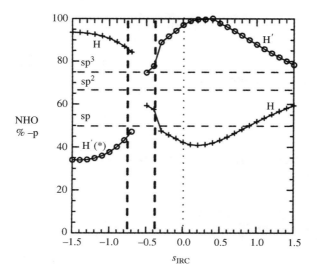

Figure 10.9   Percentage p-character (%-p) of h_C NHOs in product-type CH (plusses) and CH′ (circles) NBOs along the IRC. Horizontal dashed lines mark standard sp, sp^2, sp^3 hybrid types, and vertical lines mark the geometry of NTS_1, NTS_2 transition species (dashed; cf. Fig. 10.6) and the energetic $TS^{\ddagger}$ saddle-point (dotted). The discontinuity at $s_{IRC} \cong -0.7$ marks the transition to $CHOOSE structures with higher occupancy in the CH′ *anti*bond, indicative of essential nonbonding character [cf. (10.13)].

$CHOOSE-reactant and $CHOOSE-product NBO patterns therefore provide *complementary* descriptions of the intermediate $TS^{\ddagger}$ region, suggesting the curiously bent L-shaped geometry of the final $TS^{\ddagger}$ transition-state species and the dual (two-bond) "push–pull" rearrangements that are required to surmount the high-energy $TS^{\ddagger}$ saddle-point. Such dual-demand delocalization leads to a high activation barrier for H_2CO formation ($\Delta E^{\ddagger}_{II \rightarrow I} \cong 77\,kcal/mol$), suggesting (correctly) that H_2/CO mixtures can coexist safely at equilibrium over long periods without appreciable conversion to "more stable" H_2CO.

10.3   EXAMPLE: UNIMOLECULAR ISOMERIZATION OF FORMALDEHYDE

Let us now briefly consider the alternative *uni*molecular isomerization reaction of formaldehyde to hydroxymethylene (CHOH), whose calculated IRC reaction profile was shown in Fig. 10.3. In this case, NRT analysis of the IRC leads to bond-order variations as plotted in Fig. 10.10, indicating a simple NTS crossing very near the energetic $TS^{\ddagger}$ saddle-point at $s_{IRC} = 0$ (quite different from the corresponding Fig. 10.4 for the bimolecular formation reaction). The reaction profile is seen to involve primarily a direct switch between $b_{CH'}$ and $b_{OH'}$ bond orders, with little change of b_{CH} and b_{CO} bond orders. Accordingly, we anticipate that unimolecular hydroxymethylene isomerization is intrinsically simpler and more direct than

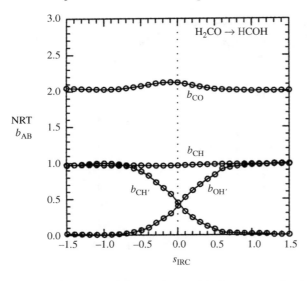

Figure 10.10 Analogous to Fig. 10.4, for the unimolecular HCOH → H_2CO reaction.

bimolecular $H_2 + CO$ reaction, with fewer contributing resonance structures and NTS transitions along the pathway.

Figure 10.11 displays the default Lewis structure (NLS) and leading donor–acceptor delocalizations ($\Delta E_{DA}^{(2)}$) in each region of the IRC (analogous to Fig. 10.6). As shown at the top of the figure and marked by the vertical dashed line, the NBO Lewis structure undergoes a single NTS transition from hydroxymethylene-like to formaldehyde-like bonding pattern near $s_{IRC} \cong -0.1$, close to the energetic TS‡ saddle-point. The figure insets identify the principal $\Delta E_{DA}^{(2)}$ delocalizations plotted on each side of the NTS.

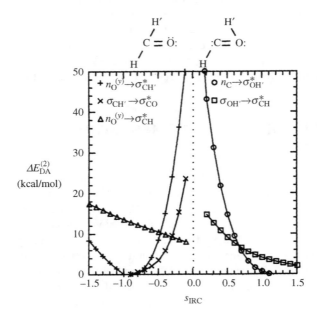

Figure 10.11 Analogous to Fig. 10.6, for unimolecular rearrangement of formaldehyde (left) to hydroxymethylene (right).

As suggested in the Lewis structure depictions at the top of the diagram, the NTS (or TS‡) structure corresponds to the migrating H$'$ being about half-way between C and O, forming a near-isosceles CH$'$O triangle (cf. Fig. 10.1d, Table 10.1). How is this low-energy isomerization pathway achieved?

Let us first seek a product-like (hydroxymethylene) perspective of the IRC pathway. As shown in Fig. 10.11, the leading $\Delta E_{DA}{}^{(2)}$ delocalizations of the HCOH region are found to be as follows:

$$n_C \rightarrow \sigma^*_{OH'} \tag{10.18}$$

$$\sigma_{OH'} \rightarrow \sigma^*_{CH} \tag{10.19}$$

corresponding to tandem push–pull attack on the target OH$'$ linkage by the vicinal carbon lone pair (donor n_C) and hydride antibond (acceptor σ^*_{CH}). Figure 10.12 exhibits NBO plots of these interactions (analogous to Fig. 10.7) for the TS‡ saddle point geometry. As shown by the orbital overlap diagrams and estimated $\Delta E_{DA}{}^{(2)}$ values, the isomerization is primarily promoted (activated) by "push" delocalization (11.18), corresponding (cf. Fig. 5.6) to a resonance admixture of hydridic H$'^-$ and "protonated carbon monoxide" character, namely,

$$n_C \rightarrow \sigma^*_{OH'} \implies \begin{array}{c} :H'^- \\ C\equiv O:^+ \\ {}^{/}\!H \end{array} \tag{10.20}$$

Much weaker is the secondary "pull" delocalization (10.19) and associated resonance mixing

$$\sigma_{OH'} \rightarrow \sigma^*_{CH} \implies \begin{array}{c} H'^+ \\ :C\equiv O: \\ :H^- \end{array} \tag{10.21}$$

which promotes cationic (rather than anionic) character of the migrating H$'$.

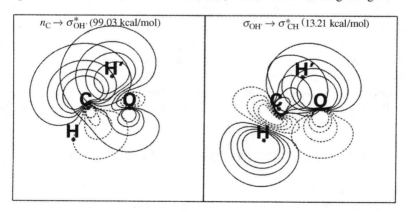

$n_C \rightarrow \sigma^*_{OH'}$ (99.03 kcal/mol) $\sigma_{OH'} \rightarrow \sigma^*_{CH}$ (13.21 kcal/mol)

Figure 10.12   Leading product-type (hydroxymethylene) push–pull delocalizations (10.8 and 10.19) (and $\Delta E_{DA}{}^{(2)}$ estimates) in TS‡ geometry for formaldehyde isomerization (cf. Fig. 10.11).

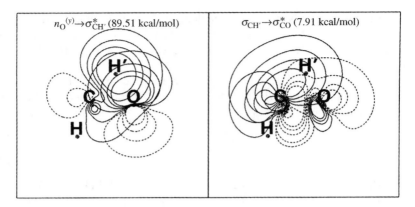

Figure 10.13   Similar to Fig. 10.12, for reactant-type (formaldehyde) push–pull delocalizations.

From a reactant-like (formaldehyde) perspective, the principal delocalizations of the $TS^{\ddagger}$ region are

$$n_O^{(y)} \rightarrow \sigma^*_{CH'} \tag{10.22}$$

$$\sigma_{CH'} \rightarrow \sigma^*_{CO} \tag{10.23}$$

corresponding to tandem push–pull attack on the sacrificial CH' bond. As shown in Fig. 10.13, the contour plots of the $TS^{\ddagger}$ NBOs are almost unrecognizably distorted from the equilibrium formaldehyde forms to which they connect continuously at large $s_{IRC} > 0$.

As is clear from the tortured forms of the $TS^{\ddagger}$ NBOs, the isomerization reaction requires strenuous reorientation and *rehybridization* of bonding hybrids at both C and O centers, whether seen from reactant or product viewpoints. From the reactant hydroxymethylene side, the carbon lone pair reorients toward the bridging H' atom while the oxygen $h_{O(H')}$ bonding hybrid reorients from $\angle_{COH'} \cong 120°$ (sp^2-like) toward 90 (p-like) hybrid directions. From the product formaldehyde side, the carbonyl O atom rehybridizes the in-plane p-type $n_O(y)$ lone pair to become a directed hybrid toward H' while the C atom reorients the $h_{C(H')}$ hybrid toward bridging geometry and increased bond polarization and geminal delocalization (10.23), giving rise to atomic charge or rehybridization profiles analogous to Figs. 10.8 and 10.9. However, further details of hybridization, polarization, and atomic charge variations along the IRC are left as student exercises.

10.4   EXAMPLE: S$_N$2 HALIDE EXCHANGE REACTION

The chemical reactions of Sections 10.1–10.3, although "elementary" by the standard ($TS^{\ddagger}$ energetic) criterion, involve *multiple* NBO push–pull delocalizations. Such reactions appear mechanistically complex from a diabatic NBO perspective, typically leading to high activation barrier and formal "forbidden" character. We now wish to

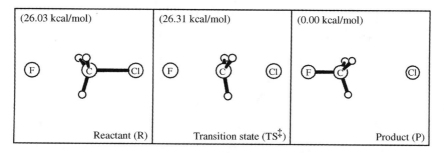

| (26.03 kcal/mol) | (26.31 kcal/mol) | (0.00 kcal/mol) |
| Reactant (R) | Transition state (TS‡) | Product (P) |

Figure 10.14   Reactant complex (left), transition state (center), and product complex (right) for S$_N$2 halide exchange reaction (10.24) (with relative energies in parentheses).

illustrate the simpler case of diabatic NBO-level elementarity in which only a *single* strong NBO delocalization (resonance shift) is required to achieve the desired product. Such "direct" NBO resonance pathways are expected to correspond to favored (low activation "allowed") mechanisms for chemical transformation.

As a prototype of NBO-level elementarity, we consider the well-known "S$_N$2 reaction" for halide exchange, namely,

$$F^- + CH_3Cl \rightarrow FCH_3 + Cl^- \tag{10.24}$$

The reactive transformation (10.24) is expected to proceed via the strong NBO 3c/4e hyperbonding interaction (cf. Section 8.3),

$$n_F \rightarrow \sigma^*_{CCl} \;(\text{resonance structure F-CH}_3 + \; :Cl^-) \tag{10.25}$$

which is maximized in the collinear F:$^-$···C-Cl geometry of the calculated TS‡ geometry. Figure 10.14 displays the optimized reactant (R), transition state (TS‡), and product (P) structures for reaction (10.24), with relative energies (compared to final FCH$_3$···Cl$^-$ product complex) in parentheses. As shown by the calculated energy values, the reaction is reasonably exothermic [$\Delta E_{rxn} = 26.03$ kcal/mol, intermediate between the values for formaldehyde reactions (10.6) and (10.8)]. However, the activation energy barriers for forward and reverse directions are, respectively,

$$\Delta E_f^{\ddagger} = 0.28 \,\text{kcal/mol} \tag{10.26}$$

$$\Delta E_r^{\ddagger} = 26.31 \,\text{kcal/mol} \tag{10.27}$$

showing that the forward reaction (10.16) is virtually *barrierless* in this case. This is in strong contrast to CH$_2$OH isomerization (10.8), which apparently involves a similar one-bond shift [from H-C to H-O in (10.8); from C-Cl to C-F in (10.24)] but lacks the directness of a *single* strong NBO donor-acceptor delocalization (resonance shift) to accomplish the desired reaction.

Figure 10.15 displays the calculated reaction energy profile along the IRC, showing the extreme asymmetry around the formal TS‡ at $s_{IRC} = 0$. The equilibrium

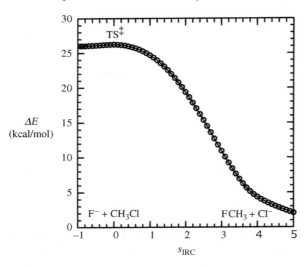

Figure 10.15 IRC reaction energy profile for halide exchange reaction (10.24), showing extreme low-barrier passageway for "S_N2" reaction mechanism [via 3c/4e hyperbonding interaction (10.25)].

reactant $F^-\cdots CH_3Cl$ complex lies near the left border of the figure at $s_{IRC} = -1$, whereas the equilibrium $FCH_3\cdots Cl^-$ complex lies far beyond the right border, $s_{IRC} > 5$. The central panel of Fig. 10.14 indicates that carbon pyramidalization ("Walden inversion") has proceeded to product-like geometry at $TS^{\ddagger}$ (apparent "late" transition state), whereas the IRC-asymmetry of Fig. 10.15 suggests "early" transition state character. How can we judge where the electronic shifts are "half complete" in this near-barrierless case?

Figure 10.16 shows the calculated NRT resonance weightings for reactant (w_R) and product (w_P) along the IRC pathway. In this case, the energetic transition state (vertical dotted line) is quite reactant-like ("early" in the sense of Hammond's

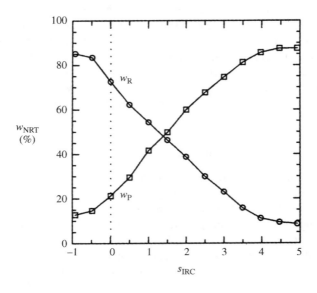

Figure 10.16 NRT resonance weights for reactant (w_R: CH_3Cl) and product (w_P: CH_3F) along the IRC for S_N2 halide exchange reaction (10.24), showing "early" (reactant-like) character of the energetic $TS^{\ddagger}$ state ($s_{IRC} = 0$) with respect to the direct NTS crossing ($w_R = w_P$) near $s_{IRC} \cong 1.5$.

postulate), whereas the electronic half-way point occurs near $s_{IRC} \cong 1.5$ according to the NTS criterion. This example illustrates the extreme allowedness of the simple 3c/4e hyperbonding resonance interaction for forward reaction ($\Delta E_f^{\ddagger} = 0.28$ kcal/ mol) and shows how to quantify early versus late character of the energetic transition state with respect to the equi-resonance NTS criterion.

In this case, the forward and reverse S_N2 reactions occur by analogous 3c/4e interaction and present analogous geometrical demands for $F \cdots C \cdots Cl$ linearity. However, Section 5.4.3 of *V&B* and the NRT website tutorial http://www.chem .wisc.edu/~nbo5/tutf_nrt.htm describe the more interesting "allowed" reaction mechanism of Diels-Alder type, where distinct forward and reverse NBO delocalizations impose *complementary* stereoelectronic constraints on the transition state.

In summary, we can see how the freedom to apply either reactant-like or product-like NBO analysis offers valuable insights into the ambivalent structural and electronic demands of transition-state species, suggesting new possibilities for rational catalyst design. More generally, the NRT method allows easy extension to *continuous* description of electronic shifts along the IRC reaction path (or indeed along *any* chosen path on the potential energy surface) involving *multiple* orbital transmutations. The examples of this chapter provide simple illustrations of how a reaction path may be dissected into constituent "elementary NBO bond-switches," thereby opening the door to improved orbital-level understanding of "reaction mechanism." Such NBO/NRT-based methods offer the promise of many future discoveries in chemical dynamics and catalysis, complementing theoretical advances that have been achieved in the domain of near-equilibrium phenomena.

PROBLEMS AND EXERCISES

10.1. Open-shell reactivity presents interesting additional complexity with regard to differential "progress" of electronic rearrangements (i.e., differential weighting of reactant- and product-like resonance species) in the two spin sets. Consider the isomerization reaction of ozone (O_3; cf. Section 3.3.2) that interchanges terminal and central O atoms,

$$O_a-O_b-O_c \rightarrow TS^{\ddagger} \rightarrow O_b-O_a-O_c$$

A sample input file to determine the open-shell $TS^{\ddagger}$ species for this reaction is shown below:

```
%chk=o3
#ub3lyp/6-311++G** scf=tight nosymm stable=opt

O(a)-O(b)-O(c) reactant, E=-225.4833192

0 1
O
O   1   1.2786
O   2   1.2786   1   116.64
```

```
--Link1--
%chk=o3
#ub3lyp/6-311++G** scf=tight nosymm guess=read
opt=(qst3,calcall)

O(a)-O(b)-O(c) reactant, E=-225.4833192

0 1
O
O  1  1.2786
O  2  1.2786  1  116.64

O(b)-O(a)-O(c) product, E=-225.4833192

0 1
O
O  1  1.2786
O  1  1.2786  2  116.64

cyclic transition species guess

0 1
O
O  1  1.3
O  1  1.3    2    60.0
```

[The first jobstep (above "--Link1--") insures the lowest-energy UHF-type solution is used as initial guess in the difficult "opt=qst3" transition-state search. Note that the fully optimized O_3 geometry of the input file differs slightly from the experimental geometry employed in I/O-3.9.]

(a) Draw a ball-and-stick model of the optimized $TS^{\ddagger}$ species with bond distances and angles to each atom. What is the calculated activation energy for this isomerization reaction?

(b) Use NBO/NRT analysis to determine the α-spin, β-spin, and total bond orders for $O_a{-}O_b$, $O_a{-}O_c$, and $O_b{-}O_c$ in the optimized $TS^{\ddagger}$ geometry (as determined above). Do the bond orders "make sense" in terms of the optimized $TS^{\ddagger}$ geometry?

(c) Evaluate the IRC for this reaction and plot your results in terms of the Arrhenius-like reaction profile (analogous, e.g., to Fig. 10.2).

(d) Perform NRT analysis for points along the IRC. Prepare plots of your results showing total b_{NRT} (analogous, e.g., to Fig. 10.4) and the relative w_R/w_P weightings (analogous, e.g., to Fig. 10.5) for each spin set. Can you discern different NTS "half-way points" in the two spin sets?

10.2.   Many important atmospheric ion–molecule reactions proceed in barrierless fashion. Some important examples in "OOCO^{+} chemistry" [*J. Chem. Phys.* **127**, 064313, 2007] include

$$O^{+} + CO_2$$

$$O_2^{+} + CO$$

$$CO^{+} + O_2$$

but many other examples could be selected. In such a case, one can select the distance R between ion and molecule as reaction coordinate (RC) and optimize remaining geometrical values, as illustrated below for $O^+ + CO_2$:

```
%chk=o3
#ub3lyp/6-311++G** scf=tight opt=z-matrix

O(+) + CO2 reaction coordinate: E(RC=5.0)=

+1 2
O
O    1   RC
C    2   co   1   ang1
O    3   cop  2   ang2  1   dihed

ang1    120.
ang2    180.
dihed   165.

RC      5.0

--Link1--
```

   (etc.)

(a) Choose an ion–molecule reaction of interest and calculate reaction profile energies $E(RC)$ for RC values 5.0, 4.0, ... (down to equilibrium ion–molecule separation). Plot $E(RC)$ versus RC to determine if there are any transition-state maxima or other interesting energetic features along the chosen reaction coordinate.

(b) Obtain NRT weightings for reactant (w_R) and product (w_P) species along the chosen RC. Plot your results to estimate the NTS ($w_R = w_P$) transition state for the reaction.

(c) Describe the primary NBO donor–acceptor interaction most "active" at the NTS, both verbally and with NBOView orbital plots. Does the optimized ion–molecule geometry at the NTS "make sense" in terms of this interaction? Why or why not?

(d) Similarly, attempt to "make NBO sense" out of any other interesting kinks or inflections along the chosen RC (if necessary, supplemented with additional points).

Chapter 11

Excited State Chemistry

Each electronic promotion to an excited state leads to new chemistry—new structural geometry, new bond orders, new charge distribution, and a new palette of donor and acceptor orbitals for intra- and intermolecular resonance phenomena. The details of each excited-state domain are reflected in the color spectrum of the associated photon (light energy) emissions and absorptions. In turn, the structural features around the chromaphoric (light-absorbing) center become participants in the sequence of chemical transformations initiated by photoexcitation of specific wavelength. Understanding the unique relationship between structural and reactive propensities of an excited state and its associated spectral excitation band is thus a prerequisite for effective chemical control of color (e.g., in dyestuffs or LED-type applications), energy storage (e.g., in photovoltaic or other "light harvesting" applications), and photochemical reaction products (e.g., in medical, environmental, or combustion engineering applications). Many current technological challenges require improved understanding and control of excited state chemistry.

Quantum chemical exploration of excited states lags far behind that of ground states. Limitations of current excited-state computational technology are of a technical nature, stemming from the greater difficulties of obtaining an accurate wavefunction (Ψ_i) and energy (E_i) for the chosen i^{th} excited state ($i > 0$); see Sidebars 11.1 and 11.2. In most respects, NBO/NRT analysis of excited-state wavefunctions parallels that for ground states, presenting few new difficulties of principle. In practice, however, excited-state wavefunctions of useful accuracy (*if* obtainable at all) typically manifest complex multiconfigurational open-shell character and strong delocalization features that may challenge NBO/NRT numerical thresholds and the limits of the localized Lewis structure concept itself. Chemical discovery in this frontier domain is indeed "exciting," but not for the faint of heart.

11.1 GETTING TO THE "ROOT" OF THE PROBLEM

We first briefly discuss the more complex "getting started" questions presented by excited-state calculations: What multistate methods are available in a chosen ESS program, and how does one specify the particular state i of interest? As described in

Discovering Chemistry With Natural Bond Orbitals, First Edition. Frank Weinhold and Clark R. Landis.
© 2012 John Wiley & Sons, Inc. Published 2012 by John Wiley & Sons, Inc.

Sidebar 11.1, variational multistate methods generally involve finding a desired *root* of the characteristic secular polynomial equation for the multistate Hamiltonian matrix **H** associated with the chosen theoretical level (method and basis set). In this simplified introduction, we focus on some current multistate options of the *Gaussian 09* (G09) program system, describing the keyword syntax for specifying the chosen root and available options (if any) for NBO/NRT analysis of the resulting wavefunction.

The principal multistate methods of Gaussian and other leading ESS programs are generally of "configuration interaction" (CI) or "complete active space multiconfiguration SCF" (CAS) type. Characteristic advantages, restrictions, and Gaussian input syntax for leading options of each type are described in Sidebar 11.2. The simple CI-singles (CIS) method, although of limited accuracy, provides the greatest range of analytic properties and analysis options for illustrative purposes. We therefore focus primarily on CIS-level wavefunctions, with selected comparative applications of more advanced CAS-type methods to illustrate capabilities and limitations of current Gaussian multistate options.

SIDEBAR 11.1	*MULTISTATE VARIATIONAL METHODS*

The well-known variational theorem of quantum mechanics establishes an *upper bound* on the true ground-state energy E_0

$$E_0 \leq \int \Phi_0 {}^* H \Phi_0 d\tau \equiv \varepsilon_0 \tag{11.1}$$

for any normalized "trial function" Φ_0

$$1 = \int \Phi_0 {}^* \Phi_0 d\tau \tag{11.2}$$

satisfying proper symmetry and boundary conditions for the chosen Hamiltonian operator H_{op}. Accordingly, the form of Φ_0 can be varied (according to the flexibility afforded by the chosen method and orbital basis set) to achieve "best" (lowest) ε_0 approximation to the true E_0. This powerful theorem underlies modern *ab initio* technology for ground-state properties.

The corresponding multistate variational theorem must be suitably generalized [Hylleraas-Undheim-MacDonald (HUM) "interleaving theorem"; see J. K. L. MacDonald, *Phys. Rev.* **43**, 830, 1933]. In this case, we consider an orthonormal *set* of n trial functions $\{\Phi_i\}$ ($i = 0, 1, 2, \ldots, n-1$) satisfying

$$\delta_{ij} = \int \Phi_i {}^* \Phi_j d\tau \tag{11.3}$$

The associated Hamiltonian "matrix elements" $\{H_{ij}\}$

$$H_{ij} = \int \Phi_i {}^* H_{op} \Phi_j d\tau = (\mathbf{H}^{(n)})_{ij} \ (j = 0, 1, 2, \ldots, n-1) \tag{11.4}$$

compose the $n \times n$ matrix $\mathbf{H}^{(n)}$ that "represents" the physical system. Successive eigenvalues $\varepsilon_i^{(n)}$ of $\mathbf{H}^{(n)}$ are obtained as the successive *roots* of the characteristic polynomial equation for the associated "secular determinant," namely,

$$\det|\mathbf{H}^{(n)} - \varepsilon\mathbf{1}| = 0 \quad (\text{for } \varepsilon = \varepsilon_0^{(n)}, \, \varepsilon_1^{(n)}, \, \varepsilon_2^{(n)}, \dots, \varepsilon_{n-1}^{(n)}) \tag{11.5}$$

According to the HUM interleaving theorem, the ordered roots for order n interleave those for order $n + 1$, namely,

$$\dots \leq \varepsilon_i^{(n+1)} \leq \varepsilon_i^{(n)} \leq \varepsilon_{i+1}^{(n+1)} \leq \varepsilon_{i+1}^{(n)} \leq \dots \tag{11.6}$$

The inexorable *de*crease of variational eigenvalues with increasing n insures that each $\varepsilon_i^{(n)}$ is a rigorous upper bound to the corresponding true energy level E_i, namely,

$$E_i \leq \varepsilon_i^{(n)} \; (i = 0, 1, 2, \dots, n - 1) \tag{11.7}$$

Systematically improved approximations for each excited-state E_i are therefore obtained simply by increasing the dimensionality n of $\mathbf{H}^{(n)}$, that is, expanding the set $\{\Phi_i\}$ toward completeness, with $\varepsilon_i \to E_i$ as $n \to \infty$. Theorem (11.7) underlies *ab initio* methods for computing excited-state properties, limited only by the number and type of excitation functions $\{\Phi_i\}$ used to construct $\mathbf{H}^{(n)}$ in the chosen approximation method.

SIDEBAR 11.2   *MULTISTATE CI AND CAS OPTIONS IN GAUSSIAN*

In the Gaussian program, the simple "CI with single-excitations" (CIS) method is particularly convenient, because it supports virtually all familiar optimization, frequency, and analysis options of ground-state HF or DFT methods. However, this convenience is offset by significant accuracy limitations, because the CIS method is intrinsically based on uncorrelated HF ground-state starting point and restrictive choice of excitations (i.e., neglect of all double and higher excitations). As a result, only a limited selection of excited states and moderate level of accuracy are accessible to CIS description.

Closely related "time-dependent" (TD) DFT-based CIS methods are available for all common DFT functionals. These methods are potentially more accurate than HF-based CIS (due to DFT-type incorporation of leading dynamic correlation effects), but their current Gaussian implementation is considerably less general. In former Gaussian versions, TD methods lacked the necessary density corrections ("density=current") for NBO/NRT or other analysis options. Many other Gaussian multistate CI options (such as CISD, QCISD, etc.) are similarly deficient with respect to analytic gradients and density corrections for excited-state roots.

A more accurate but computationally expensive set of multistate Gaussian options is provided by the SAC-CI keyword, based on coupled-cluster (CC) techniques for incorporating electron correlation and size-consistency corrections for higher-level excitations. The SAC-CI options cover a wide range of excitation and ionization phenomena, including provision for spin state and symmetry control with analytic derivatives and density corrections for specific target roots. Thus, SAC-CI can offer benchmark-like accuracy for assessing lower-level CI/CAS methods, but involves considerable increase in computational cost and input-keyword complexity.

"Complete active space" (CAS) SCF methods employ a more complete CI expansion, incorporating *all* possible configurations from an "active space" of N electrons and M

orbitals, with self-consistent optimization of each orbital. In principle, this leads to the *best possible* multiconfigurational wavefunction for the chosen active space, but in practice, the active space must be chosen judiciously. Even for small N, M values, numerical convergence of the iterative CAS(N,M) search is notoriously difficult when the initial orbital "guess" is formulated in terms of canonical MOs. Fortunately, CAS convergence characteristics are found to improve markedly if NBOs are chosen as the starting orbitals [A. V. Nemukhin and F. Weinhold, *J. Chem. Phys.* **97**, 1095–1108, 1992], and standard Gaussian options now make such "CASNBO" calculations rather routine.

Gaussian input syntax for various CI/CAS methods can be illustrated for the simple case of vertical (fixed geometry) excitation to the first excited ($i = 1$) state of nitric oxide (NO). A sample Gaussian input file to evaluate and analyze the CIS wavefunction for this state at fixed $R_{NO} = 1.2$ Å is shown below:

```
#CIS(root=1)/6-311++G** scf=tight density=current pop=nboread

1st-excited state of doublet radical NO at DIST=1.2

0 2
N
O  1  dist

dist  1.2

$nbo file=NO_xs fixdm $end
```

The "root=1" parameter of the CIS keyword specifies the excited state $i = 1$ (consistent with the notation in Sidebar 11.1), and "density=current" requests that density corrections for this excited state (rather than the default ground state) be stored for subsequent NBO analysis ("pop=nboread"). The "fixdm" keyword in the $nbo keylist applies an additional correction to prevent unphysical (negative or Pauli-violating) orbital populations from the CIS-corrected density as approximated by the Gaussian program.

For the DFT-based TD approximation in G09, the route (card may be replaced by

```
TD(root=1) B3LYP/6-311++G** scf=tight
```

Current G09 also supports the "density = current" keyword for NBO analysis of TD excited states.

For the SAC-CI method, a simplified route card (neglecting symmetry) might be specified as

```
#SAC-CI=(full,doublet=(nstate=6),
targetstate=(spinstate=doublet,root=2))
6-311++G** nosymm density=current pop=nboread
```

Note that "root=2" identifies the first excited "targetstate" of interest (i.e., SAC-CI counts the ground state as root=1).

Finally, for the CAS/NBO procedure, one must first select active space N, M parameters and source NBOs for the state of interest. For example, we might select N=11, M=8 to include a full-valence active space with CIS-level NBOs as input "guess" orbitals. The two-step Gaussian input file for such CIS-seeded CASNBO(11,8) calculation is shown below:

```
%chk=no_cas
#CIS(root=1)6-311++G** scf=tight density=current pop=nboread

    (z-matrix as above)

$nbo aonbo=cs $end

--Link1--
%chk=no_cas
#CAS(11,8,nroot=2)/6-311++G** guess=read density=current
pop=nboread

    (z-matrix as above)

$nbo file=no_cas fixdm $end
```

The $nbo "aonbo=cs" keyword requests storing the CIS-level NBOs in the shared checkpoint file where "guess=read" will read them as initial guess for the CAS/NBO job. Note that CAS identifies the 1st excited state as "nroot=2" whereas CIS uses "root=1" for this state. Note also that the Gaussian open-shell CAS implementation *fails* to provide relevant spin density information to NBO, forcing "spin-averaged" NBO description of reduced accuracy. This restriction strongly detracts from the potential usefulness of CAS calculations for excited-state analysis. However, illustrative use of this method allows one to see how one can still obtain useful NBO-based descriptors of the excited state despite loss of spin information.

11.2 ILLUSTRATIVE APPLICATIONS TO NO EXCITATIONS

As simple illustrative applications, let us first consider low-lying states of nitric oxide (NO), recognized as *Science* magazine's (1992) "Molecule of the Year" for its central role in numerous atmospheric and physiological phenomena. Figure 11.1 displays potential curves for low-lying excited states of NO as calculated at the CIS/6-311++G** level, showing the complex excitation features that are predicted to lie within 10 eV of the ground state. As seen in the figure, the two lowest states (X) appear as near-degenerate potential curves of rather ordinary single-well character, whereas the next five states (A–E) exhibit a confusing spaghetti-tangle of oscillatory maxima and minima (barriers and wells). What's going on here?

Before looking at details of CIS-level excitation features, we might ask how these compare with experimental reality. As shown in Table 11.1, such comparisons are far from reassuring. The agreement with experimentally inferred excitation (T_e), vibration (v_e), and geometry (R_e) values is so poor that even the presumed state associations

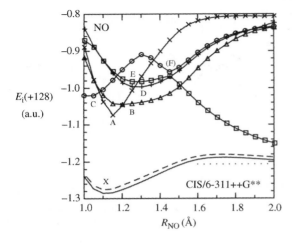

Figure 11.1 Low-lying states of nitric oxide (NO) in the range $R_{NO} =$ 1.0–2.0 Å, calculated at CIS/6-311++G** level from stable UHF configuration at each R_{NO}. The horizontal dotted line marks the asymptotic dissociation limit to ground state $N(^4S) + O(^3P)$ atoms (−129.20629 a. u.). Successive minima are labeled (X, A–E) in accordance with presumed spectroscopic identification. (The outer "F" well of the C double-minimum state, if real, may appear to be an additional "state" according to spectroscopic criteria; cf. Table 11.1.)

are questionable. The calculated CIS-level dissociation energy (ca. 2.2 eV) is also far below the inferred experimental value (ca. 5.3 eV).

Further evidence of theoretical uncertainty is provided by comparison calculations at higher levels. Figure 11.2 displays corresponding calculations at SAC-CI level, with successive curves (X, A–E) symbolically marked to suggest possible associations with CIS curves in Fig. 11.1. The calculated SAC-CI dissociation energy is now far too *high* (cf. dotted line in Fig. 11.2) and the low-lying excitation energies too *low* compared to experimental values (cf. Table 11.1). The pattern of crossings and dissociation limits beyond ca. 1.5 Å also disagree qualitatively in the two theoretical descriptions. Thus, neither theoretical level should be considered realistic, and the

Table 11.1 Calculated (CIS/6-311++G**) excitation energy (T_e), vibrational frequency (v_e), and bond distance (R_e) for low-lying excited states of nitric oxide (with identifying state labels corresponding to Fig. 11.1a), compared with experimental values for presumed spectroscopic association (G. Herzberg, *spectra of diatomic dolecules*, D. Van Nostrand, New York, 1950, pp. 558).

State	T_e (eV)		v_e (cm^{-1})		R_e (Å)		
	CIS	Exp	CIS	Exp	CIS	Exp	Herzberg
X (−)	0	0	2216	1904	1.118	1.151	$X^2\Pi$
A (x)	5.73	5.45	6817	2371	1.154	1.064	$A^2\Sigma^+$
B (Δ)	6.53	5.69	1598	1038	1.179	1.385	$B^2\Pi$
C (o_{in})	7.15	6.47	2850	2347	1.025	1.075	$C^2\Sigma^+$
D (+)	7.81	6.58	1626	2327	1.280	1.065	$D^2\Sigma^+$
E (°)	8.21	7.52	1358	2374	1.284	1.066	$E^2\Sigma^+$
(F) (o_{out})	8.96	?	4527	?	1.451	?	?

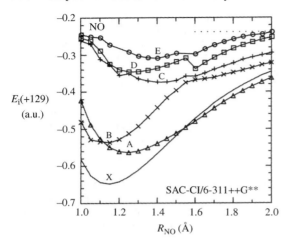

Figure 11.2 Similar to Fig. 11.1, for SAC-CI method.

extreme differences warn of severe sensitivity of excitation details with respect to theoretical level. Sidebar 11.3 describes related CAS/NBO-type calculations that provide still another picture of the NO excitation spectrum (with characteristic CAS-type limitations on analysis options). Under these circumstances, we can only hope that deeper analysis assists understanding of possible reasons for the sensitivity to theory level and success or failure of one method versus another. Accordingly, our present mission is to illustrate NBO-level description of the CIS-level excitation features portrayed in Fig. 11.1, while acknowledging the likely inadequacies of this level for quantitative purposes.

Let us first consider the two ground-state levels (solid and dashed curves of Fig. 11.1), which exhibit a characteristic idiosyncrasy of UHF and CIS-level description. The expected ground-state configuration of NO, namely,

$$(n_N)^2(n_O)^2(\sigma_{NO})^2(\pi_x)^2(\pi_y)^2(\pi_x^*)^{\uparrow} \tag{11.8a}$$

can equivalently be written in separated α/β spin configurations as

$$[(n_N)^{\uparrow}(n_O)^{\uparrow}(\sigma_{NO})^{\uparrow}(\pi_y)^{\uparrow}(p_{N(x)})^{\uparrow}(p_{O(x)})^{\uparrow}][(n_N)^{\downarrow}(n_O)^{\downarrow}(\sigma_{NO})^{\downarrow}(\pi_x)^{\downarrow}(\pi_y)^{\downarrow}] \tag{11.8b}$$

[because $(\pi_x)^2(\pi_x^*)^{\uparrow}$ is equivalent to $(\pi_x)^{\downarrow}(p_{N(x)})^{\uparrow}(p_{O(x)})^{\uparrow}$]. This configuration should yield a doubly degenerate ground state of $^2\Pi$ symmetry (because the odd electron might equally be placed in the π_y^* orbital). However, in "broken symmetry" UHF theory (cf. Sidebar 3.2), only one of the two equivalent configurations is chosen as starting point, and the π_x, π_y orbitals optimize to slightly *a*symmetric forms, requiring a slight "excitation energy" for the $\pi_x^* \rightarrow \pi_y^*$ configuration of CIS theory. [Only the ground-state CIS curve displays this artifact, because corresponding $\pi_x \rightarrow \pi_y$ replacements for other $^2\Pi$ states differ by two (or more) substitutions from the ground UHF configuration, and hence never appear in CIS-level description.] Such symmetry breaking artifacts are intrinsic to UHF/CIS description, and should remind us that CIS-level wavefunctions are at best useful only for crude qualitative purposes.

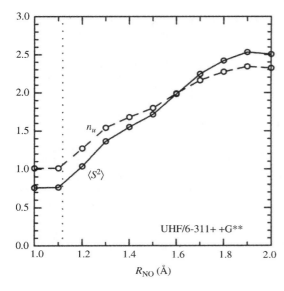

Figure 11.3 Total squared spin angular momentum ($\langle S^2 \rangle$, a.u.; solid) and corresponding number of unpaired electrons (n_u; dashed) at various interatomic separations (R_{NO}) along the ground-state NO potential in UHF/6-311++G** description, with vertical dotted line marking the calculated equilibrium geometry.

The characteristic symmetry breaking of UHF/CIS theory is also displayed by significant deviations from expected doublet spin symmetry, even in the near-neighborhood of equilibrium. The number of unpaired electrons (n_u) of spin 1/2 can be related to the expectation value of total squared spin angular momentum ($\langle S^2 \rangle$) by the following equation:

$$\langle S^2 \rangle = (n_u/2)(n_u/2 + 1) \tag{11.9}$$

For a doublet radical ($n_u = 1$), this should lead to $\langle S^2 \rangle = 0.75$, but UHF-level wavefunctions often deviate significantly from exact spin symmetry. Figure 11.3 displays calculated UHF $\langle S^2 \rangle$ and n_u values over the range of R_{NO} bond distances in Fig. 11.1, showing the significant "spin contamination" that affects UHF/CIS excited-state behavior beyond the equilibrium region.

The R-dependent n_u behavior displayed in Fig. 11.3 corresponds to the expected spin-configurational changes associated with electronic promotion and bond formation as the atoms approach. These configurational changes are shown in greater detail in Fig. 11.4, which plots valence NAO occupancies (from NPA or NEC output) during NO bond formation. As seen in Fig. 11.4, the 2s → 2p$_z$ "promotion" effect of σ-bond formation begins at longer range (ca. 1.8 Å for 2p$_{O(z)}$), whereas the onset of shorter-range π-bond interactions (occupancy shifts in 2p$_x$, 2p$_y$ NAOs) occurs near 1.4 Å. The spin pairings associated with successive bond formation lead to progressive spin diminution (cf. Fig. 11.3) toward the low-spin molecular limit near R_e, consistent with the NEC variations of Fig. 11.4.

The longer-range region of UHF/CIS excitations is also characterized by interesting "barrier" features, such as the ca. 0.7 eV barrier that separates the ground-state attractive well from the limiting long-range dissociation to ground-state atoms (dotted line in Fig. 11.1). Such barriers lead to characteristic "predissociation"

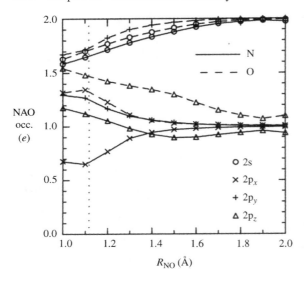

Figure 11.4   Occupancies of principal valence NAOs in NO bond formation, showing expected atom-like configurations for N $[(2s)^2(2p_x)^1(2p_y)^1(2p_z)^1]$ and O $[(2s)^2(2p_x)^1(2p_y)^2(2p_z)^1]$ at long range ($R_{NO} \cong 2.0$ Å), "promotion" to σ-bonding configuration (increased $2p_z$, reduced 2s occupancy) at intermediate range ($R_{NO} < 1.8$Å), and π-bonding interactions ($2p_x$, $2p_y$ occupancy shifts toward electronegative O) at shorter range ($R_{NO} < 1.4$ Å). (See figure insets for atom and orbital labeling.) A vertical dotted line marks the ground-state equilibrium geometry.

phenomena that are beyond the scope of this discussion [see Herzberg, p. 420ff]. A still more interesting barrier feature is seen near 1.3 Å along the C (circles) curve, separating the *two* distinct equilibrium minima (near 1.02, 1.45 Å) in this excited state. The double-well character of the C/F curve leads to the interesting phenomenon of "bond-stretch isomers" [well known in excited states of the even simpler H_2 molecule; see E. R. Davidson, *J. Chem. Phys.* **35**, 1189, 1961; W.-D. Storer and R. Hoffmann, *J. Am. Chem. Soc.* **94**, 1661, 1972], in which two distinct isomeric minima on the *same* potential energy surface differ only by a change of bond length. The outer "F" minimum apparently results from an avoided crossing with the E (squares) curve near 1.45 Å, as shown in greater detail in Fig. 11.5. Figure 11.5

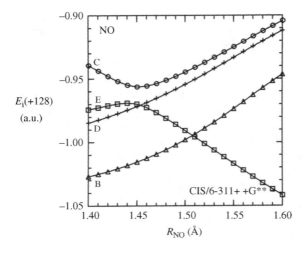

Figure 11.5   Blow-up of "avoided crossing" region in Fig. 11.1.

also displays two actual (*un*avoided) crossings, one of the D/E curves (near 1.45 Å) and the other of the B/D curves (near 1.50 Å). In polyatomic molecules, such curve-crossings correspond to "conical intersections" that are often the site of radiation-less transitions and unusual vibrational excitation (heating) effects. All such effects reflect the complex multiconfigurational character of molecular excited states and warn of the inherent challenges in describing these states in terms of usual MO-type single-configuration concepts.

Let us now turn to NBO-based description of the NO excited states, highlighting selected points of special interest. Table 11.2 summarizes key NBO/NRT descriptors, including bond order b_{NO}, atomic charge Q_N, and NLS spin configurations, for CIS equilibrium species in low-lying states (X, A–F) of nitric oxide. For each spin NLS, we employ a symbolic Lewis structural diagram with nonbonding electron "bar" symbols positioned to distinguish p_x (upper bar), p_y (lower bar) or σ-hybrid (side bar) type 1-c NBO. For example, the structures "ı$\bar{\text{N}}$=$\bar{\text{O}}$ı" or "ı$\underline{\text{N}}$=$\underline{\text{O}}$ı" differ only in orientation of the off-axis nonbonding electrons in p_x or p_y directions, respectively (so that the π_{NO} NBO is π_y in the first case and π_x in the second). Sidebar 11.4 provides additional details of $NRTSTR keylist input that leads to the calculated b_{NO} entries of Table 11.2. As seen in the table, the excited-state species vary widely in bond order (from 1.8 to 3.5), charge distribution (from -0.3 to $+0.3$), and NLS spin configuration (from single-bond to quadruple-bond forms). The accuracy of single-configuration NLS description is also seen to vary significantly from state to state, with ground-state-like values ($>99\%$) for some species (A, C, E) but significantly lower accuracy (93–95%) for others (B, D, F). How can we understand these variations in terms of simple NBO-based concepts?

As shown in Table 11.2, the NLS description of the ground X-state corresponds to the expected form (Eqs. 11.8a and 11.8b), with double-bonded (ı$\underline{\text{N}}$=$\underline{\text{O}}$ı) α NLS and triple-bonded (ı$\underline{\text{N}}$≡$\underline{\text{O}}$ı) β NLS. From this starting configuration, we can describe each excited species concisely in terms of the formal NBO excitations shown in the final column of Table 11.2. Each excitation denotes a one-electron promotion between distinct NBO forms (such as "$n_N \rightarrow \pi^*$" promotion from σ-type n_N to π-type π_x^* or π_y^*), or a polarization shift from bonding to nonbonding form (such as "$\pi_{CO} \rightarrow p_O$" relocalization from two-center π_{CO} to one-center p_O). As shown in Table 11.2, each excited-state species carries a unique NBO-excitation "signature" that suggests its energy ordering, structure, and chemical properties.

Among the CIS excitations of Table 11.2, the C-state (inner well) is clearly distinguished by one-e promotion to a *Rydberg* bond orbital,

$$\sigma_{3s} \cong 0.79\,(3s)_N + 0.62\,(3s)_O \qquad (11.10)$$

built primarily from 3s-type NAOs (of "Rydberg" character, beyond the formal valence shell). Figure 11.6 shows two-dimensional contour and three-dimensional surface view of the strange ("pac-man") σ_{3s} Rydberg bond spin NBO, which augments the three usual valence bond spin NBOs to give a *quadruply* bonded α NLS of unusually short bond length.

Table 11.2 Calculated bond length (R_e), NRT bond order (b_{NO}), atomic charge at N (Q_N), and α/β natural lewis structure (NLS) representations (and percentage accuracy) for low-lying states of nitric oxide (CIS/6-311++G** level) (cf. Fig. 11.1a for state labels).

State	R_e(Å)	b_{NO}	Q_N	NLS (% ρ_{NLS})		NBO excit.
				α	β	
X(−)	1.118	2.495	+0.268	׀N̈=Ö׀ (99.80%)	׀N≡Ö׀ (99.93%)	—
A(x)	1.154	2.458	−0.274	׀N̲–Ö׀ (99.68%)	׀N≡Ö׀ (99.92%)	$\pi_x^\uparrow \to p_{o(x)}^\uparrow$
B(Δ)	1.179	2.116	+0.310	׀N̲–Ö׀ (95.72%)	N̄–Ö׀ (92.51%)	$n_N^\uparrow \to \pi_x^{*\uparrow}$ $\pi_x^\uparrow \to p_{o(x)}^\uparrow$
C(o$_{in}$)	1.025	(3.5)	+0.221	׀N≡Ö׀ (98.91%)	׀N≡Ö׀ (99.94%)	$\pi_y^{*\uparrow} \to \sigma_{3s}^\uparrow$
D(+)	1.280	2.179	+0.206	׀N̲=Ö׀ (93.52%)	N̄–Ö׀ (94.45%)	$n_N^\downarrow \to \pi_x^{*\downarrow}$
E(⊙)	1.284	2.020	+0.104	׀N̲=Ö׀ (99.70%)	N̲–Ö̲׀ (98.04%)	$n_N^\downarrow \to \pi_y^{*\downarrow}$ $\pi_x^\downarrow \to p_{o(x)}^\downarrow$
(F)(o$_{out}$)	1.451	1.805	−0.008	N̄–Ö׀ (94.00%)	N̲–Ö׀ (93.43%)	$n_N^\downarrow \to p_{N(x)}^\uparrow$ $\pi_x^\downarrow \to p_{o(x)}^\downarrow$

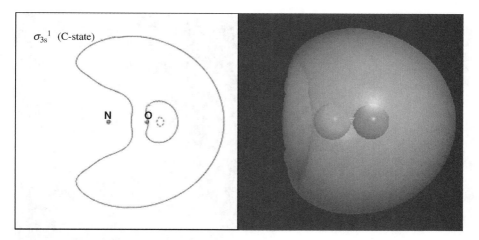

Figure 11.6 Contour and surface plots for σ_{3s} Rydberg-type NBO in quadruply bonded C-state inner well (1.025 Å; cf. Table 11.2) of NO (CIS/6-311++G** level). (See the color version of this figure in Color Plates section.)

A surprising feature of Table 11.2 is the charge distribution of the A-state, which exhibits *negative* charge at N despite the nominal $\pi_x^\uparrow \rightarrow p_{O(x)}^\uparrow$ repolarization of a π-bond toward O. The explanation apparently lies in the anomalous π-bond polarizations of the β-spin set, which are conspicuously *reversed* compared to π_x, π_y bonds (of either spin) in other states. Figure 11.7 illustrates the difference between π-bond NBOs of ordinary (X state) versus reversed-polarity (A state) form. The reversed π-polarity suggests reversed propensities for the direction of nucleophilic or electrophilic π-attack in the A-state, according to CIS theory.

Finally, we briefly describe the variations of multiconfigurational character within a *single* state, taking as an illustrative example the C-state potential (circles) of Fig. 11.1. Table 11.3 compares the bond order, charge distribution, and dominant NLS bonding pattern at three distances (1.2, 1.4, and 1.6 Å), showing the strong variations of electronic character that accompany oscillatory features along the C-state potential curve. In each spatial region, a single-configuration NLS description is of reasonably high accuracy (98–99%), but these configurations differ surprisingly from region to region in both α and β spin sets. The accuracy of single-configuration NLS description is expected to drop significantly near crossings and avoided crossings (e.g., the B, D, and F wells), as shown by the $\%\rho_{NLS}$ values in Table 11.2.

The α NLS (symbolically denoted as ⊪N≡O⊪) at 1.40 Å is particularly noteworthy. Unlike other double-bonded forms in Tables 11.2, 11.3, both 2-c bonds are here of π type (π_x, π_y). The usual σ_{NO} bond is replaced by filled *on*-axis 1-c orbitals (the former σ-bonding hybrids) that are mutually opposed for strong *repulsive* interaction. This configuration therefore leads to a steeply repulsive region on the C-state potential (ca. 1.30–1.45 Å), which forms the apparent "inner wall" of the F

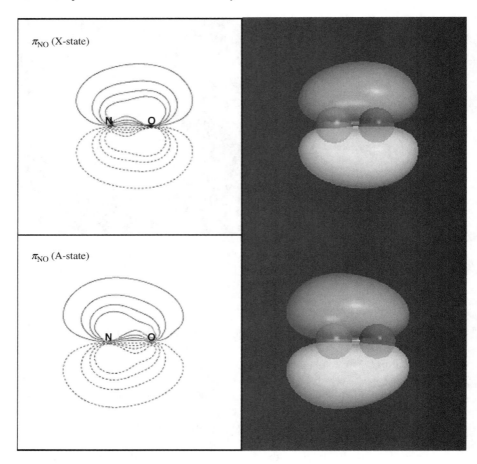

Figure 11.7   Contour and surface plots comparing π_{NO} NBO for normal-polarized ground-state (X; upper panels) versus reversed-polarized excited state (A; lower panels) π-bonds of NO (CIS/6-311++G** level). (See the color version of this figure in Color Plates section.)

potential well. The electronic configuration therefore switches conspicuously from one side of the F potential minimum to the other. The CIS-level description suggests pronounced anharmonicities and other spectral anomalies in F-well vibrations, with strong shifts of bond-order, polarity, and electronic configuration accompanying each transit between inner and outer turning points of the vibrational motion.

Deeper electronic insights into the origins of unusual IR or UV-VIS spectral features could be gained from more detailed NBO analysis of C-state or other potential curves in Fig. 11.1. However, the preceding examples may adequately suggest the surprising excited-state complexities that arise even in simple diatomic species. These diatomic examples also illustrate generic NBO/NRT techniques (cf. Sidebar 11.4) for comparing different states, or different methods for the same state, in more complex polyatomic species, as considered in the following sections.

Table 11.3 Similar to Table 11.2 for representative distances (1.2, 1.4, and 1.6 Å) along the C-state potential curve.

$R_e(\text{Å})$	b_{NO}	Q_N	NLS (% ρ_{NLS})		NBO excit.
			α	β	
1.20	2.033	+0.268	ꞮN̲=O̲Ꞓ (99.04%)	ꞮN=OꞒꞒ (98.01%)	$\pi_y^{\downarrow} \rightarrow p_{o(y)}^{\downarrow}$
1.40	2.474	−0.274	ꞒꞒN̲=OꞒꞒ (98.50%)	ꞮN−O̲Ꞓ (98.80%)	$\pi_y^{*\uparrow} \rightarrow \sigma_{NO}^{*\uparrow}$
					$\pi_x^{\downarrow} \rightarrow p_{o(x)}^{\downarrow}$
					$\pi_y^{\downarrow} \rightarrow p_{o(y)}^{\downarrow}$
1.60	2.027	+0.310	ꞮN̲̄−O̲Ꞓ (99.49%)	ꞮN−O̲Ꞓ (98.24%)	$\pi_x^{\uparrow} \rightarrow p_{N(x)}^{\uparrow}$
					$\pi_x^{\downarrow} \rightarrow p_{o(x)}^{\downarrow}$
					$\pi_y^{\downarrow} \rightarrow p_{o(y)}^{\downarrow}$

SIDEBAR 11.3   *CAS/NBO FOR NITRIC OXIDE EXCITATIONS*

The excited states of NO can also be approximated by complete active space (CAS) calculations (Sidebar 11.2) employing NBOs as starting orbitals ("CAS/NBO method"). As an illustrative application we consider full-valence CAS(11,8) active space (with $N = 11$ electrons, $M = 8$ orbitals) to optimize and analyze the "nroot=6" (fifth excited root) of NO, as shown in the Gaussian input file below:

```
%chk=no_cas
#cis(root=5)/6-311++g** density=current pop=nboread

NO, initial NBOs from cis root=5 (1.45 A)
    56     1.4500    -129.14323

0 2
N
O  1  dist

dist  1.45

$nbo fixdm aonbo=cs $end

--Link1--
%chk=no_cas
#cas(11,8,nroot=6,fulldiag)/6-311++g** density=current
guess=read sleazy
```

```
NO, cas nroot=6

0 2
N
O  1  dist

dist  1.45

--Link1--
%chk=no_cas
#cas(11,8,nroot=6,fulldiag)/6-311++g** density=current
guess=read pop=nboread opt

NO, cas nroot=6 (optimizes to -129.1432292 at R = 1.4473)

0 2
N
O  1  dist

dist  1.45

$nbo file=no_cas fixdm archive $end
```

As shown in this complex three-link job, a simpler CIS(ROOT=5) calculation was first performed at the approximate geometry ($R = 1.45$ Å) suggested by an earlier scan job. The NBOs from this calculation were then occupancy sorted and stored ("aonbo=cs") in the checkpoint file, where they were picked up ("guess=read") to initialize the CAS/NBO calculation in the second link. The precarious initial CAS(11,8)/6-311++G** convergence was first performed with relaxed numerical thresholds ("sleazy"), then this poorly converged CAS solution was used in the final link (with another "guess=read") to perform geometry optimization ("opt") and NBO analysis of the resulting equilibrium species.

The NBO results expose a characteristic weakness of Gaussian CAS calculations. Due to the spin-free GUGA (graphical unitary group approach) method employed for energy evaluations, all spin-dependent density information is *lost*, thereby sacrificing considerable detail of the open-shell density and NBOs. The "spin-averaged" character of the CAS description is signaled by absence of the usual α and β spin NBO output sections, and by the unusually low accuracy (92.49% in this case) of the resulting ROHF-like natural Lewis structure.

As shown in the leading entries of the final NBO summary output,

```
Natural Bond Orbitals (Summary):

          NBO                 Occupancy
-------------------------------------------
Molecular unit  1   (NO)
   1. BD ( 1) N  1- O  2        1.94856
   2. CR ( 1) N  1              1.99996
   3. CR ( 1) O  2              1.99995
```

```
4. LP ( 1) N  1            1.99423
5. LP ( 2) N  1            1.06164
6. LP ( 1) O  2            1.99554
7. LP ( 2) O  2            1.43656
8. LP ( 3) O  2            1.43656
9. LP*( 3) N  1            1.06164
```

five NBOs are found to be of near-double occupancy but four others have occupancy 1.06–1.44, indicative of extreme (~*tetra*radical) spin unpairing as expected in this highly excited state. These four partially occupied orbitals evidently correspond to the four unmatched spin-orbitals in the CIS F-state NLS structures of Table 11.2, but additional details of the CAS open-shell density can only be dimly perceived through the veil of spin-free description. Where any feasible alternative exists, use of spin-free CAS calculations is not recommended for analysis purposes, and such calculations are not considered further in this book.

SIDEBAR 11.4 *NBO/NRT ANALYSIS OF NO EXCITATION SPECIES*

The CIS/6-311++G** excited states of NO (Table 11.2) present several challenges to default NBO/NRT analysis. Unlike the starting UHF-level description of the ground-state X $^2\Pi$ species, the CIS-level excited-state description offers no simple one-e "effective Hamiltonian" (analogous, e.g., to the Fock or Kohn–Sham operator) to assess orbital energetics. Hence, second-order $E^{(2)}$ stabilization energies and related energy-type NBO descriptors disappear from CIS-level output.

Analysis of CIS-level density for a chosen root (specified by "density=current" keyword) also requires Gaussian calculation of additional density corrections. These corrections are of approximate perturbative form, often leading to inconsistent mathematical features of the density matrix such as pathological "negative" or "Pauli-violating" orbital populations that are recognized as fatal errors by the NBO program. The FIXDM ("fix density matrix") keyword corrects the worst such pathologies, and should always be included for CIS-level NBO/NRT analysis, but this keyword is only a partial fix for inherent errors of CIS density corrections. Similar remarks apply to SAC-CI and related methods.

For NRT bond order calculations reported in Tables 11.2 and 11.3, we employ $NRTSTR keylists to force consistent inclusion of alternative single-, double-, and triple-bonded resonance structures in each state. A sample Gaussian input file to compute the final entries of Table 11.3 for the C-state at $R = 1.60$ Å is given below:

```
%chk=no_cis
#uhf/6-311++g** scf=(qc,tight)
density=current stable=opt

NO, stable uhf ground state for cis

0 2
N
O  1  dist

dist  1.6
```

```
--Link1--
%chk=no_cis
#cis(root=6,nstates=10)/6-311++g** scf=tight
density=current guess=read pop=nboread

NO, C-state

0 2
N
O  1  dist

dist  1.60

$nbo file=no_cis_c fixdm nrt $end
 $NRTSTRA
   STR1         ! single
      LONE 1 2   2 3 END
      BOND S 1 2 END
   END

   STR2         !double
     LONE 1 2   2 2 END
     BOND D 1 2 END
   END
   STR3          !atoms
     LONE 1 4   2 2 END
   END
$END
$NRTSTRB
   STR1        ! single
     LONE 1 3   2 1 END
     BOND S 1 2 END
   END
   STR2         ! double
     LONE 1 1   2 2 END
     BOND D 1 2 END
   END
   STR3          ! triple
     LONE 1 1   2 1 END
     BOND T 1 2 END
   END
   STR4         ! atoms
     LONE 1 1   2 4 END
   END
$END
```

Note that the first job employs STABLE=OPT to obtain the most stable UHF solution, which then becomes (through GUESS=READ) the starting point for subsequent CIS (root=6) calculation and analysis.

11.3   FINDING COMMON GROUND: NBO VERSUS MO STATE-TO-STATE TRANSFERABILITY

A characteristic feature of NBOs is their high *transferability* compared to MOs. For example, a carbon–carbon π-bond NBO (π_{CC}) from any unsaturated hydrocarbon (e.g., butadiene, benzene, or buckminsterfullerene) appears visually indistinguishable from that of ethylene. In contrast, π-type MOs tend to differ *qualitatively* from one molecule to another, often varying confusingly even with small geometric changes in a single molecule.

Figure 11.8 compares frontier MOs of acrolein ($CH_2{=}CHCH{=}O$) in planar and twisted geometries, showing the rather confusing changes of form that accompany a ground-state torsional distortion ($\Delta E = 8.48\,kcal/mol$). The visual changes are also confirmed by CMO keyword output, where, for example, the LUMO varies from strongly mixed $[\varphi_{LUMO} \cong 0.69\pi^*_{CC} - 0.68\pi^*_{CO}]$ to nearly pure π^*_{CO} character during torsional motions. Figure 11.9 shows the corresponding torsional variations of π_{CC}, $n_O^{(\pi)}$, and π^*_{CO} NBOs (principal components of SOMO, HOMO, and LUMO, respectively), illustrating the near-equivalence to corresponding ethylene (π_{CC}) or formaldehyde ($n_O^{(\pi)}$, π_{CO}) NBOs. Whereas the NBOs of larger molecule exhibit high transferability from small-molecule precursors, those of MOs tend to exhibit surprising variability that may challenge extrapolation even for small geometry distortions.

However, we might inquire whether similar NBO versus MO transferability differences extend to different *states* of the same species. Such state-to-state transferability would allow each spectral excitation to be identified with specific orbital "quantum jumps" based on a transferable set of orbitals (of similar form but distinct occupancies) that are *common* to states of an excitation manifold. Such an orbital-based description of spectral excitations affords considerable conceptual economy, but requires selection of the transferable orbital set (if any) that most aptly serves as "common ground" for such simplification.

Do MOs or NBOs better meet the criterion of state-to-state transferability? We can address this question for the simple case of acrolein, using the lowest-lying "n to π^*" triplet state as an example. Figures 11.10 and 11.11 exhibit MOs and NBOs of the vertical (ground singlet geometry) ^3A excited state of acrolein, reordered as necessary to pair α- and β-spin orbitals and allow direct comparisons with corresponding left panels of Figs. 11.8 (MOs) and 11.9 (NBOs).

In NBO language, the transition is aptly described as "$n_O^{(\pi)} \to \pi^*_{CO}$" excitation, removing a β-spin electron from $n_O^{(\pi)}$ and replacing it (with reversed spin) in π^*_{CO}. If we write the relevant portion of the singlet ground-state configuration as

$$\ldots (\pi_{CO})^2 (\pi^*_{CO})^0 (n_O^{(\pi)})^2 \tag{11.11}$$

the excited triplet configuration (written in separated $[\alpha][\beta]$ form) becomes

$$\ldots [(\pi_{CO})^\uparrow \, (\pi^*_{CO})^\uparrow (n_O^{(\pi)})^\uparrow] \, [(\pi_{CO})^\downarrow (n_O^{(\pi)})^0] \tag{11.12}$$

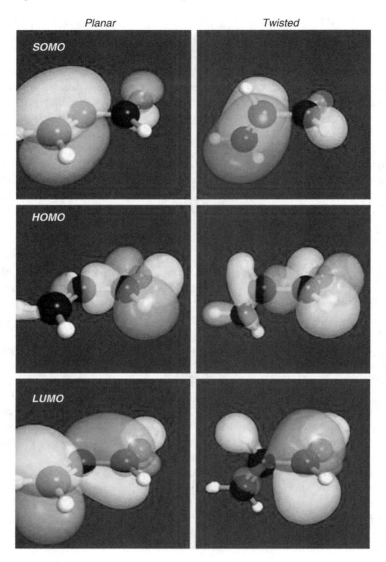

Figure 11.8   Frontier MOs of ground-state acrolein (SOMO = second occupied MO; HOMO = highest occupied MO; LUMO = lowest unoccupied MO), showing qualitative variations of form with torsions from planar (left) to twisted (right) geometry. (See the color version of this figure in Color Plates section.)

This can equivalently be written as

$$\ldots [(n_O'^{(\pi)})^\uparrow (n_C'^{(\pi)})^\uparrow (n_O^{(\pi)})^\uparrow] [\pi_{CO})^\downarrow (n_O^{(\pi)})^0] \tag{11.13}$$

because occupied π_{CO}, π^*_{CO} spin-orbitals (with zero net bond order) are equivalent to occupied $n_C'^{(\pi)}$, $n_O'^{(\pi)}$ nonbonding NBOs. The lower-left panel of Fig. 11.11 shows

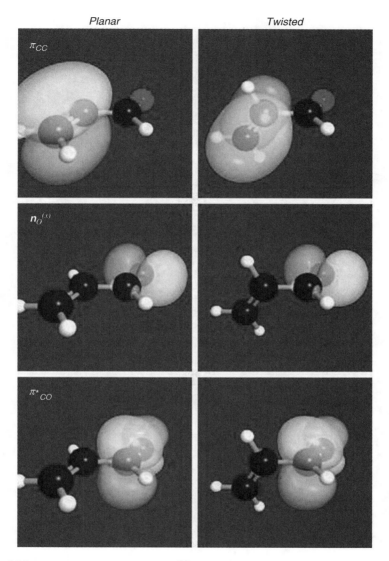

Figure 11.9   Similar to Fig. 11.8, for π_{CC}, $n_O^{(\pi)}$, π^*_{CO} NBOs that are leading contributors to MOs of Fig. 11.8, showing near-transferable NBO forms in planar and twisted geometry. (See the color version of this figure in Color Plates section.)

the new $n_C'^{(\pi)}$ NBO, whereas the new $n_O'^{(\pi)}$ resembles $n_O^{(\pi)}$ (middle-left panel), but rotated by 90°. Remaining NBOs of the two configurations are seen to be closely matched, so the NEC-based "$n_O^{(\pi)} \to \pi^*_{CO}$" designation describes the actual excitation quite concisely and accurately.

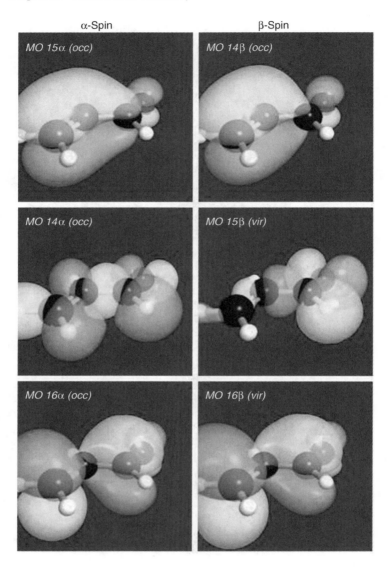

Figure 11.10   Frontier MOs of lowest triplet excited state of acrolein in vertical ground-state geometry (cf. left panels of Fig. 11.8 for ground singlet state), showing significant variations in state-to-state MO forms. (See the color version of this figure in Color Plates section.)

The corresponding MO-based configurational description is considerably less apt. One can see resemblance between singlet and triplet MOs that are uninvolved in excitation, but the formally vacated MO 15β spin-orbital (middle right in Fig. 11.10) deviates strongly from its supposed occupied counterpart

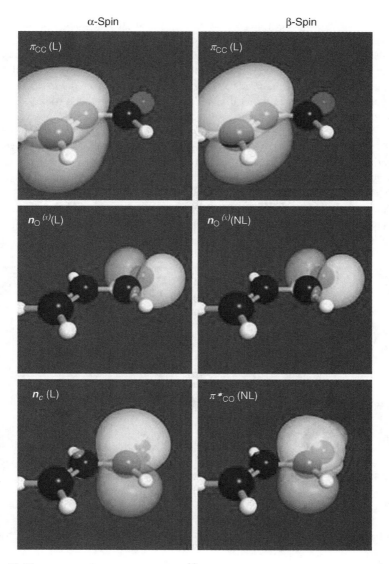

Figure 11.11   Similar to Fig. 11.10, for π_{CC}, $n_O^{(\pi)}$, π^*_{CO} NBOs (cf. left panels of Fig. 11.9 for ground singlet state), showing high state-to-state NBO transferability. (See the color version of this figure in Color Plates section.)

MO 14α, and neither closely resembles the corresponding parent singlet MO ("HOMO" of Fig. 11.8). The qualitative visual comparisons suggest that a Koopmans-type MO-based description of acrolein excitation is significantly less accurate than the corresponding "$n_O^{(\pi)} \rightarrow \pi^*_{CO}$" NBO description. The superior transferability of NBOs compared to MOs is expected to become still more evident when strong conjugative coupling between adjacent π-bonds (as in acrolein) is absent.

Despite the widespread and somewhat unjustified influence that has been accorded to Koopmans' theorem [see, e.g., critical discussion by E. Heilbronner, in R. Daudel and B. Pullman (eds.), *The World of Quantum Chemistry* D. Reidel, Dordrecht, 1974], there is generally *no* reason to assume that ground-state MOs provide the "best" or "only" basis for describing electronic excitation. Even if MOs were judged to exhibit superior state-to-state transferability, their capricious ground-state forms (Fig. 11.8) make it unlikely that such transferability could serve as a productive route to conceptual understanding of excited-state properties. Sidebar 11.5 provides additional details of the unphysical "delocalization" commonly associated with MO-based description of both ground and excited states.

State-to-state NBO transferability suggests how familiar NBO/NRT methodology may be applied consistently to analysis of an entire excitation manifold. Some simple applications to acrolein excited states are illustrated in ensuing sections.

SIDEBAR 11.5   *PHYSICAL AND UNPHYSICAL DELOCALIZATION IN NBO AND MO THEORY*

In place of the usual direct LCAO-MO (AO $\rightarrow$ MO) transformation, we may consider the intermediate sequence of localized transformations

$$\text{AO} \rightarrow \text{NAO} \rightarrow \text{NBO} \rightarrow \text{NLMO} \rightarrow \text{MO} \qquad (11.14a)$$

to clarify the meaning of "delocalization" in MO and NBO theory. Except for the initial basis AOs (which are commonly taken as nonorthogonal), each orbital basis in (11.14a) provides a complete orthonormal set that can be used to *exactly* describe the wavefunction or density. This allows the "completely delocalized" limit of MO theory to be characterized in terms of intermediate localized (NAO, NBO) or semi-localized (NLMO) matrix representations of the one-electron Hamiltonian F_{op} (Fock or Kohn–Sham operator), namely,

$$\mathbf{F}^{(\text{NAO})} \rightarrow \mathbf{F}^{(\text{NBO})} \rightarrow \mathbf{F}^{(\text{NLMO})} \rightarrow \mathbf{F}^{(\text{MO})} \qquad (11.14b)$$

where, for example, for the NAO basis set $\{\varphi_i^{(\text{NAO})}\}$,

$$\left(\mathbf{F}^{(\text{NAO})}\right)_{i,j} = \langle \varphi_i^{(\text{NAO})}|F_{op}|\varphi_j^{(\text{NAO})}\rangle = \int \varphi_i^{(\text{NAO})*} F_{op} \varphi_j^{(\text{NAO})} d\tau \qquad (11.15)$$

and similarly for other bases. Diagonalization of any of the matrices (11.14b) must lead to the *same* final MO eigenvalues $\varepsilon_i = (\mathbf{F}^{(\text{MO})})_{ii}$ and MOs, expressed as linear combinations of the respective localized basis functions. We assume a rudimentary knowledge of matrix diagonalization (e.g., for a 2×2 matrix) in the following discussion.

Each transformation in (11.14a) brings the Fock matrix $\mathbf{F}$ to increasingly diagonal form (with fewer and smaller off-diagonal F_{ij} elements), culminating in the final diagonal $\mathbf{F}^{(\text{MO})}$ matrix with $F_{ij} = \varepsilon_i \delta_{ij}$. The initial strong mixing of hybridization and bond formation

(NAO → NBO transformation) results in the localized NBO Fock matrix, whose sparse off-diagonal structure leads to the simple patterns of donor–acceptor mixing. As a result, weak non-Lewis (NL) "tails" are attached to parent Lewis (L) NBOs to form *semi*-localized NLMOs (Section 5.4)

$$\varphi_i^{(\text{NLMO})} \cong \varphi_i^{(\text{NBO})} + \sum_j C_{ij} \varphi_j^{(\text{NBO})} \tag{11.16}$$

that are readily approximated by second-order perturbation theory (Section 5.2). The NLMOs tend to differ almost imperceptibly from NBOs and are easily associated with the "valence bonds" of the classical Lewis structure diagram.

However, it is important to recognize (cf. Section 5.4) that the semi-localized NLMOs are filled to capacity and hence *unitarily equivalent* to "textbook" canonical MOs (see *V&B*, p. 115ff). This means that the NBO donor–acceptor mixings of (11.16) are the only delocalizations of physical significance, and that the subsequent NLMO → MO mixings produce *no effect on the density or any other measurable property* of the determinantal wavefunction. (Canonical MOs may provide a genuine alternative starting point for perturbed systems or for multiconfigurational approximations, but these are separate issues.) Most of the confusing "delocalization" of MOs is physically illusory.

Let us consider some details of the $\mathbf{F}^{(\text{NLMO})}$ matrix diagonalization that reveals spurious MO delocalization. The NLMO Fock matrix elements are known to vanish between L (occupied) and NL (virtual) blocks, but nonvanishing F_{ij} elements *within* the L or NL block lead to mixing of NLMOs to form canonical occupied or virtual MOs. In a 2×2 approximation, the mixing of NLMOs $\varphi_i^{(\text{NLMO})}$ and $\varphi_j^{(\text{NLMO})}$ to form final MOs $\varphi_i^{(\text{MO})}$, $\varphi_j^{(\text{MO})}$ can be expressed in terms of the in-phase and out-of-phase linear combinations

$$\varphi_i^{(\text{MO})} = (1 - \lambda^2)^{1/2} \varphi_i^{(\text{NLMO})} + \lambda \varphi_j^{(\text{NLMO})} \tag{11.17}$$

$$\varphi_j^{(\text{MO})} = (1 - \lambda^2)^{1/2} \varphi_j^{(\text{NLMO})} - \lambda \varphi_i^{(\text{NLMO})} \tag{11.18}$$

with mixing coefficient λ (reducing to $\lambda = 1$ in the degenerate limit)

$$\lambda \cong |F_{ij}/(F_{jj} - F_{ii})| \tag{11.19}$$

According to (11.17) and (11.18), the final MOs appear "completely delocalized" when λ is of order unity, but "localized" (NLMO-like) for sufficiently small $\lambda \to 0$.

As a simple example of meaningless MO mixing, we consider the twisting of an acrolein molecule from anti ($\theta = 180°$) to syn ($\theta = 0°$) conformation through a perpendicular ($\theta = 90°$) transition geometry. If we focus on π-type MOs, we can examine the dihedral dependence of off-diagonal NLMO coupling elements as displayed in Figure 11.12 for occupied $F_{\pi-\pi'}$ (solid) and virtual $F_{\pi^*-\pi'^*}$ (dotted) MO mixings,

$$F_{\pi-\pi'} = \langle \pi_{\text{CO}} | F_{\text{op}} | \pi_{\text{CC}} \rangle \tag{11.20}$$

$$F_{\pi^*-\pi'^*} = \langle \pi_{\text{CO}}^* | F_{\text{op}} | \pi_{\text{CC}}^* \rangle \tag{11.21}$$

The figure shows that both $F_{\pi-\pi'}$ and $F_{\pi^*-\pi'^*}$ couplings vary *strongly* with twist angle, leading to vivid changes of MO morphology with *no* physical significance.

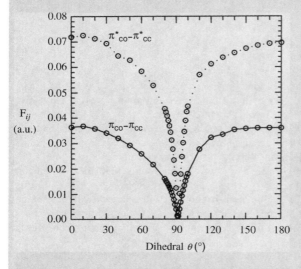

Figure 11.12 Dihedral variations of NLMO Fock matrix elements $F_{ij} = (\mathbf{F}^{(NLMO)})_{ij}$ for occupied $\pi_{CO}-\pi_{CC}$ (solid) or virtual $\pi^*_{CO}-\pi^*_{CC}$ (dotted) couplings in twisting of acrolein (B3LYP/6-311++G** level).

Figure 11.13 displays the associated π^*_{CO}, π^*_{CC} variations of %-NLMO composition for the acrolein LUMO (MO 16), showing how LUMO composition varies from strongly localized (π^*_{CO}-like) near 90° to highly delocalized ($\pi^*_{CO}-\pi^*_{CC}$ mixture) in less twisted geometry, with additional 10–20% contributions from other NLMOs throughout the angular range. Although the forms of the NLMOs change somewhat during twisting (reflecting the torsional dependence of conjugative and hyperconjugative delocalization), these changes are evidently dwarfed by morphological changes in the MOs that are devoid of physical significance.

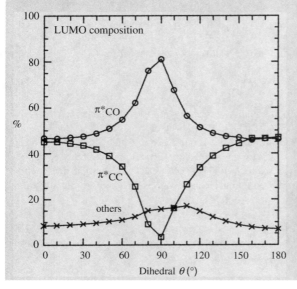

Figure 11.13 Dihedral variation of LUMO composition for acrolein (cf. dotted curve of Fig. 11.12), showing rapidly varying percentage contributions of π^*_{CO} (circles), π^*_{CC} (squares), and other (x's) NLMOs to canonical MO 16. Similarly vivid (but meaningless) variations are exhibited by CMO keyword analysis of practically all valence-level MOs.

Section 4.11 (p. 561ff) of *V&B* describes the instructive example of PtH$_4{}^{2-}$, showing additional details of each step in the sequence (11.14a). As shown in that example, the conceptual benefits of ignoring superfluous NLMO → MO transformation are complemented by the *many* simplifying features of NAO → NBO → NLMO transformations, each of which yields to simple perturbative modeling. By focusing on NBO → NLMO delocalization effects that are physically substantive and ignoring superfluous NLMO → MO mixings, one dispels the mystical ambiguity that surrounds current conceptions of MO theory and regains the powerful link to Lewis structural concepts.

11.4   NBO/NRT DESCRIPTION OF EXCITED-STATE STRUCTURE AND REACTIVITY

Let us now take up specific aspects of excited-state structure and reactivity, using the lowest-lying $^3(n \rightarrow \pi^*)$ vertical triplet excitation of acrolein as an example. I/O-11.1 shows Gaussian z-matrix and $NBO keylist input for the analyses to be described below.

In the vertical (trans conformer) geometry of the ground singlet species, the open-shell NBO Lewis structures of the excited triplet species are found to be represented by

$$
\begin{array}{cc}
\underset{\alpha\ (97.7\%\text{-}\rho_L)}{
\begin{array}{c}
\mathrm{H}\quad\ \ \mathrm{H} \\
\diagdown\ \ \diagup \\
\mathrm{C}=\mathrm{C} \\
\diagup\quad\ \diagdown \\
\mathrm{H}\quad\ \ \ \underset{\underset{\mathrm{H}}{|}}{\mathrm{C}}-\bar{\mathrm{Q}}\mathrm{I}
\end{array}}
&
\underset{\beta\ (97.6\%\text{-}\rho_L)}{
\begin{array}{c}
\mathrm{H}\quad\ \ \mathrm{H} \\
\diagdown\ \ \diagup \\
\mathrm{C}=\mathrm{C} \\
\diagup\quad\ \diagdown \\
\mathrm{H}\quad\ \ \underset{\underset{\mathrm{H}}{|}}{\mathrm{C}}=\mathrm{OI}
\end{array}}
\end{array}
\qquad (11.22)
$$

――――――――― *I/O-11.1* ―――――――――

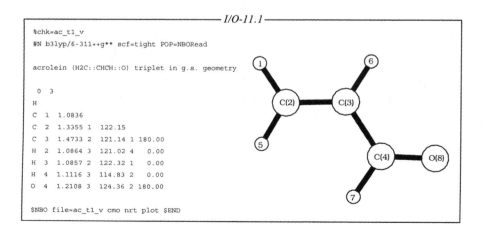

```
%chk=ac_t1_v
#N b3lyp/6-311++g** scf=tight POP=NBORead

acrolein (H2C::CHCH::O) triplet in g.s. geometry

 0  3
H
C   1   1.0836
C   2   1.3355  1   122.15
C   3   1.4733  2   121.14  1  180.00
H   2   1.0864  3   121.02  4    0.00
H   3   1.0857  2   122.32  1    0.00
H   4   1.1116  3   114.83  2    0.00
O   4   1.2108  3   124.36  2  180.00

$NBO file=ac_t1_v cmo nrt plot $END
```

As shown in the Lewis structural diagrams, the α NLS has the $\bar{\text{C}}-\bar{\text{O}}\text{I}$ "trivalent carbanion" and β NLS the $\text{C}=\text{OI}$ "hypovalent oxygen" (LP*) expected from formal $^3(n_O \rightarrow \pi^*_{CO})$ excitation, but both structures show evidence of significant NL delocalization (ca. 2.3%-ρ_{NL} density).

The general principles of NBO donor–acceptor interactions (Chapter 5) suggest the strong delocalizations to be expected for the triplet-state Lewis structures (11.22). In the alpha spin set, the allylic-like $\text{C}=\text{C}-\bar{\text{C}}\text{I}$ pattern immediately suggests strong $n_C \rightarrow \pi^*_{CC}$ delocalization, and the anionic $-\bar{\text{O}}\text{I}$ oxo pattern suggests enhanced $n_O^{(y)} \rightarrow \sigma^*_{CH}$ and $n_O^{(y)} \rightarrow \sigma^*_{CC}$ vicinal delocalizations, as found in the perturbative $\Delta E^{(2)}$ estimates for α spin:

$$n_C(\text{NBO } 13\alpha) \rightarrow \pi^*_{CC}(\text{NBO } 110\alpha): 36.4\,\text{kcal/mol} \tag{11.23}$$

$$n_O^{(y)}(\text{NBO } 15\alpha) \rightarrow \sigma^*_{CH}(\text{NBO } 115\alpha): 10.0\,\text{kcal/mol} \tag{11.24}$$

$$n_O^{(y)}(\text{NBO } 15\alpha) \rightarrow \sigma^*_{CC}(\text{NBO } 113\alpha): 7.3\,\text{kcal/mol} \tag{11.25}$$

Similarly, the $\text{C}=\text{C}-\text{C}=\text{OI}$ pattern of the beta spin set suggests unusually strong vicinal delocalizations into the vacated $n^*_O{}^{(y)}$ (LP*) acceptor orbital ($\sigma_{CH} \rightarrow n^*_O{}^{(y)}$, $\sigma_{CC} \rightarrow n^*_O{}^{(y)}$) as well as the usual conjugative π–π^* interaction of vicinal π-bonds, as found in the leading few $\Delta E^{(2)}$ values

$$\sigma_{CH}(\text{NBO } 13\beta) \rightarrow n^*_O{}^{(y)}(\text{NBO } 15\beta): 17.1\,\text{kcal/mol} \tag{11.26}$$

$$\pi_{CC}(\text{NBO } 3\beta) \rightarrow \pi^*_{CO}(\text{NBO } 116\beta): 15.1\,\text{kcal/mol} \tag{11.27}$$

$$\sigma_{CC}(\text{NBO } 5\beta) \rightarrow n^*_O{}^{(y)}(\text{NBO } 15\beta): 7.6\,\text{kcal/mol} \tag{11.28}$$

The strong allylic-type delocalization (11.23) is also identified as a significant 3c/4e hyperbonding interaction by the 3CHB search (automatically activated by the NRT keyword).

The NHO directionality and bond-bending table also give hints of incipient angular deformations in nonplanar torsional geometries. The α-spin formyl carbon C4(H7) hybrid shows significant out-of-plane bond-bending strain ($\sim$4°), anticipating the expected pyramidalization of a trivalent amino-like Lewis structure pattern. The corresponding β-spin C4(H7) hybrid exhibits even stronger in-plane strain ($\sim$6°), indicative of the strong inverse-hyperconjugative leverage exerted by (11.26).

NRT analysis gives a still more complete picture of triplet structural and reactive propensities. The α-spin density is dominated by the expected two strong allylic-like resonance structures ($\mathbf{I}_\alpha$, $\mathbf{II}_\alpha$) arising from donor–acceptor interaction (11.23) (cf. Figs. 5.3 and 5.6),

$$\tag{11.29}$$

$$\mathbf{I}_\alpha\ (47.7\%) \qquad\qquad \mathbf{II}_\alpha\ (42.7\%)$$

The leading few β-spin resonance structures are similarly those expected from leading donor–acceptor interactions (11.26 and 11.27),

$$\text{(11.30)}$$

The calculated (half-)bond orders for each spin

$$\text{(11.31)}$$

are combined to give the total triplet-state NRT bond orders

(11.32)(triplet excited state; vertical)

These may be compared with corresponding values for the ground singlet state

(11.33)(singlet ground state)

to obtain the singlet–triplet bond-order *changes* (Δb_{NRT}) for vertical excitation

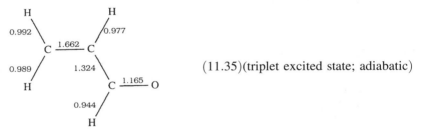

$$(11.34)(\text{singlet–triplet } \Delta b_{\text{NRT}}; \text{ vertical})$$

The vertical Δb_{NRT} values (11.34) provide interesting predictors of incipient structural and reactivity changes in the nascent triplet species. Consistent with usual bond-order–bond-length relationships and the large skeletal changes in (11.34), the initial forces on the triplet species are expected to *lengthen* the weakened C(2)–C(3) and C(4)–O(8) bonds ($\Delta b_{1,3} = -0.219$, $\Delta b_{4,8} = -0.362$) and *shorten* the C(3)–C(4) bond ($\Delta b_{3,4} = +0.104$). These structural predictions are confirmed by experiment (see the extensive studies of O. S. Kokareva, V. A. Bataev, V. I. Pupyshev, and I. A. Gudunov, *Int. J. Quantum Chem.* **108**, 2719–2731, 2008) and by theoretical geometry parameters for the fully optimized triplet species, as summarized in Table 11.4.

The final geometric adjustment to the altered resonance pattern further shifts the adiabatic triplet b_{NRT} values in the direction expected for resonance structure $\mathbf{II}_\alpha$:

$$(11.35)(\text{triplet excited state; adiabatic})$$

Table 11.4 Optimized singlet and triplet geometry of acrolein (*trans* conformer, C_s symmetry; B3LYP/6-311++G** level), with net shifts $\Delta_{S \to T}$.

Property	Singlet	Triplet	$\Delta_{S \to T}$
Bond lengths (Å)			
C(2)–C(3)	1.3355	1.3788	+0.0433
C(3)–C(4)	1.4733	1.3948	−0.0785
C(4)–O(8)	1.2108	1.3110	+0.1002
C(2)–H(1)	1.0836	1.0812	−0.0024
C(2)–H(5)	1.0854	1.0835	−0.0019
C(3)–H(6)	1.0857	1.0858	+0.0001
C(4)–H(7)	1.1116	1.0963	−0.0153
Bond angles (°)			
O(8)–C(4)–H(7)	120.81	111.56	−9.25
O(8)–C(4)–C(3)	124.36	125.99	+1.63

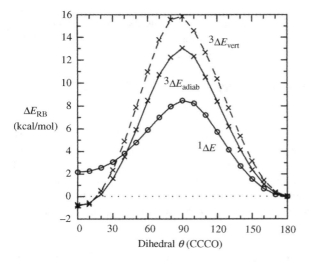

Figure 11.14 Acrolein rotation barrier ΔE_{RB} (kcal/mol) in ground singlet ($^1\Delta E$; solid, circles) and excited triplet ($^3\Delta E_{adiab}$; solid, x) states, shown with respect to trans equilibrium geometry in each state. The "vertical triplet" barrier ($^3\Delta E_{vert}$; dashed, x) for vertical (fixed singlet geometry) excitation at each dihedral angle θ is shown for comparison.

The adiabatic singlet–triplet Δb_{NRT} shifts are therefore seen to be given by

$$(11.36)(\text{singlet–triplet } \Delta b_{NRT}; \text{adiabatic})$$

Although the NBO/NRT algorithms make no use of structural information, the final triplet b_{NRT} values exhibit the expected qualitative correlations with optimized bond lengths that are well known for ground-state species.

The enhanced double bonding at C(3)–C(4) (cf. $\mathbf{II}_\alpha$) is also expected to significantly increase the triplet barrier to internal rotation. Figure 11.14 compares the singlet barrier curve (solid; circles) with the corresponding triplet curve (solid; x's), showing the significant barrier increase in the latter case. The figure also displays an idealized "vertical triplet" barrier curve (dashed; x's) in which the triplet-state geometry is held *identical* to that of the singlet state at each θ. Of course, the unphysical (inflexible) $^3\Delta E_{vert}$ barrier is far too high compared to the adiabatically relaxed $^3\Delta E_{adiab}$ barrier, but the model calculation emphasizes the *electronic* origin of the increased triplet barrier (rather than, e.g., increased "steric crowding" due to a change in triplet geometry).

Other $NBO and $DEL keyword options could be used to investigate the electronic origins of the pronounced OCH angle reduction, the reversed conformational preference from trans to cis geometry, the expected pyramidalization in nonplanar torsional geometry, the expected acidity increase at the formyl proton, and other interesting features of the excited triplet state. However, the foregoing examples may adequately suggest how NBO/NRT descriptors can provide a useful picture of forces acting on the nascent excitation species and its subsequent structural and reactive evolution, allowing familiar ground-state resonance "arrow-pushing" concepts to be successfully extended to excited-state processes.

11.5   CONICAL INTERSECTIONS AND INTERSYSTEM CROSSINGS

Light absorption leading to electronic excitation is commonly accompanied by radiative reemission as the electron decays back to the ground potential energy surface. However, in certain cases (dependent on excited-state radiative lifetime and potential features to be described below) the system returns to the ground electronic surface *without* optical emission, a so-called "radiationless transition." Such non-radiative transition processes are important features of reaction pathways on both ground and excited surfaces.

The key potential feature required for facile nonradiative decay is a *conical intersection* (CI), an accessible low-energy molecular geometry in which the potential energy surfaces undergo degenerate *crossing*. At the special CI geometry, such as $CI(S_0/S_1)$ for the lowest allowed singlet excitation $S_0 \to S_1$, no optical photon is required to interconvert "ground" S_0 and "excited" S_1 surfaces, and radiationless transitions occur with high probability. For a spin-forbidden (e.g., S_0/T_1) transition, the analogous degeneracy feature is called an *intersystem crossing* (ISC). Figure 11.15 suggests how simple curve-crossing in a diatomic species (where only bond length R is needed to specify the degenerate crossing point) becomes the common apex of a "double funnel" in a polyatomic species, perhaps forming the low-energy terminus of a higher-dimensional "seam" of degeneracies. The special CI geometry therefore provides a "portal" for facile radiationless transitions between distinct states and bonding patterns. The multistate and multidimensional aspects of conical intersections challenge both visualization and computation.

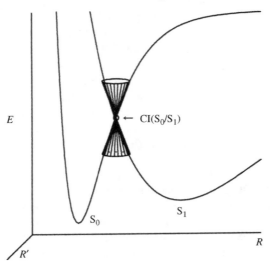

Figure 11.15   Schematic conical intersection $CI(S_0/S_1)$ of molecular potential energy surfaces S_0, S_1, suggesting how a simple curve-crossing in the diatomic $E(R)$ case is broadened to a "double funnel" in the polyatomic $E(R,R')$ case.

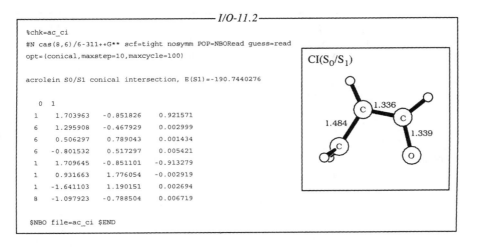

——— *I/O-11.2* ———

```
%chk=ac_ci
#N cas(8,6)/6-311++G** scf=tight nosymm POP=NBORead guess=read
opt=(conical,maxstep=10,maxcycle=100)

acrolein S0/S1 conical intersection, E(S1)=-190.7440276

   0  1
   1     1.703963    -0.851826    0.921571
   6     1.295908    -0.467929    0.002999
   6     0.506297     0.789043    0.001434
   6    -0.801532     0.517297    0.005421
   1     1.709645    -0.851101   -0.913279
   1     0.931663     1.776054   -0.002919
   1    -1.641103     1.190151    0.002694
   8    -1.097923    -0.788504    0.006719

$NBO file=ac_ci $END
```

CI(S_0/S_1)

1.336

1.484

1.339

The Gaussian program CAS keyword (Sidebar 11.2) provides the "OPT= CONICAL" search for a conical intersection in the neighborhood of a chosen starting geometry. For acrolein, an important conical intersection between singlet S_0/S_1 states (first discovered by M. Reguero, M. Olivucci, F. Bernardi, and M. A. Robb, *J. Am. Chem. Soc.* **116**, 2013–2114, 1994) can be obtained from the Gaussian input file shown in I/O-11.2. Unlike other special geometrical features, the CI(S_0/S_1) point typically exhibits no distinguishing stability or gradient properties that might suggest its electronic character (as deduced, for example, from familiar bonding principles governing near-equilibrium species). As shown in I/O-11.2, the CI(S_0/S_1) geometry for acrolein resembles the *s-cis* conformer with twisted vinyl group, suggesting the importance of π-diradical interactions. Given the nascent CI-geometry emerging from a radiationless transition (somewhat analogous to the vertical geometry produced by optical transition), we seek NBO/NRT descriptors for the structural and reactive propensities that guide subsequent evolution on the emergent potential energy surface.

Direct NBO analysis of the Gaussian-calculated CAS wavefunction incurs the severe limitations of the spin-averaged CAS description of electron density (Sidebar 11.2). Table 11.5 summarizes the accuracies (%-ρ_L) and frontier NBO occupancies of low-lying singlet and triplet states of *s-cis* acrolein, showing the distinctly inferior quality (93–95%-ρ_L) of CAS-type spin-averaged description of electronic excitation. Still worse limitations of this type are expected for the distorted CI(S_0/S_1) geometry of I/O-11.2. Without proper spin-dependent density matrices, the CAS wavefunctions of the current Gaussian program implementation are therefore scarcely suitable for modern chemical analysis.

However, given the distinctive geometry (I/O-11.2) of the CI(S_0/S_1) species, we are free to use more informative wavefunction methods to describe bonding propensities of the nascent species. In particular, for the ground-state S_0 species of principal interest we can conveniently revert to single-configuration methods as employed in previous chapters. Because DFT methods often exhibit spurious numerical behavior

Table 11.5 CAS(8,6)/6-311++G** energetics (ΔE), NLS accuracy (%-ρ_L), and frontier NBO occupancies for low-lying singlet S_0, S_1, S_2, and triplet T_1 states of acrolein in equilibrium *s-cis* geometry, with descriptive NBO excitation(s) for each state. The spin-averaged CAS quantities are only loosely comparable to spin-dependent descriptors provided by other methods.

	S_0	S_1	S_2	T_1
ΔE (eV)	0.00	3.03	6.44	5.60
%-ρ_L	98.85	95.08	93.11	95.36
NBO occ. (*e*)				
$\pi_{C(2)C(3)}$	1.855	1.791	—[a]	—[a]
$\pi_{C(4)O}$	1.990	—[a]	1.948	1.767
$n_O^{(y)}$	1.909	1.041	1.030	1.929
$n_{C(2)}^{(z)}$	—[a]	—[a]	0.999	0.980
$n_{C(3)}^{(z)}$	—[a]	—[a]	1.043	0.892
$n_{C(4)}^{(z)}$	—[a]	1.019	—[a]	—[a]
$n_O^{(z)}$	—[a]	1.851	—[a]	—[a]
$\pi^*_{C(2)C(3)}$	0.094	0.311	—[a]	—[a]
$\pi^*_{C(4)O}$	0.060	—[a]	0.897	0.346
NBO excitation(s)	—[a]	$n_O^{(y)} \to \pi^*_{CO}$	$\pi_{CC} \to \pi^*_{CC}$	$\pi_{CC}^{\downarrow} \to \pi^*_{CC}^{\uparrow}$
			$n_O^{(y)} \to \pi^*_{CO}$	

[a] Not present in this NLS.

in multiradical species of "unusual" geometrical and electronic character, we employ *ab initio* UHF/6-311++G** wavefunctions for a zeroth-order picture of the bonding pattern. As expected, the broken-symmetry UHF wavefunction of the twisted CI geometry exhibits considerable open-shell diradical character and spin contamination ($\langle S^2 \rangle = 1.237$). Figure 11.16 summarizes various NBO/NRT descriptors of the CI-S_0 species, showing the natural atomic charges (Fig. 11.16a), spin density (Fig. 11.16b), and NRT bond orders (Fig. 11.16c).

The NRT bond orders of Figs. 11.16c point to significant oxetene (cyclic enol ether) character of the CI-S_0 species, with pronounced diradical character and high polarity of the strained C(2)O(8) long-bond. The α-spin NRT description is dominated by the cyclic "enol-like" oxetene pattern $CI_{1\alpha}$, with weaker admixture of the corresponding "keto-like" pattern $CI_{2\alpha}$, namely,

$$\tag{11.37}$$

$CI_{1\alpha}$ (87.8%) $CI_{2\alpha}$ (5.4%)

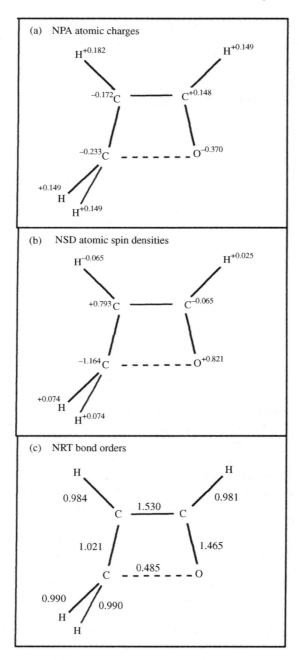

Figure 11.16 NBO/NRT descriptors for CI-S_0 conical intersection of acrolein on ground-state S_0 surface (UHF/6-311++G**//CAS(8,6)/6-311++G** level), showing (a) natural atomic charges, (b) natural spin densities, and (c) NRT bond orders.

The β-spin NRT description differs dramatically at all four skeletal centers, with leading structures

$$
\begin{array}{cc}
\text{H}\diagdown \quad \diagup\text{H} & \text{H}\diagdown \quad \diagup\text{H} \\
\text{C—C} \quad \longleftrightarrow & \text{C=C} \qquad \beta\text{-spin} \\
\text{H}\diagup\text{C}_{|} \quad \text{O}_{|} & \text{H}\diagup\text{C}_{|} \quad \text{O}_{|} \\
\quad\text{H} & \quad\text{H} \\
\mathbf{CI}_{1\beta}\ (78.4\%) & \mathbf{CI}_{2\beta}\ (13.9\%)
\end{array}
\tag{11.38}
$$

As shown in Fig. 11.16b, *all four* skeletal centers exhibit large spin density values (|NSD| > 0.5), indicative of extreme $\text{C}^{\downarrow}\text{C}^{\uparrow}\text{C}^{\downarrow}\text{O}^{\uparrow}$ (*tetra*radical) spin-polarized character. The overall atomic charges (Fig. 11.16a) and bond orders (Fig. 11.16c) also reflect the formal resonance averaging of α/β-spin charges and bond orders in (11.37) and (11.38).

The tortured character of the nascent C(2)–O(8) "half-bond" is also manifested in the default Lewis structures, namely,

$$
\begin{array}{cc}
\text{H}\diagdown \quad \diagup\text{H} & \text{H}\diagdown \quad \diagup\text{H} \\
\text{C=C} & \text{C}^{+}\text{—C} \qquad \text{CI–S}_0 \\
\text{H}\diagup\text{C}^{+} \quad \text{O}_{|}^{-} & \text{H}\diagup\text{C}_{|}^{-} \quad \text{O}_{|} \\
\quad\text{H} & \quad\text{H} \\
\alpha\text{-NLS}\ (99.1\%\text{-}\rho_{\text{L}}) & \beta\text{-NLS}\ (98.5\%\text{-}\rho_{\text{L}})
\end{array}
\tag{11.39}
$$

As seen in (11.39), the α-NLS represents C(2), O(8) interaction as "zwitterionic" $[\text{C}(2)^{-}\text{O}(8)^{+}]$, whereas the α-NRT $\mathbf{CI}_{1\alpha}$ structure in (11.37) prefers the covalent C(2)–O(8) representation. Consistent with its marginally covalent character, the α-NRT $b_{\text{C}(2)\text{O}(8)}$ bond order is only 2% covalent (98% ionic), and the highly ionic and spin-polarized character of C(2)–O(8) interaction is further exhibited in NPA charges (Fig. 11.16a) and spin densities (Fig. 11.16b). Figure 11.17 displays three-dimensional surface plots of the in-plane $n_{\text{O}}^{(y)}$–n_{C}^{*} α-NBOs leading to C(2)–O(8) bonding and the out-of-plane $n_{\text{O}}^{(z)}$–π^{*}_{CC} α-NBOs leading to keto-enol resonance, with estimated second-order stabilizations. Despite the strained CI geometry, the interacting NBOs of Fig. 11.17 retain quite recognizable forms.

On the basis of these NBO/NRT descriptors, the emergent CI-S$_0$ species is expected to exhibit strong electronic propensity for ring closure toward the cyclic oxetene isomer, in competition with simple acyclic relaxation to ground-state acrolein. More detailed computational exploration of the potential energy surface near the nascent CI-S$_0$ geometry [Robb et al.] supports this conclusion, which is also consistent with available experimental data.

Finally, we may briefly examine the corresponding spin-forbidden ISC(S$_0$/T$_1$) transition portal between ground S$_0$ and lowest excited triplet T$_1$ surfaces, as shown in I/O-11.3. In this case, inclusion of the "SLATERDET" (Slater determinant) option in the CAS keyword allows OPT=CONICAL to include both singlet and triplet surfaces in the search for low-lying degeneracies. [Note that both CI (I/O-11.2) and

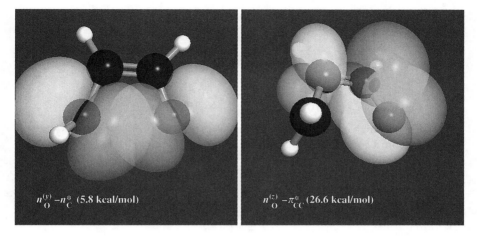

$n_O^{(y)} - n_C^*$ (5.8 kcal/mol) $n_O^{(z)} - \pi_{CC}^*$ (26.6 kcal/mol)

Figure 11.17 Leading (a) in-plane and (b) out-of-plane α-NBO interactions (and estimated second-order stabilization energies) for CI-S_0 conical intersection of acrolein (UHF/6-311++G** level in geometry of I/O-11.2). (See the color version of this figure in the Color Plates section.)

ISC (I/O-11.3) searches employed restricted MAXSTEP parameters (and increased MAXCYCLE parameters) to preserve fragile CAS convergence from step to step of geometry optimization.] As seen in I/O-11.3, the twisted ISC(S_0/T_1) geometry resembles that of the CI(S_0/S_1) species (I/O-11.2) with altered skeletal bond lengths and slightly inverted pyramidalization at C(2).

Once again we can employ more informative UHF calculations to analyze the ISC(S_0/T_0) species. For the ISC-S_0 species, the nominal α-NLS, β-NLS bond patterns

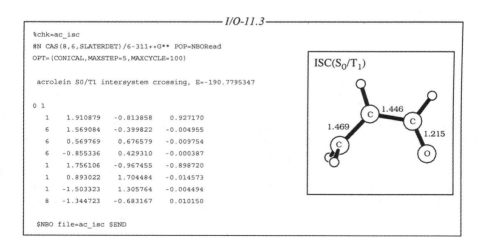

I/O-11.3

```
%chk=ac_isc
#N CAS(8,6,SLATERDET)/6-311++G** POP=NBORead
OPT=(CONICAL,MAXSTEP=5,MAXCYCLE=100)

acrolein S0/T1 intersystem crossing, E=-190.7795347

0 1
    1    1.910879    -0.813858     0.927170
    6    1.569084    -0.399822    -0.004955
    6    0.569769     0.676579    -0.009754
    6   -0.855336     0.429310    -0.000387
    1    1.756106    -0.967455    -0.898720
    1    0.893022     1.704484    -0.014573
    1   -1.503323     1.305764    -0.004494
    8   -1.344723    -0.683167     0.010150

$NBO file=ac_isc $END
```

ISC(S_0/T_1)

are found to be identical to those of the CI-S_0 species, namely,

$$
\begin{array}{cc}
\underset{\alpha\text{-NLS (98.5\%-}\rho_L)}{
\begin{array}{c}
\text{H} \diagdown \quad \diagup \text{H} \\
\text{C}{=}\text{C} \\
\text{C}^+ \quad {}_{|}\underset{..}{\text{O}}{}_{|}^- \\
| \diagdown \\
\overset{}{\text{H}}\text{H}
\end{array}}
&
\underset{\beta\text{-NLS (98.9\%-}\rho_L)}{
\begin{array}{c}
\text{H} \diagdown \quad \diagup \text{H} \\
\text{C}^+{-}\text{C} \\
{}_{|}\text{C}^- \quad \text{O}{}_{|} \\
| \diagdown \\
\overset{}{\text{H}}\text{H}
\end{array}}
\end{array}
\tag{11.40}
$$

again reflecting the rather extreme spin-unpairing of enol-like (α) versus keto-like (β) Lewis structural patterns in the two spin sets [as suggested also by the strong spin contamination ($\langle \mathbf{S}^2 \rangle = 1.162$) of the open-shell singlet UHF solution]. The composite NRT bond orders for this species

$$
\begin{array}{c}
\text{H} \qquad\qquad\qquad \text{H} \\
0.949 \diagdown \qquad\qquad \diagup 0.953 \\
\text{C} \;\underline{1.452}\; \text{C} \\
1.307 \diagup \qquad\qquad \diagdown 1.542 \\
\text{C} \qquad\qquad \text{O} \\
0.982 \Big| \;\diagdown 0.965 \\
\text{H} \quad \text{H} \qquad \text{ISC-}S_0
\end{array}
\tag{11.41}
$$

suggest the strong shifts in skeletal bonding [particularly at C(2)–C(3)] and complete *absence* of cyclic oxetene-like character in this case. Thus, despite their somewhat superficial geometrical resemblance, the radiationless ISC(S_0/T_1) and CI(S_0/S_1) transitions yield nascent photoproducts of distinct electronic and vibrational character on the ground S_0 potential energy surface. Composite NRT bond orders for the initial ISC-T_1 species

$$
\begin{array}{c}
\text{H} \qquad\qquad\qquad \text{H} \\
0.985 \diagdown \qquad\qquad \diagup 0.971 \\
\text{C} \;\underline{1.266}\; \text{C} \\
1.035 \diagup \qquad\qquad \diagdown 1.749 \\
\text{C} \qquad\qquad \text{O} \\
0.990 \Big| \;\diagdown 0.990 \\
\text{H} \quad \text{H} \qquad \text{ISC-}T_1
\end{array}
\tag{11.42}
$$

provide a corresponding picture of altered skeletal bonding (and absence of cyclization) at the entrant portal of the triplet surface.

From the starting Lewis structural representations (11.40), NBO donor–acceptor delocalizations of expected form lead to the NRT resonance mixings and bond-order shifts depicted in (11.41). Figure 11.18 shows three-dimensional surface plots of the leading NBO delocalizations in each spin set, displaying the expected resemblances to analogous keto-enol ($n_O \rightarrow \pi^*_{CC}/n_C \rightarrow \pi^*_{CO}$) interactions in ground-state equilibrium species, despite the rather tortured ISC geometry.

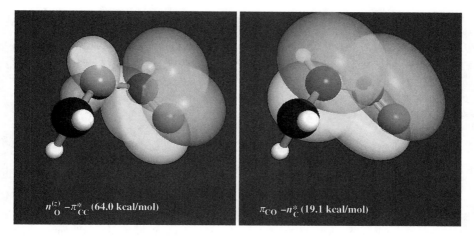

$n_O^{(z)} - \pi_{CC}^*$ (64.0 kcal/mol) $\pi_{CO} - n_C^*$ (19.1 kcal/mol)

Figure 11.18 Leading (a) α-spin and (b) β-spin π-type NBO donor–acceptor interactions (and estimated second-order stabilization energies) for ISC-S_0 intersystem crossing of acrolein (UHF/6-311++G^{**} level in geometry of I/O-11.3). [In the ISC-S_0 species, the in-plane α-spin $n_O^{(y)} - n_C^*$ interaction (cf. Fig. 11.17a for the CI-S_0 species) is negligibly weak (0.27 kcal/mol).] (See the color version of this figure in the Color Plates section.)

In this and many cases, we shall not be surprised to discover that the electronic propensities of highly distorted excited-state species exhibit strong NBO/NRT analogies to those studied in earlier chapters. The simple example of acrolein suggests how Lewis structural concepts continue to yield rich explanatory dividends as NBO-based tools are employed to penetrate ever deeper into the excited-state domain.

PROBLEMS AND EXERCISES

[The following problems are expressed in terms of spin-forbidden singlet–triplet $(S_0 \rightarrow T_1)$ transitions so that you can conveniently use single-configuration UHF/UB3LYP methods to examine NBO/NRT descriptors of each state. To bring additional "excitement" to the problem, consider using multiconfigurational (CIS, CAS, SAC-CI, etc.) methods for the corresponding spin-allowed singlet–singlet $(S_0 \rightarrow S_1)$ transitions.]

11.1. The characteristic "color" [wavelength $\lambda = hc/\Delta E(S_0 \rightarrow T_1)$] of a spectroscopic $S_0 \rightarrow T_1$ transition is expected to vary with the nature of the substituent (X) adjacent to a carbonyl chromophore. Consider the $^3n \rightarrow \pi^*$ transition for a series of substituted XCHO aldehydes,

$$\begin{array}{c} H \\ \diagdown \\ C = \bar{O}| \\ \diagup \\ X \end{array} \qquad X = H, CH_3, NH_2, OH, F$$

Optimize the ground-state S_0 geometry and calculate $\Delta E(S_0 \to T_1)$ and λ_X for the vertical $S_0 \to T_1$ transition in each XCHO species, using the B3LYP/6-311++G** method for each state. Plot your calculated λ_X versus electronegativity of the attached X atom (Ξ_X, Table 4.2) and look for evidence of a discernible correlation.

(a) The energy $E(S_0)$, $E(T_1)$ of each state can be expressed as usual (cf. Sections 5.1–5.3) in terms of Lewis (E_L) and non-Lewis (E_{NL}) contributions,

$$E(S_0) = E_L(S_0) + E_{NL}(S_0)$$

$$E(T_1) = E_L(T_1) + E_{NL}(T_1)$$

The transition energy $\Delta E(S_0 \to T_1) = E(T_1) - E(S_0)$ is therefore similarly expressed as

$$\Delta E(S_0 \to T_1) = \Delta E_L(S_0 \to T_1) + \Delta E_{NL}(S_0 \to T_1)$$

Evaluate $\Delta E_L(S_0 \to T_1)$, $\Delta E_{NL}(S_0 \to T_1)$ for each X. Characterize the contribution of each component ΔE_L, ΔE_{NL} to the overall variation in λ_X, and comment on whether L or NL contributions seem to have greater effect on the spectroscopic color shift.

(b) In the simple $n_O{}^{(y)} \to \pi^*{}_{CO}$ description of the $S_0 \to T_1$ transition, one could expect that leading ground-state delocalizations involving the spectroscopic orbitals (such as $n_O{}^{(y)} \to \sigma^*{}_{CX}$, or $n_X \to \pi^*{}_{CO}$) would *lower* S_0 [thereby increasing $\Delta E(S_0 \to T_1)$ and reducing λ_X], whereas the corresponding excited-state delocalizations should lower T_0 and increase λ_X. Try to identify the leading donor–acceptor delocalizations in each state (by $DEL deletions, $\Delta E^{(2)}$ values, or other method of your choice) that seem to dominate ΔE_{NL} and the spectroscopic color shift.

(c) Summarize your conclusions concerning the most important electronic characteristics of X that lead to red-shifting or blue-shifting of λ_X. Can you suggest *new* substituents X_B, X_R that should lead to still shorter (bluer) or longer (redder) wavelengths than any considered above? Explain your predictions carefully, then check their validity with full $\Delta E(S_0 \to T_1)$ calculations for each substituent.

11.2. Following the logic of Problem 11.1, compare the $S_0 \to T_1$ transition of formaldehyde (X = H) with the corresponding $^3n_O \to \pi^*{}_{CO}$ transition in (a) CO, (b) CO_2, and (c) ketene ($CH_2C{=}C{=}O$). Try to identify the important electronic difference (Lewis or non-Lewis; perhaps only a single NBO donor–acceptor interaction.) that principally accounts for the observed variations in wavelength λ in each species (a)–(c).

11.3. Similarly, extend your analysis to one or more of conjugated cyclic ketones, such as semiquinone, 1,4-benzoquinone (quinone), and 1,2-benzoquinone

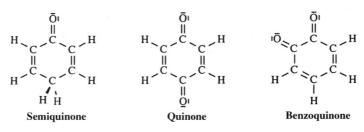

Semiquinone   Quinone   Benzoquinone

The quinones exhibit interesting pharmocological activity and appear as key species in photosynthesis and other biologically important electron transfer processes. Their excited states also exhibit highly creative NLS solutions to the symmetry problem of "which end to choose" for excitation. Optimized S_0 geometries for each species are shown in z-matrices below:

```
%chk=semiq
#b3lyp/6-311++g** scf=tight pop=nboread

 semiquinone (S0), E=-307.5313161

    0  1
    6     1.1113    -0.0000     0.0000
    6     0.3296     1.2559     0.0000
    6    -1.0096     1.2563     0.0000
    8     2.3351     0.0007     0.0000
    1     0.9049     2.1753     0.0000
    1    -1.5566     2.1950     0.0000
    6    -1.8188    -0.0006     0.0000
    1    -2.4978    -0.0013     0.8660
    1    -2.4986    -0.0013    -0.8660
    6    -1.0080    -1.2565     0.0000
    1    -1.5548    -2.1954     0.0000
    6     0.3312    -1.2556     0.0000
    1     0.9066    -2.1748     0.0000

 $NBO file=semiq $END

--Link1--
%chk=quinone
#b3lyp/6-311++g** scf=tight pop=nboread

 quinone (C6H4O2), E(S0) = -381.561889167

    0  1
    8     0.0000     2.6622     0.0000
    6     0.0000     1.4425     0.0000
    6     0.0000     0.6698     1.2686
    6     0.0000    -0.6698     1.2686
    1     0.0000     1.2558     2.1811
    1     0.0000    -1.2558     2.1811
    6     0.0000    -1.4425     0.0000
    8     0.0000    -2.6622     0.0000
    6     0.0000    -0.6698    -1.2686
    6     0.0000     0.6698    -1.2685
    1     0.0000    -1.2558    -2.1811
    1     0.0000     1.2558    -2.1811

 $NBO file=qui_s0 $END

--Link1--
%chk=benzq
#b3lyp/6-311++g** scf=tight pop=nboread

 1,2-benzoquinone (S0), E=-381.5499171
```

```
0  1
8     1.7241     1.3894     0.0000
6     0.6755     0.7838     0.0000
6    -0.6379     1.4529     0.0000
6    -1.7744     0.7313     0.0000
1    -0.6358     2.5368     0.0000
1    -2.7363     1.2334     0.0000
6    -1.7744    -0.7313     0.0000
1    -2.7363    -1.2334     0.0000
6    -0.6379    -1.4529     0.0000
1    -0.6358    -2.5368     0.0000
6     0.6755    -0.7838     0.0000
8     1.7241    -1.3894     0.0000
```

$NBO file=benzq_s0 $END

(a) Describe how quinone T_1 makes use of 3c/4e *long*-bonding (Section 8.3; denoted as "$\hat{\sigma}$-type" to distinguish from ordinary 3c/4e ω-bonding) to escape the symmetry dilemma. Show plots of the unusual $\hat{\sigma}_{OO}$ and $\hat{\sigma}^*_{OO}$ NBOs and their representative hyperconjugative interactions with the cyclic carbon σ-skeleton.

(b) Similarly, describe the unusual O–O "bonding" features in the β-NLS of triplet benzoquinone. Show two-dimensional and three-dimensional plots of the principal NBO donor–acceptor delocalizations that stabilize the β-NLS. From these plots, explain why it "makes sense" to employ an *out*-of-phase ($\hat{\sigma}^*_{OO}$-type) orbital as the filled (Lewis-type) donor in this case.

11.4. A simple z-matrix for rigid rotation of glyoxal (CHOCHO) is shown below. By altering TAU from 0 to 180 (e.g., in 10 degree increments), one obtains an approximation to the ground-state torsional potential energy surface $E_{S0}(\tau)$. By changing the spin multiplicity and evaluating the vertical (Franck–Condon) triplet energy at each point, one obtains the corresponding $E_{T1}(\tau)$ torsional potential. The difference

$$\Delta E_{S \to T}(\tau) = E_{T1}(\tau) - E_{S0}(\tau)$$

represents the torsional shift in spectroscopic $S_0 \to T_1$ excitation energy. Evaluate $\Delta E_{S \to T}(\tau)$ for glyoxal, determine the dihedral angles τ_B, τ_R for most blue-shifted and red-shifted excitation energy, and describe the principal NBO interactions that seem to account for the torsional effect on spectroscopic excitation. Can you see a connection between hyperconjugative interactions that govern the potential curves and the shifts in optical properties that accompany torsional distortion? Explain briefly.

```
%chk=gly_rr
#b3lyp/6-311++g** scf=tight POP=NBORead

glyoxal O::CHCH::O rigid-rotation by TAU

  0  1
  O
  C    1    1.2200
  C    2    1.4600    1    120.00
  O    3    1.2200    2    120.00    1      tau
  H    2    1.0800    3    120.00    1    180.00
  H    3    1.0800    2    120.00    4    180.00

tau   180.

$NBO gly_s180 $END
```

11.5. In the ROHF approximation (see Sidebar 2.1), electrons are viewed as being maximally paired, with spin density localized in the minimal number of singly occupied orbitals. This means that only *one* atom could exhibit nonzero spin density in a doublet species (one unpaired spin), *two* atoms in a triplet species (two unpaired spins), and so forth. Moreover, the nonvanishing spin density at any atom could only be of *positive* sign in this approximation. As noted elsewhere in this book (e.g., Section 7.3), the truth is quite different. As an example of a simple doublet species, consider the ubisemiquinone radical (formally obtained by removing one of the H atoms from semiquinone in Problem 11.3), with z-matrix as shown below:

```
%chk=usq
#b3lyp/6-311++g** scf=tight pop=nboread

 semiquinone radical ("ubisemiquinone"), E(D0)=-306.9134552

   0  2
   6    -1.0470    -0.0000     0.0000
   6    -0.2897     1.2388     0.0000
   6     1.0855     1.2237     0.0000
   8    -2.3001    -0.0000     0.0000
   1    -0.8552     2.1632     0.0000
   1     1.6422     2.1543     0.0000
   6     1.7824    -0.0000     0.0000
   1     2.8663    -0.0000     0.0000
   6     1.0854    -1.2237     0.0000
   1     1.6422    -2.1543     0.0000
   6    -0.2897    -1.2388     0.0000
   1    -0.8553    -2.1631     0.0000

 $NBO file=usq_d0 $END
```

In ROHF approximation, 100% of the spin density ($+1$) should be localized at C_7. What is the actual NSD percentage found at C_7? How many other atoms have significant spin density (say, 10% or greater) of the expected total? Do any atoms exhibit *negative* spin densities? Use simple resonance arguments (or carry out NRT analysis) to rationalize why delocalized spin-density "ripples" of *oscillating* sign should be a *ubiquitous* feature of conjugated radicals.

(a) As further examples of delocalized spin density distributions, consider the excited triplet species from vertical $^3\pi \to \pi^*$ excitation of the following acyclic and cyclic conjugated species:

Butadiene   *Hexatriene*

Benzene   *Naphthalene*

At what site(s) are the two unpaired electrons most strongly localized in each species? Do you find evidence of significant ($>10\%$) spin delocalization onto more than two centers? Do you find evidence of *negative* spin density at any center?

(b) For each species considered in (a), try to rationalize the principal features of the NSD in terms of the α/β NLS bonding patterns, principal NBO donor–acceptor interactions, and/or NRT resonance weightings in each spin set.

11.6. Phenolphthalein is a familiar laboratory *indicator*,

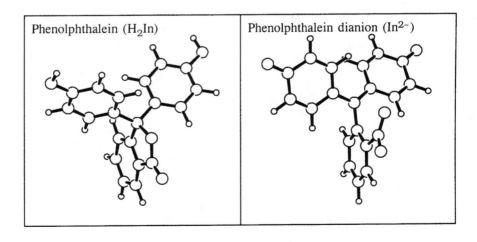

Acidic (colorless)   *Basic (pink)*

a substance (like many in nature) that undergoes color changes in response to a change in pH. Formally, phenolphthalein (H_2In) is a weak diprotic acid that is visually "colorless" (with excitation wavelength too short to be seen by the human eye) in its intact form in neutral or acidic media, but becomes pink-colored (due to absorption at the blue end of the human visual spectrum) in its ionized In^{2-} form in basic media, namely,

$$H_2In \rightleftarrows In^{2-} + 2\,H^+$$
colorless   pink

Optimized ground-state S_0 geometries of H_2In and In^{2-} are shown in the ORTEP views and z-matrices below:

Phenolphthalein (H_2In)	Phenolphthalein dianion (In^{2-})

```
%chk=phph
#b3lyp/6-311++g** scf=tight POP=NBORead

phenolphthalein H2In (colorless), E(S0)=-1071.7538974

    0   1
    6    -0.1401     2.1046     1.7766
    6    -0.3207     2.5870     0.3903
    6    -0.2626     1.5016    -0.4714
    6    -0.4722     1.6768    -1.8354
    1    -0.4621     0.8381    -2.5216
    6    -0.7118     2.9697    -2.3025
    1    -0.8773     3.1301    -3.3620
    6    -0.7518     4.0645    -1.4285
    1    -0.9434     5.0560    -1.8218
    6    -0.5609     3.8815    -0.0624
    1    -0.6037     4.7077     0.6374
    8    -0.1228     2.7148     2.8105
    8    -0.0070     0.7416     1.7217
    6    -0.0005     0.2383     0.3395
    6     1.3747    -0.3919     0.1154
    6     1.7575    -1.4587     0.9376
    6     3.0065    -2.0518     0.8100
    1     1.0726    -1.8241     1.6934
    1     3.2858    -2.8740     1.4623
    6     3.9035    -1.5869    -0.1561
    6     3.5408    -0.5248    -0.9799
    1     4.2456    -0.1655    -1.7198
    6     2.2879     0.0667    -0.8362
    1     2.0360     0.9022    -1.4767
    6    -1.1501    -0.7615     0.1826
    6    -2.3166    -0.6427     0.9472
    6    -3.3904    -1.5031     0.7580
    1    -2.3812     0.1183     1.7142
    1    -4.2867    -1.4172     1.3603
    6    -3.3199    -2.5002    -0.2156
    6    -2.1664    -2.6308    -0.9880
    1    -2.0987    -3.4105    -1.7411
    6    -1.0912    -1.7685    -0.7826
    1    -0.1955    -1.8969    -1.3785
    8    -4.4062    -3.3192    -0.3605
    1    -4.2319    -3.9761    -1.0427
    8     5.1437    -2.1340    -0.3349
    1     5.2894    -2.8328     0.3115

 $NBO file=h2in $END

 --Link1--
 %chk=in_2m
 #b3lyp/6-311++g** scf=tight POP=NBORead

phenolphthalein In(2-) (pink), E(S0)=-1070.5928249
```

```
-2 1
6      0.9311     2.0778     1.8788
6      0.6485     2.3340     0.3718
6      0.1155     1.3566    -0.4917
6     -0.2167     1.7177    -1.8086
1     -0.6376     0.9612    -2.4630
6     -0.0099     3.0090    -2.2856
1     -0.2743     3.2575    -3.3096
6      0.5371     3.9721    -1.4368
1      0.7049     4.9855    -1.7932
6      0.8588     3.6260    -0.1280
1      1.2874     4.3434     0.5629
8      1.6592     2.9318     2.4412
8      0.3644     1.0733     2.3594
6     -0.0888    -0.0899    -0.1289
6      1.0495    -0.9407    -0.1146
6      1.0498    -2.2222     0.5258
6      2.1253    -3.0668     0.4807
1      0.1849    -2.4953     1.1184
1      2.1081    -4.0143     1.0119
6      3.3487    -2.7371    -0.2303
6      3.3583    -1.4066    -0.8123
1      4.2689    -1.0996    -1.3189
6      2.2828    -0.5627    -0.7363
1      2.3486     0.4171    -1.1960
6     -1.4396    -0.5453    -0.0352
6     -1.8477    -1.8703    -0.3870
6     -3.1584    -2.2724    -0.3525
1     -1.0973    -2.5597    -0.7565
1     -3.4379    -3.2726    -0.6722
6     -4.2348    -1.3918     0.0627
6     -3.8030    -0.0498     0.4117
1     -4.5732     0.6427     0.7395
6     -2.4932     0.3439     0.3466
1     -2.2233     1.3513     0.6426
8     -5.4402    -1.7539     0.0959
8      4.3240    -3.5267    -0.3147
```

$NBO file=in_2m_s0 $END

(a) Calculate the $S_0 \rightarrow T_1$ vertical excitation energy and wavelength for both H_2In and In^{2-}. Do you see evidence for dependence of $S_0 \rightarrow T_1$ excitation energy on state of ionization?

(b) Locate the principal site(s) of triplet excitation in H_2In, In^{2-} species from the NSD values. Do you see evidence for delocalized (non-ROHF-type) spin-density waves and negative spin densities in these species? Describe briefly.

(c) Try to identify the principal NBO donor–acceptor interactions (or other electronic features) that seem to best account for the pH-dependent shifts in excitation wavelength.

(d) As in Problem 11.1(c), suggest modifications of the H_2In structure by chemical substitution that might further shift the wavelength difference between H_2In and In^{2-} and explain your reasoning.

Appendix A

What's Under the Hood?

The "engine" of NBO analysis is the NBO program (currently, NBO 5.9), composed of over 350 subprograms [subroutines (SR) and functions (FN)] and 50,000 lines of Fortran code. To really understand what is going on, you should understand Fortran, the original FORmula TRANslation programming language for scientific applications. However, about a third of the program source code consists of comment lines (beginning with "C" in column 1) that explain in plain English the gist of what is going on. It pays to know where the Fortran source code is located and how to scan its contents with a text editor, even if you cannot understand a word of Fortran.

The "owner's manual" for the NBO program is the *NBO Manual*. It is divided into three main sections: Section A is for casual users, interested only in rudimentary NBO analysis (default options); Section B is for more serious users, interested in keyword options for advanced analysis (NRT, STERIC, NCS, etc.); and Section C is for programmers, interested in deeper details of program construction (e.g., those needed for interfacing to a new ESS host program). Section C of the *NBO Manual* is the authoritative guide to "what's under the hood" in the NBO program engine.

In addition to a general description of overall program logic and data-flow between NBO and the host ESS program, Section C contains a comprehensive summary of all ~350 subprograms of the main NBO program, giving the arguments (the data passed to and from other subprograms) and a brief explanation of each subprogram. Subprograms are further classified and grouped into categories: "Group I" are base NAO/NBO algorithms underlying default options; "Group II" are those of $DEL energetic analysis. Additional groups include supplemental modules (such as NRT, STERIC, NEDA, and other post-NBO3 options) and general utility programs (called by many modules). Within each group, subprograms are further organized into distinct modules under main "driver" subroutines (SR NAODRV for NAO generation, SR NBODRV for NBO determination, and so forth), more or less in the order called to perform the task of the module.

A general overview of the subprograms of Groups I and II (all called by the highest-level SR RUNNBO) is shown in the accompanying flow chart, which indicates the logical relationship of the routines to one another and to the order of discussion in Sections C.5 and C.6 of the *NBO Manual*. The sequence of execution is

Discovering Chemistry With Natural Bond Orbitals, First Edition. Frank Weinhold and Clark R. Landis.
© 2012 John Wiley & Sons, Inc. Published 2012 by John Wiley & Sons, Inc.

generally from top to bottom and from left to right, with subprograms of equal precedence shown at an equal vertical level.

NBO Flow Chart for Group I, II Subprograms

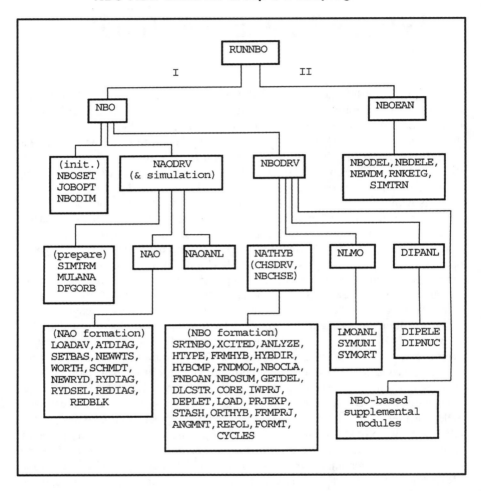

Each program module is in turn related to a particular subsection and keyword(s) of Section B, such as default NPA/NBO analysis (Section B-2), NRT analysis (Section B-72), STERIC analysis (Section B-100), and so forth. For any particular subroutine or function of the NBO program, you can look up its index entry under SUBROUTINE or FUNCTION and find its detailed description in Section C of the *NBO Manual*. At the beginning of each keyword subsection in Section B, you will generally find an original literature citation that describes the underlying NBO algorithm and documents its numerical application. For example, you can find close connections between the original NPA paper (R. B. Weinstock, A. E. Reed, and F. Weinhold, *J. Chem. Phys.* **83**, 735–746, 1985; Appendix) and the coding of SR NAODRV and supporting subroutines, beginning near line 4423 of the NBO 5.9 source program.

Section B also contains additional information on individual keyword options, including syntax of varied $DEL options (p. B-16ff), rules for constructing $CHOOSE (p. B-14ff) and $NRTSTR (p. B-77ff) keylist input, details of ARCHIVE (.47) file structure (p. B-62ff), general print-level control with the PRINT keyword (p. B-10), and other guides to intelligent usage of the program. A serious NBO user should read Section B thoroughly even if there is no aspiration to penetrate Section C.

Section C also provides useful information on program limits and thresholds (consult individual THRESHOLD entries in the Index), default logical file numbers of user-requested external files (p. C-10ff), internal details of the NBO direct access (read–write) file (p. C-16ff); and other program control variables or error indices that may be referenced in error messages (cf. Appendix D). Keeping the *NBO Manual* close at hand and knowing how to troubleshoot with Sections B and C can solve many problems before it becomes necessary to contact nbo@chem.wisc.edu for further assistance.

Appendix B

Orbital Graphics: The NBOView Orbital Plotter

Conventional molecular orbitals (MOs) are of famously variable morphology, even in closely related molecular environments. Seeing the graphical MO plot is usually a unique visual experience, of little pedagogical value in anticipating what to look for in the next molecule of interest. In contrast, the graphical forms of NBOs are highly transferable and predictable, with subtle variations that provide richly rewarding clues to chemical behavior. Thus, it is extremely important that a chemistry student is able to accurately visualize the graphical forms of NBOs and other localized orbitals, including overlays of interacting donor and acceptor NBOs and their subtle shifts from one molecule to another. NBO orbital graphics goes far beyond the usual cartoon-like depictions.

NBO checkpointing options (p. B-127ff of the *NBO Manual*) make it easy to replace the MOs in the Gaussian checkpoint file with NBOs or any other chosen localized set. Thus, any graphical utility that is designed to display the MOs stored in the checkpoint file can be "tricked" into displaying NBOs instead. However, this workaround makes it difficult to compare, e.g., NBOs with their constituent NHOs, or PNBOs with NBOs, because the MOs can only be replaced by one localized set at a time. Furthermore, graphical MO utilities are often restricted to display one MO at a time, so that NBO donor–acceptor interactions cannot be conveniently depicted.

The *NBOView* program provides a convenient and flexible utility for displaying one-dimensional orbital profiles, two-dimensional contour diagrams, or photo-like three-dimensional surface plots of up to *eight* simultaneous orbitals and/or the total electron density. The orbitals to be individually or simultaneously displayed may be selected freely from orthogonal (NAO, NHO, NBO, and NLMO) or pre-orthogonal (PNAO, PNHO, PNBO, and PNLMO) localized natural sets, as well as from the ordinary basis AOs or delocalized MOs. The many examples shown throughout this book illustrate *NBOView* usage with default program settings. However, the programmable parameters for two-dimensional contour plots (controlling number, interval, and line-type of contour lines) or three-dimensional surface plots (controlling color, opacity, reflectivity, lighting sources, camera angle, and other details of the

Discovering Chemistry With Natural Bond Orbitals, First Edition. Frank Weinhold and Clark R. Landis.
© 2012 John Wiley & Sons, Inc. Published 2012 by John Wiley & Sons, Inc.

sophisticated optical model) allow a broad range of alternative visual effects. *NBO-View* can also automate the successive view frames for movie-like animated rotation of the three-dimensional visual display, as seen on the homepage of the *NBO website* (www.chem.wisc.edu/~nbo5). Full instructions for using the *NBOView* program are given in the online manual (www.chem.wisc.edu/~nbo5/v_manual.html) where other illustrations and details can be found.

Input for the *NBOView* program is generated by including the PLOT keyword as input to the *NBO 5* program. (Note that PLOT files generated by the older "NBO 3.1" version, as included in the standard Gaussian program distributions, are not acceptable to *NBOView*.) The PLOT files will be saved with the stem-name specified by the FILE keyword (e.g., FILE=myjob) and numerical extensions in the range 31–46 (e.g., myjob.31, myjob.32, ..., myjob.41, myjob.46). You should therefore avoid requesting these logical file numbers for other file I/O (Appendix C).

For pedagogical purposes, the pre-orthogonal "P" versions of natural localized orbitals (e.g., PNAOs, PNHOs, or PNBOs) are generally preferred, because they exhibit the visual *orbital overlap* that conveys a powerful intuition of the strength of orbital interaction. The "What Are NBOs?" website link (www.chem.wisc.edu/~nbo5/web_nbo.htm) provides discussion and *NBOView* illustrations of the important difference between pre-orthogonal (visualization) orbitals versus the orthonormal (physical) orbitals of NAO/NBO/NLMO theory. Throughout this book, graphical illustrations of "NBO donor–acceptor overlap" employ pre-orthogonal PNBOs, whereas numerical matrix elements (e.g., for second-order perturbative estimates and other purposes) are based on the NBOs. Although only a seeming technicality, and not belabored in the running text, the distinction between "visualization orbitals" and "physical orbitals" [i.e., those that could be eigenfunctions of a physical (hermitian) Hamiltonian operator] should be kept in mind.

Appendix C

Digging at the Details

In this appendix, we briefly summarize general features of *matrix output keywords* that provide virtually unlimited detail on operators or orbitals of interest.

In principle, the NBO program can provide the complete matrix representation of leading one-electron operators "OP" likely to be of interest to the user, such as

> Allowed 1e operators (OP):
> K = kinetic energy operator
> V = one-electron potential energy operator (nuclear-electron attraction)
> F = one-electron Hamiltonian (Fock or Kohn–Sham operator)
> S = unit ("overlap") operator
> DI = dipole moment operator (three Cartesian components)
> DM = density operator

If we symbolize the operator (OP) as O_{op}, *all possible* information about O_{op} is given by its matrix elements $(\mathbf{O})_{i,j}$ in a chosen orbital basis set (BAS) [a complete set of orbitals, symbolized as $\{b_i\}$]. Specifically, each $(\mathbf{O})_{i,j}$ (the matrix element in row i and column j) is evaluated as

$$(\mathbf{O})_{i,j} = \langle b_i | O_{op} | b_j \rangle = \int b_i{}^*(\mathbf{r}) \, O_{op} b_j(\mathbf{r}) \, \mathrm{d}^3\mathbf{r} = \int b_i{}^* \, O_{op} \, b_j \, d\tau$$

i.e., as the interaction integral between orbitals $b_i(\mathbf{r})$ and $b_j(\mathbf{r})$, integrated over all space.

The orbital basis "BAS" of interest to the user can be selected from any of the following:

> Allowed orbital basis sets (BAS):
> AO = basis functions ("atomic orbitals") of the wavefunction
> NAO = natural atomic orbitals
> NHO = natural hybrid orbitals
> NBO = natural bond orbitals
> NLMO = natural localized molecular orbitals
> MO = molecular orbitals (for HF, DFT, or CAS methods)

Discovering Chemistry With Natural Bond Orbitals, First Edition. Frank Weinhold and Clark R. Landis.
© 2012 John Wiley & Sons, Inc. Published 2012 by John Wiley & Sons, Inc.

For selected operators, BAS could also be selected as one of the preorthogonal sets: PNAO, PNHO, PNBO, or PNLMO; see p. B-2 of the *NBO Manual* for a complete listing.

To obtain numerical matrix elements for the chosen operator (OP) and basis (BAS), the user merely inserts a composite keyword of the form "OPBAS" into the $NBO ... $END keylist. For example, to obtain matrix elements of the dipole operator (OP = DI) in the NBO basis (BAS = NBO), the keylist entry would be "DINBO"

$NBO DINBO $END

Similarly, to obtain both the Fock operator in the NHO basis ("natural Hückel matrix") and the overlap matrix in the PNHO basis, the keyword entries would be "FNHO" and "SPNHO"

$NBO FNHO SPNHO $END

For unmodified keywords, as shown above, the *full* $(\mathbf{F})_{i,j}$, $(\mathbf{S})_{i,j}$ matrices would be printed in the output file. To restrict output, e.g., to only diagonal $(\mathbf{F})_{5,5}$, $(\mathbf{F})_{14,14}$ and off-diagonal $(\mathbf{F})_{5,14}$ elements, one would use a *bracket-list* of the form

$NBO FNHO <5 5/14 14/5 14> $END

For other ways to control or redirect output to an external file in machine-readable format, see the full discussion in Section B.2.4 of the *NBO Manual.*

The NBO program also allows you to obtain complete details of the matrix transformation from one basis set "BAS1" to another "BAS2" by including a conjoined keyword of the form "BAS1BAS2" in the $NBO keylist. For example, the transformation from the AO (BAS1 = AO) to the NBO (BAS2 = NBO) basis would be specified by the "AONBO" keyword, whereas the transformation from NBOs to MOs would similarly be specified by "NBOMO," and so forth.

Consult Section B.2.4 of the *NBO Manual* for further details on restricting output (e.g., to valence shell or Lewis orbitals only), redirecting output to an external file, or "checkpointing" AO-based transformations (e.g., AONBO or AONLMO) to the host checkpoint file for CAS-type (Sidebar 11.2) or graphical (Appendix B) applications. The AOINFO keyword also provides additional detail (probably more than you wish to see) on orbital exponents and contraction coefficients of the Gaussian-type basis AOs that underlie the host ESS wavefunction calculation.

Although matrix output keywords can consume lots of paper, they are the ultimate resource for the building blocks of NBO analysis and its extensions in specialized cases.

Appendix D

What If Something Goes Wrong?

Constructive troubleshooting usually begins by consulting the FAQ (frequently asked questions) section of the *NBO website* http://www.chem.wisc.edu/~nbo5/faq.htm

Problems appear with varying degree of seriousness. Let us mention typical cases in the ascending order of concern.

Least serious are common "WARNING" messages that accompany natural population analysis. A message about population inversion (occupancy order not matching energy order) or low core-orbital occupancy is for informational purposes only, signaling that something unusual or interesting may be happening, but ordinarily of no real concern.

More problematic are cases in which NBO analysis failed to give an "expected" result, or some kind of unphysical erratic behavior is observed. The most severe problems of this type are as follows:

- *Linear dependence.* Apparent linear dependence of basis functions (due to finite accuracy of machine arithmetic) leads to insidious numerical pathologies that can infect all aspects of wavefunction calculation and analysis. The problems arise most frequently when diffuse ($+/++$) functions and/ or short interatomic distances are present. The host ESS program may attempt to "correct" linear-dependence pathologies by removing entire blocks of basis functions (look for messages to this effect), and the NBO program similarly checks for instabilities and removes additional troublesome functions from the AO basis as necessary. However, any sign of linear independence (even if "corrected") is a danger sign. Consider altering the basis whenever ESS or NBO warning messages about linear dependence appear in your output.

- *DFT instabilities.* DFT functionals are parameterized to give good results for near-equilibrium geometries of "ordinary" molecules, but they may become increasingly erratic if presented with highly unusual or unphysical densities, as required, e.g., in certain extreme limits of $DEL and STERIC evaluations.

Discovering Chemistry With Natural Bond Orbitals, First Edition. Frank Weinhold and Clark R. Landis.
© 2012 John Wiley & Sons, Inc. Published 2012 by John Wiley & Sons, Inc.

In case of doubt, follow the *caveats* in the *NBO Manual* (e.g., p. B-20) and check against Hartree–Fock or other non-DFT methods for consistency.

- *Unexpected NBO structure.* You can check whether the NLS returned by default NBO analysis is really "best" by using $CHOOSE keylist input to evaluate alternative structures that NBO may have overlooked. In rare cases, NBO may "skip over" a superior Lewis structure because the default occupancy threshold (starting at 1.90e, with 0.1e decrements) is too coarse-grained. In such cases, you can use the DETAIL keyword to see why NBO made the decisions it did, and you can then use the THRESH keyword to alter the occupancy threshold and perhaps find a structure of improved %-ρ_L (e.g., by "recognizing" a lone pair that fell just below the default threshold). Such cases, although highly unusual, are not unprecedented, and the structure of highest %-ρ_L should be considered *the* NLS even if the default NBO search dropped the ball.

Most vexing are cases where the program goes into an infinite trance and/or halts midway through execution, with or without an error message. Here are suggestions for common cases:

- *NRT hang.* Any infinite stall or abrupt program termination without an error message normally portends memory conflict or overrun, particularly in NRT jobs that end with a "hypervalency detected" message. In such cases, restart the job with NRTFDM keyword included and, if possible, increase available memory (%mem) allocation. (NRT tries to "turn on" NRTFDM if apparent hypervalency is detected in mid-task, but this often requires additional memory beyond that judged necessary by initial memory check; including NRTFDM allows more accurate assessment of memory demands and a more graceful exit if memory resources are found inadequate.)

- *Other NRT problems.* NRT is a memory hog, and error messages about "too many resonance structures" often require an iterative approach. Begin by reducing basis size as far as possible, then reduce NRTMEM (try NRTMEM=1) and increase NRTTHR (try NRTTHR=20) until you can get the NRT job to complete. Based on what you find, judiciously increase NRTMEM and provide explicit $NRTSTR entries for the corresponding number of reference structures. By then, you can probably begin reducing NRTTHR and increasing the basis back to full size.

- *"Unphysical population" halt.* Problems with unphysical (negative or Pauli-violating) populations can often be cured by including the FIXDM keyword.

- *Unrecognized keywords.* If NBO output for a requested option is missing (and not listed among the requested job options near the top of NBO output), check above the NBO banner to see if the keyword was listed as "unrecognized." This could arise from misspelling, a problem in keyword syntax, or an older NBO version that does not support the requested keyword.

- *Other inscrutable error messages.* Most error messages should be intelligible in context. If not, you might try using a text editor to search the source code for the message text and condition that caused the failure. Be sure that the error message actually comes from NBO code (rather than the host ESS) before seeking assistance from nbo.chem.wisc.edu.

Appendix E

Atomic Units (a.u.) and Conversion Factors

Electronic structure theorists generally employ "atomic units" (a.u.) in which the three defining *base units* are the fundamental natural constants e (electronic charge), m_e (electronic mass), $\hbar$ (Planck's constant/2π) rather than arbitrarily chosen macroscopic objects (e.g., the Pt–Ir bar in Paris that defines "unit mass" in conventional SI units). These units have many advantages, not least that they bring the electronic Schrödinger equation to its intrinsically simplest form, expressed in pure numbers only, so that it can be solved once for all, independent of remeasured physical quantities. The atomic units are also sensibly proportioned and "sized" such that the key atomic properties tend to have values of order unity; for example, the hydrogenic 1s orbital radius turns out to be exactly 1 a.u. of length. By working out the combination of e, m_e, and $\hbar$ whose practical units match those of a desired physical property (such as energy $= m_e e^4/\hbar^2$, length $= \hbar^2/m_e e^2$, and so forth), one obtains the corresponding "atomic unit" of that property, which is usually designated simply as "a.u." rather than assigned a special symbol and name for each property.

Tables E.1–E.4 provide conversion factors from a.u. to SI units and a variety of practical (thermochemical, crystallographic, and spectroscopic) units in common usage. An abbreviated exponential notation is employed in which 6.02214 (23) means 6.02214×10^{23}. Throughout this book, we follow a current tendency of the quantum chemical literature by expressing energy changes in common thermochemical units (kcal/mol), structural parameters in crystallographic Angstrom units (Å), vibrational frequencies in common spectroscopic wavenumber units (cm^{-1}), and so forth, thereby facilitating communication between theoretical and experimental practitioners.

Discovering Chemistry With Natural Bond Orbitals, First Edition. Frank Weinhold and Clark R. Landis.
© 2012 John Wiley & Sons, Inc. Published 2012 by John Wiley & Sons, Inc.

Table E.1   Conversion Factors from Atomic to SI Units

Atomic unit (base units)	SI value	Name (symbol)
Mass (m_e)	9.10939 (−31) kg	Mass of the electron
Charge (e)	1.602188 (−19) C	Electronic charge
Angular momentum ($\hbar$)	1.05457 (−34) J/(s rad)	Planck's constant/2π
Energy ($m_e e^4/\hbar^2$)	4.35975 (−18) J	Hartree (H)
Length ($\hbar^2/m_e e^2$)	5.29177 (−11) m	Bohr; Bohr radius (a_0)
Time ($\hbar^3/m_e e^4$)	2.41888 (−17) s	Jiffy
Electric dipole moment ($\hbar^2/m_e e$)	8.47836 (−30) C m	2.541765 Debye (D) units
Magnetic dipole moment ($e\hbar/2m_e$)	9.27402 (−24) J/T	Bohr magneton (μ_B)

Table E.2   Energy Conversion Table for Non-SI Units

Unit	Value in non-SI units					
	a.u.	kcal/mol	eV	cm^{-1}	Hz	K
a.u.	1	6.27510 (2)	2.72114 (1)	2.19475 (5)	6.57968 (15)	3.15773 (5)
kcal/mol	1.59360 (−3)	1	4.33641 (−2)	3.49755 (2)	1.04854 (13)	5.03217 (2)
eV	3.67493 (−2)	2.30605 (1)	1	8.06554 (3)	2.41799 (14)	1.16044 (4)
cm^{-1}	4.55634 (−6)	2.85914 (−3)	1.23984 (−4)	1	2.99792 (10)	1.43877
Hz	1.51983 (−16)	9.5371 (−14)	4.13567(−15)	3.33564 (−11)	1	4.79922 (−11)
K	3.16683 (−6)	1.98722 (−3)	8.61739 (−5)	6.95039 (−1)	2.08367 (10)	1

Table E.3   Fundamental Constants, in Atomic and SI Units

Physical constant	Symbol	Value (a.u.)	Value (SI)
Rydberg constant	R_∞	2.29253 (2)	1.09737 (−23)/m
Planck constant	h	6.28319 (=2π)	6.62608 (−34) J s
Speed of light	c	1.37036 (2)	2.99792 (8) m/s
Proton mass	m_p	1.83615 (3)	1.67262 (−27) kg
Atomic mass unit	amu	1.82289 (3)	1.66054 (−27) kg
Fine structure constant	α	7.29735 (−3)	7.29735 (−3)

Table E.4   Other Constants and Conversion Factors

Quantity (symbol)	SI value or equivalent
Avogadro's number (N_0)	6.02214 (23)/mol
Kilocalorie (kcal)	4.18400 (3) J
Kelvin (K)	C−273.15
Boltzmann constant (k)	1.38066 (−23) J/K
Faraday constant ($\mathcal{F}$)	9.64853 (4) C/mol

Index

absolutely localized molecular orbital (ALMO)   226
acceptor (Lewis acid)   219
   non-Lewis orbital   94, 96–7, 104, 107, 109, 196–7, 210, 219–21, 239, 245, 252, 278
acetylene, HCCH   76, 173
acrolein, $H_2C=CHCHO$ 269–289
agostic interaction   206–7
Alabugin, I.V.   102
allylic-type resonance   60, 85, 100, 117, 278
aluminum trichloride, $AlCl_3$ 179
   dimer, Al_2Cl_6 179
amide resonance   56–60, 77, 97–127, 146–152
aminoborane, H_2NBH_2 184
aminomethane, H_2NCH_3
ammonia, NH_3 74–5, 177, 190, 193–4, 196–200, 203
ammonia-boron trifluoride complex, H_3N:BF_3 178, 191, 193
antibond (BD*)   81, 97, 102, 104–5, 167, 183, 219, 221, 239
anticooperativity   106, 221–223
antiperiplanar influence, see stereoelectronic effects
antisymmetry   11, 95, 130, 136–144, 155, 224
AO, see basis atomic orbital
archive (.47) file, see NBO Keywords, ARCHIVE
aromaticity   92, 129
Arrhenius reaction profile   231, 235
arrow-pushing mnemonic   103–4, 210, 282
atoms   34–50
   in NBO output   23, 34
   Bohr-Rutherford model   14, 136

bonded vs. free-space character   35
   in molecules   34–51
   NAO-based definition   34
   periodicity   4, 10, 19, 27, 30, 128, 136, 201
atomic charge   39–44
   natural   40, 91, 140, 185, 191
   atomic polar tensor   41
   Bader QTAIM   41
   dipole-fitting   40–41
   Mulliken   21, 42–43, 226
atomic configuration   10–31, 34–43, 77, 193, 201–204
atomic radii   136, 197
atomic spin-orbitals   10, 18
atomic unit (a.u.)   17, 25, 104
Aufbau principle   19–21, 84, 201
Autschbach, J.   161

B3LYP, see density functional theory (DFT)
back-bonding   196–8, 200, 203–4
Badenhoop, J.K.   139
Bader, R.W.F.   41–2
Bader charges   41–2
Baerends, E.J.   226
ball-and-stick model   135, 139, 250
banana bond   60
basis atomic orbital (AO)   9–10, 13–19, 21–2, 27, 29–31, 34–5, 37, 42–3, 50, 61–2, 130–1
   contributions of   41
   effective core potential (ECP)   30
   gauge-including   155–157
   occupied   32
   Slater type   22
   vs. NAOs   21–2

Discovering Chemistry With Natural Bond Orbitals, First Edition. Frank Weinhold and Clark R. Landis.
© 2012 John Wiley & Sons, Inc. Published 2012 by John Wiley & Sons, Inc.